ADVANCES IN

GEOPHYSICS

VOLUME 15

Contributors to This Volume

H. U. Dütsch
Bijan Esfandiari
Floyd A. Huff
Carl A. Moore
Donald J. Williams

Advances in
GEOPHYSICS

Edited by

H. E. LANDSBERG

Institute for Fluid Dynamics and Applied Mathematics
University of Maryland, College Park, Maryland

J. VAN MIEGHEM

Royal Belgian Meteorological Institute
Uccle, Belgium

Editorial Advisory Committee

BERNARD HAURWITZ R. STONELEY
ROGER REVELLE URHO A. UOTILA

VOLUME 15

1971

Academic Press • New York and London

COPYRIGHT © 1971, BY ACADEMIC PRESS, INC.
ALL RIGHTS RESERVED
NO PART OF THIS BOOK MAY BE REPRODUCED IN ANY FORM,
BY PHOTOSTAT, MICROFILM, RETRIEVAL SYSTEM, OR ANY
OTHER MEANS, WITHOUT WRITTEN PERMISSION FROM
THE PUBLISHERS.

ACADEMIC PRESS, INC.
111 Fifth Avenue, New York, New York 10003

United Kingdom Edition published by
ACADEMIC PRESS, INC. (LONDON) LTD.
24/28 Oval Road, London NW1 7DD

LIBRARY OF CONGRESS CATALOG CARD NUMBER: 52-12266

PRINTED IN THE UNITED STATES OF AMERICA

CONTENTS

LIST OF CONTRIBUTORS vii

Geochemistry and Geology of Helium

CARL A. MOORE and BIJAN ESFANDIARI

1. Introduction...	2
2. Radioactivity...	3
3. Origin and Abundance of Elements.............................	9
4. Origin of Helium...	11
5. Helium-Generating Potential of Rocks and Fluids...............	14
6. Distribution and Occurrence of Helium	29
7. Migration of Helium	32
8. Accumulation or Entrapment of Helium	41
9. Natural Gas Analysis	46
10. Conclusions..	53
References	54

Evaluation of Precipitation Records in Weather Modification Experiments

FLOYD A. HUFF

1. Introduction...	60
2. Effects of Natural Precipitation Variability in Evaluating Cloud Seeding Experiments	64
3. Use of Precipitation Climatology in Assessing Potential Benefits of Weather Modification...	70
4. Space and Time Variability Relationships	78
5. Other Relevant Climatological Studies	92
6. Precipitation Measurement Requirements	97
7. Statistical Evaluation Techniques	111
8. Downwind Seeding Effects	119
9. Design of a Precipitation Modification Experiment	123
References	133

Charged Particles Trapped in the Earth's Magnetic Field

Donald J. Williams

1. Introduction	137
2. Magnetic Field Considerations	146
3. Particle Survey	156
4. Sources, Losses, and Transport	170
5. Concluding Remarks and Future Directions	206
List of Symbols	207
References	208

Photochemistry of Atmospheric Ozone

H. U. Dütsch

1. Observed Distribution of Total Ozone	219
2. The Classical Photochemical Theory: "Pure Oxygen Atmosphere"	244
3. Ozone Photochemistry in a "Moist" Atmosphere	271
4. Possible Importance of Nitrogen Oxides to Ozone Photochemistry	303
5. Ozone as a Tracer	306
6. Interrelation between Ozone Photochemistry and Stratospheric Dynamics	312
List of Symbols	315
References	315
Author Index	323
Subject Index	330

LIST OF CONTRIBUTORS

Numbers in parentheses indicate the pages on which the authors' contributions begin.

H. U. DÜTSCH, *Federal Institute of Technology, Laboratory of Atmospheric Physics, Zurich, Switzerland* (219)

BIJAN ESFANDIARI, *Department of Geology, University of Tehran, Tehran, Iran* (2)

FLOYD A. HUFF, *Illinois State Water Survey, Urbana, Illinois* (60)

CARL A. MOORE, *School of Petroleum and Geological Engineering, University of Oklahoma, Norman, Oklahoma* (2)

DONALD J. WILLIAMS, *Space Environment Laboratory, Environmental Research Laboratories, National Oceanic and Atmospheric Administration, Boulder, Colorado* (137)

ADVANCES IN

GEOPHYSICS

VOLUME 15

GEOCHEMISTRY AND GEOLOGY OF HELIUM

Carl A. Moore and Bijan Esfandiari

School of Petroleum and Geological Engineering, University of
Oklahoma, Norman, Oklahoma, and Department of Geology,
University of Tehran, Tehran, Iran

	Page
1. Introduction	2
Statement of the Problem	2
2. Radioactivity	3
2.1. Introduction	3
2.2. Radioactive Nuclides	4
2.3. Alpha Radioactivity	4
3. Origin and Abundance of Elements	9
3.1. Origin of the Elements	9
3.2. Equilibrium Theory	9
3.3. Cosmic Abundances of the Elements	10
4. Origin of Helium	11
4.1. Definition and Uses of Helium	11
4.2. Theories of Helium Origin	12
4.3. Cosmogenic Helium	12
4.4. Radiogenic Helium	13
4.5. Association of Helium in Natural Gases	13
5. Helium-Generating Potential of Rocks and Fluids	14
5.1. Procedure	14
5.2. Equipment	15
5.3. Theory of Specific Radiation Activity	16
5.4. Helium-Generating Potential of Igneous Rocks	17
5.5. Geochemical Controls of Uranium and Thorium Fractionation	20
5.6. Helium-Generating Potential of Sedimentary Rocks	20
5.7. Types of Uranium Concentration in Sedimentary Rocks	22
5.8. Helium-Generating Potential of Metamorphic Rocks	23
5.9. Helium-Generating Potential of Fluids	25
6. Distribution and Occurrence of Helium	29
6.1. Extraterrestrial Helium	29
6.2. Terrestrial Helium	29
6.3. Crustal Erosion	30
6.4. Helium in Crustal Rocks	30
6.5. Helium in the Hydrosphere	32
6.6. Helium in the Atmosphere	32
7. Migration of Helium	32
7.1. Migration of Alpha Emitters	33
7.2. Migration of Helium at High Pressures	33
7.3. Migration of Helium at Low Pressures	34
7.4. Diffusional Migration of Helium	35
7.5. Subsurface Conditions	36
7.6. Migration of Helium	36

7.7. Diffusivity Process in Sediments 38
7.8. Simplification of Diffusivity Factors for Analysis 38
7.9. Mathematical Models 39
8. Accumulation or Entrapment of Helium 41
  8.1. Ability to Hold Helium 41
  8.2. Effect of Radiation 45
  8.3. Conditions Favoring Helium Accumulation 45
  8.4. Retentivity and Diffusivity of Rocks 45
9. Natural Gas Analyses 46
  9.1. Helium–Argon Ratio 46
  9.2. Helium–Crude Oil Solution 47
  9.3. Noble Gas Abundances in Natural Gases 51
  9.4. Helium Distribution in Natural Gases 51
10. Conclusions 53
References 54

1. Introduction

Statement of the Problem

There are many interesting problems in the geochemistry of the rare gases, one of which is their scarcity in terrestrial materials as compared with their abundance in the solar and stellar atmospheres. Helium is second to hydrogen in being the most abundant element in the universe. The cause or causes which may govern the scarcity on earth will be discussed in this review.

Coexistence of helium with some natural gases presents another geochemical problem. A number of theories to explain the origin of helium in natural gases have been proposed. On a broad scale, these theories assume that either: (1) all or part of the helium is derived from the disintegration of radionuclides, or (2) it is mostly primordial, being that helium which was trapped during the degassing stage of the formation of the earth. These theories have many shortcomings which have been pointed out by others workers [3,11,19]. A helium-generating mechanism to overcome these shortcomings is presented here.

A considerable amount of work has been done on the material balance of helium between the earth, the atmosphere, and space, but little on the mode of migration of the helium within the earth's crust to the atmosphere. Several possible modes of migration will be considered and conclusions drawn for that type of migration which is the most probable.

Natural gas analyses from many areas reveal a wide variation in helium content. As a general rule, the helium content increases with increasing nitrogen content. However, there seems to be no definite relationship between helium and nitrogen.

Radioactivity of the natural environment, of the host rocks, and of certain fluids has a direct bearing on the amount of helium found in source areas. However, assigning the correct relative amount of helium from each contributing source is a difficult task. Variations in the helium content of natural gases may be linked with the entrapment, factors of escape, retention, and diffusion, and other complex factors.

The exact role of rock geometry, including the communication passages caused by porosity, permeability, and fracture systems, is essentially unknown. We shall consider the individual role of each of these geometric factors and their importance.

Since alpha particles may be converted to helium atoms within a few microns of a given surface, fractured basement rocks are good sources of helium. Such fracturing creates passageways for the helium which may be entrapped in the crystalline structure of the rocks. Following release, the helium atoms are able to move into the passageways which are the migratory channels. Such channels, including fractures, may extend through the sedimentary section. Thus, any helium which is present in the basement rocks may find its way into the passageways of the sediments where it may migrate until temporarily trapped by the geometry of the enclosing rocks.

Solubility of radioactive minerals such as uraninite (UO_2) in petroleum and in brines is still poorly known. A favorable solubility factor, however, could account for a locally high helium content where the helium concentration is about eight mole percent of the natural gas. Circulating or migrating fluids, e.g., ground water or petroleum, might partially dissolve the uranium and thorium content of the host rocks and actually become an alpha-emitting fluid, and hence will be a source of helium.

Subsequent migration of oil and gas into traps is an acute problem in the geochemistry of helium. The uranium and thorium which are dissolved in petroleum can produce helium within the petroleum itself. Because of its very insoluble nature, the helium may escape from this fluid and be trapped with natural gases in the same reservoir.

2. Radioactivity

2.1. Introduction

The nuclear charge, or atomic number Z, is the integral number of protons in the nucleus. The total number of neutrons and protons in a nucleus is the mass number A. Nuclei of the same Z but different A are different forms of the same element and are called isotopes. The mass of an electron or a beta particle is only 1/1850 the mass of a proton. When a nucleus decays by beta emission, Z increases by one unit because one negative charge is removed, and A remains constant. Alpha particles are the nuclei of helium He^{2+}, and thus have a mass of four atomic mass units and a double positive charge. Therefore, in a nucleus decaying by alpha emission, Z will decrease by two and A by four.

Radioactive processes are spontaneous nuclear reactions, or spontaneous transmutations, characterized by the radiation which is emitted. These occur at random, at a measurable rate, but the exact moment when a given atom will decay cannot be predicted. This spontaneous transmutation will cause a nuclide to convert into another nuclide in an attempt to form a more stable configuration under the release of energy in the form of different radiation. If the probability of transmutation is nonexistent or is very small the nuclide is weakly radioactive and considered to be stable. If the probability is large, the nuclide is strongly radioactive. Where more than one consecutive change occurs before stability is reached, a radioactive chain or series, is formed. Radioactive nuclides are commonly called radionuclides.

2.2. Radioactive Nuclides

There are about sixty naturally occurring radioactive nuclides. More than a thousand species [58] have been prepared artificially in various nuclear transmutations and in disintegration processes from naturally stable nuclides. In fact, virtually every known element and several previously unknown elements have been produced in radioactive forms.

Several types of radiations are emitted from radioactive nuclei during their decay process. Most nuclides that are found in nature are either alpha or beta emitters, and their radiation often is accomplished through gamma radiation. Our attention here is focused on alpha radiation because of its role in the generation of helium.

2.3. Alpha Radioactivity

Alpha decay is energetically possible in all the members of the entire upper third of the periodic system. It becomes prominent in the nuclide region above $Z=82$ which is lead. For most heavy radionuclides above $A=212$, systematic trends in the alpha-decay properties indicate that at constant Z the alpha-disintegration energy decreases [59]. All elements with Z greater than 80 which are found in nature have radioactive isotopes, and above $Z=82$ no stable elements exist. Consequently the alpha half-life increases in an essentially linear fashion with increasing A.

The alpha half-lives depend to a marked degree on the alpha decay energies. Only when the alpha-decay energy exceeds a given value does the alpha lifetime become short enough to be detectable. Alpha half-lives

TABLE I. Natural radionuclides

Parent nuclide			Manner of decay	Decay product			Half-life
Atomic number	Element	Mass number		Atomic number	Element	Mass number	
Single primary radionuclides							
57	lanthanum	138	70% K, L cap	56	barium	138	1.0×10^{11} yr
			30% beta	58	cerium	138	5.1×10^{15} yr
58	cerium	142	alpha	56	barium	138	2.2×10^{15} yr
60	neodymium	144	alpha	58	cerium	140	1.25×10^{11} yr
62	samarium	147	alpha	60	neodymium	143	long
74	tungsten	180(?)	alpha	72(?)	hafnium(?)	176(?)	5.9×10^{11} yr
78	platinum	190	alpha	76	osmium	186	10^{15} yr
78	platinum	192	alpha	76	osmium	188	2×10^{17} yr, or more than 2×10^{18} yr
83	bismuth	209	alpha(?)	81	thallium	205	
Radioactive families: The ^{238}U (uranium) family							
92	uranium	238	alpha	90	thorium	234	4.51×10^{9} yr
92	uranium	234	alpha	90	thorium	230	2.48×10^{5} yr
90	thorium	230	alpha	88	radium	226	8.0×10^{4} yr
88	radium	226	alpha	86	radon	222	3.8229 days
86	radon	222	alpha	84	polonium	218	3.05 min
84	polonium	218	99.98% alpha	82	lead	214	
			0.02% beta	85	astantine	218	
85	astantine	218	99.9% alpha	83	bismuth	214	1.5 to 2 sec
			0.1% beta	86	radon	218	

(*Continued*)

TABLE I.—(Continued)

Parent nuclide			Manner of decay	Decay product			Half-life
Atomic number	Element	Mass number		Atomic number	Element	Mass number	
83	bismuth	214	0.04 % alpha	81	thallium	210	19.7 min
			99.96 % beta	84	polonium	214	
86	radon	218	alpha	84	polonium	214	0.019 sec
84	polonium	214	alpha	82	lead	210	1.64×10^{-4} sec
83	bismuth	210	99+ % beta	84	polonium	210	5.013 days
			5×10^{-5} % alpha	81	thallium	206	
84	polonium	210	alpha	82	lead	206	138.401 days
82	lead	206	stable				
Radioactive families: The ^{232}Th (thorium) family							
90	thorium	232	alpha	88	radium	228	1.39×10^{10} yr
90	thorium	228	alpha	88	radium	224	1.910 yr
88	radium	224	alpha	86	radon	220	3.64 days
86	radon	220	alpha	84	polonium	216	51.5 sec
84	polonium	216	alpha	82	lead	212	0.158 sec
83	bismuth	212	63.8 % beta	84	polonium	212	60.5 min
			36.2 % alpha	81	thallium	208	
84	polonium	212	alpha	82	lead	208	3.04×10^{-7} sec
Radioactive families: The ^{235}U (actinium) family							
92	uranium	235	alpha	90	thorium	231	7.1×10^{8} yr
91	protactinium	231	alpha	89	actinium	227	3.43×10^{4} yr

Z	element	A	decay	A'	daughter	Z'	half-life
89	actinium	227	98.8% beta	227	thorium	90	22 min
			1.2% alpha	223	francium	87	18.17 days
87	francium	223	99+% beta	223	radium	88	
			6 × 10⁻³% alpha	219	astantine	85	
90	thorium	227	alpha	223	radium	88	18.17 days
88	radium	223	alpha	219	radon	86	11.68 days
85	astantine	219	3% beta	219	radon	86	
			97% alpha	215	bismuth	83	
86	radon	219	alpha	215	polonium	84	0.9 min
84	polonium	215	99+% alpha	211	lead	82	3.92 sec
			5 × 10⁻⁴% beta	215	astantine	85	
85	astantine	215	alpha	211	bismuth	83	1.83 × 10⁻³ sec
83	bismuth	211	99.7% alpha	207	thallium	81	10⁻⁴ sec
			0.3% beta	211	polonium	84	
84	polonium	211	alpha	207	lead	82	0.52 sec

Radioactive Families: The ²³⁷U (neptunium) family

Z	element	A	decay	A'	daughter	Z'	half-life
93	neptunium	237	alpha	233	protactinium	91	2.20 × 10⁶ yr
92	uranium	233	alpha	229	thorium	90	1.62 × 10⁵ yr
90	thorium	229	alpha	225	radium	88	7340 yr
89	actinium	225	alpha	221	francium	87	10.0 days
87	francium	221	alpha	217	astantine	85	4.8 min
85	astantine	217	alpha	213	bismuth	83	0.018 sec
83	bismuth	213	98% beta	213	polonium	84	47 min
			2% alpha	209	thallium	81	
84	polonium	213	alpha	209	lead	82	4.2 × 10⁻⁶ sec
83	bismuth	209	alpha(?)	205	thallium	81	2 × 10¹⁷ yr, or more than 2 × 10¹⁸ yr

decrease gradually from 1.39×10^{10} yr for ^{232}Th, to hours and minutes for the heaviest nuclides. Among the heavy nuclides, several with a relatively low A value may decay by an orbital electron capture that often competes with alpha disintegration. Table I shows different alpha-emitting radionuclides, their manner of decay, decay product, and their half-lives.

The investigation of the collateral radioactive series, transuranium elements, and neutron-deficient heavy elements has revealed the existence of approximately 70 new alpha emitters as contrasted with only 29 alpha-active nuclides known to exist in nature [58].

Most of the alpha particles emitted from a radioactive source have very nearly straight-lined tracks of identical length [36]. Over their tracks they are capable of ionizing the matter that they traverse. The length of the track is called the range of the alpha particles, and the range is defined as the distance in dry air at standard conditions traveled by the alpha particles from their source to a point at which they no longer produce ionization. At this point the alpha particle is converted to a molecule of helium gas.

Because energy is consumed during ionization, the velocity of an alpha particle will decrease toward the end of its track. The mean range of an alpha particle passing through air R is listed in Table II, based on data from Rankama [58].

TABLE II. Mean range R of alpha particles in dry air at 15 °C and 1 atm for the natural alpha-emitters

Uranium family	R (cm)	Actinium family	R (cm)	Thorium	R (cm)
^{238}U	2.65	^{235}U	2.82	^{232}Th	2.60
^{234}U	3.21	^{227}Th	4.60	^{228}Th	4.00
^{230}Th	3.11	^{223}Ra	4.29	^{224}Ra	4.30
^{226}Ra	3.30	^{215}Po	6.46	^{216}Po	5.64
^{210}Po	3.84	^{211}Bi	5.36	^{212}Bi	4.73

A comparison of the half-life values of Table I with the range of alpha particles emitted from radionuclides reveals that an inverse relationship exists between the half-life and the range of the alpha particles which they emit. In gaseous media, the range of an alpha particle depends on the nature of the medium through which it travels, while in solids, its range is in the order of a few hundredths of a millimeter.

The relation of the range of an alpha particle in air to its range in the medium investigated yields a quantity called the relative stopping power of the medium. The range of alpha particles in solids may be estimated by means of an empirical rule formulated by Bragg and Kleeman (see Faul [25]). This equation gives the relation between range in air and the length of path L in any particular medium:

(2.1) $\qquad L = R(P_a/\rho), \qquad (A/A_a^{1/2}) = 3.2 \times 10^{-4}(R/\rho)\, A^{1/2}$

where R is the range in air, P_a is the density of air, ρ is the density of the particular medium, A_a is the mean mass number of air, and A is the mean mass number of the medium.

3. Origin and Abundance of Elements

3.1. Origin of the Elements

The question of the origin of the elements and their isotopes constituting the universe is an ancient scientific problem. With the gradual improvement of nuclear experimental data, the theories of the origin of the elements have advanced markedly [24]. Recognition of the structure of the nuclei of elements as aggregates of protons and neutrons has resulted in theories explaining their origin and their relative abundances by a synthesis, or buildup. These theories postulate that the nucleus starts with either or both of the basic building blocks, protons and neutrons. An alternative is that nuclei now in existence resulted from the breakup of a primordial nuclear fluid, with fission and evaporation processes playing a leading role.

The synthesis approach does not attempt an answer to the perhaps even more intriguing problem of the origin of protons and neutrons. For that aspect of the problem of origins, there are practically no experimental data [28].

3.2. Equilibrium Theory

Several theories have been proposed to account for the mode of formation of the chemical elements. One, which may be termed the equilibrium theory, proposes that the relative abundances of the elements are the result of a "frozen" thermodynamic equilibrium between atomic nuclei at some high temperature and density. Through suitable assumptions about temperature, pressure, and density, good agreement with observed abundances is obtained for elements with up to $Z = 40$ [48]. For elements with higher atomic numbers, however, these assumptions lead to impossibly low abundances.

Therefore, a theory has been proposed which considers the relative abundances of the elements as resulting from nonequilibrium processes. According to this latter theory the light nuclei were built up by thermonuclear processes and the remaining nuclei by successive neutron capture, with intervening beta disintegrations. This theory predicts the general trend of the observed data, but fails to explain some of the detailed features, particularly in explaining the nonexistence of nuclei of atomic weights five and eight [48]. The difficulty can be overcome by postulating the fusion of three ^{4}He nuclei to give ^{12}C. The complexities of the abundance data, however, as established by Suess and Urey [48], showed that no single process can satisfactorily account for these complexities.

3.3. Cosmic Abundances of the Elements

On the basis of data from the composition of meteorites and of solar and stellar matter, Goldschmidt [33] compiled the first adequate table of cosmic abundances of elements and isotopes. The data on hydrogen and helium were discovered largely from an examination of the sun and stars, and the figures for most of the other elements were based on their relative abundances in meteoric material.

In general, the agreement between the relative abundances determined in different regions of the universe seems reasonably good [48]. Variations in the abundances of hydrogen, helium, lithium, carbon, and nitrogen in different parts of the universe are due to the participation of these elements in thermonuclear transformations.

The relative abundance of the different elements, especially the lighter ones, varies considerably, as shown in Fig. 1. An element may be a hundred or a thousand times more, or less, abundant than its immediate neighbor in the periodic table. Nevertheless, when the data are carefully analyzed, numerous regularities are found [33]. These may be summed up as follows:

(1) The abundances show a rapid exponential decrease for elements with the lower atomic numbers (to about $Z = 30$), followed by an almost constant value for the heavier elements.

(2) Elements with an even atomic number are more abundant than those with odd atomic numbers on either side. This regularity is known as the Oddo–Harkins rule.

(3) The relative abundances for elements with higher atomic numbers than nickel ($Z = 28$) vary less than those elements with lower atomic numbers.

(4) Only ten elements (hydrogen, helium, carbon, nitrogen, oxygen, neon, magnesium, silica, sulfur, and iron), all with Z less than 27, show an appreciable abundance. Of these, hydrogen and helium far outweigh the other eight elements.

The regularities displayed in Fig. 1 suggest that the absolute abundances of the elements depend on nuclear rather than chemical properties and are related to the inherent stability of the nuclei. An element is uniquely characterized by the number of protons Z in its nucleus, but the number of neutrons N

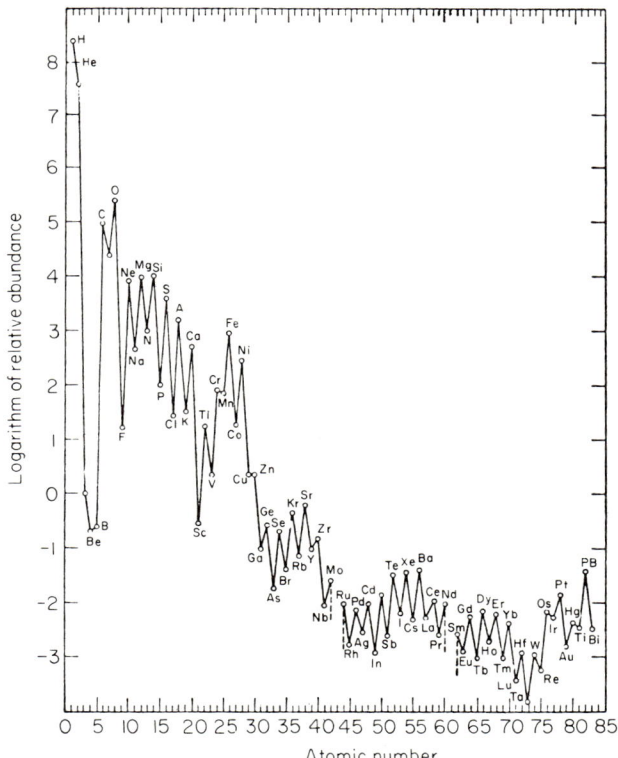

Fig. 1. Relative abundances of the elements, referred to Si = 10,000 (atoms per 10,000 atoms of Si) plotted against atomic number Z (after Ahrens [2]). Derived from terrestrial and solar data.

associated with these protons can vary. As a result, an element may have several isotopes differing in mass number or atomic weight A $(= N + Z)$, and stability, but not appreciably in chemical properties.

4. Origin of Helium

4.1. Definition and Uses of Helium

The gases of the helium group consist of helium, neon, argon, krypton, xenon, and radon. They have been referred to as rare gases, noble gases, inert gases, helium group, group 0 (zero), or VIII elements. Helium is the lightest member of the rare gas family. It is an inert, monoatomic, colorless, odorless, tasteless, inactive gas that is a good conductor of heat and electricity. It is the only substance that remains a gas below $-252.70\,°\text{C}$. It liquefies at −268.8°C., and freezes only when placed under pressure and cooled to −272.0°C. When liquid helium is cooled to within 2°C. of absolute zero, it assumes a different physical state—a superfluid almost completely devoid of viscosity. In this state it is capable of flowing through the smallest openings and becomes a superconductor of heat and electricity.

Helium owes its utility to its unique physical properties. It diffuses more rapidly, flows through a hole faster, transmits sound at a higher velocity than any other gas except hydrogen, and it conducts heat and electricity better than any other gas except neon [17]. Compared to other gases, helium has a much lower solubility in water than other gases, a lower refractive index, and a lower temperature of liquification. These properties are useful for many industrial and scientific applications, such as cryogenics, space flights, leak detection, arc welding, superconductivity, and medicine, to name a few.

4.2. Theories of Helium Origin

A lack of helium compounds under natural conditions makes it necessary to study the origin and occurrence of helium in association with the substances which accompany it in nature. The most widely accepted theory on the origin of terrestrial helium claims the production of this inert gas by alpha emitters present in terrestrial materials. However, the possibility of partial entrapment of cosmogenic or primordial helium during the degassing stages of the earth should be considered.

4.3. Cosmogenic Helium

Cosmogenic or primordial helium was first detected in the sun's spectrum by Lockyer in 1868 (see [38]). A look at the cosmic abundance of elements in Fig. 1. reveals that helium is a major constituent of most stars. Hydrogen is the most abundant element in the universe and constitutes 76% of its total mass, while helium, the second most abundant, makes up 23% of the total mass of the universe.

Twelve to fifteen billion years ago the solar system was vastly different. At that time, it is assumed to have been a rotating mass of turbulent hydrogen gas [28]. Conversion of gravitational potential energy into thermal kinetic energy caused the triggering of exothermic nuclear reactions. These reactions caused the fusion of hydrogen into cosmogenic helium and from there on into more complicated nuclides. The fusion or burning of hydrogen "fuel" into helium "ash" occurs in all stars, particularly in regions of the highest density and temperature. This can be written as

(4.1) $$4\ ^1H \rightarrow\ ^4He + 26.7\ \text{MeV}$$

4.4. Radiogenic Helium

Radiogenic helium is an end product of the decay of a radioactive parent element. Any alpha emitter is a source of radiogenic helium. The following three reactions demonstrate the helium production from a parent element:

	Total half-lives (years)
$^{238}U \rightarrow {}^{206}Pb + 8\,{}^4He$	4.5×10^9
$^{235}U \rightarrow {}^{207}Pb + 7\,{}^4He$	7.1×10^8
$^{232}Th \rightarrow {}^{208}Pb + 6\,{}^4He$	1.4×10^{10}

All the natural alpha emitters which are the sources of radiogenic helium are listed in Table I.

Henceforth in the following discussions, "helium" designates radiogenic helium. For cosmogenic or primordial helium, the proper prefix will be used.

4.5. Association of Helium in Natural Gases

Rogers [62], reviewing the occurrence of helium, and considering the different possibilities regarding the origin of helium contained in natural gases, concluded that either: (1) the helium is derived from deposits of uranium or thorium, probably disseminated through the strata not far beneath the horizons at which the helium-bearing gas occurs; or (2) the helium is derived from sources at great depth.

In favor of the first theory, Rogers pointed out: (1) the existence of such deposits is not unreasonable from the geologist's point of view; (2) if it is admitted that a considerable quantity of helium can escape from the mineral in which it forms, and can migrate upward, the size of the parent deposit should be finite; and (3) such deposits would account for the areal and also the stratigraphic distribution of helium.

In Dobbin's opinion [22], the occurrence of many of the rich helium gases in reservoirs lying just above basement rocks, is due to highly radioactive elements in the basement rocks.

Keesom [38] suggested that helium might have originated from the disintegration of the hypothetical element 87. To this, the concept of a generation of helium by short-lived radioactive elements now totally extinct may be added.

Pierce [55] pointed out that although there have been only a few studies on the geologic occurrence of helium since Rogers [62], there is an increasing volume of data pointing toward a general acceptance of Rogers' assumption that most of the helium in natural gas is radiogenic. Data on the distribution and composition of the natural gases suggest that three-fourths of the helium in the Panhandle gas fields migrated into the field from the Palo Duro basin.

Allegedly the helium there was derived from uraniferous rocks that are faulted against the gas-producing reservoir rocks along the western boundary of the Panhandle gas fields. In addition sedimentary rocks, and especially the asphaltic cement in the sandstones, could be a significant source of the helium in these fields.

Some relatively low concentrations of helium in the natural gases of the Panhandle fields may have migrated from the Anadarko basin to the East, where source rocks of the natural gases probably contained traces of uranium and thorium. Radioactivity from these traces yielded alpha particles, or helium, which migrated into the Panhandle reservoir with the hydrocarbon gas.

5. Helium-Generating Potential of Rocks and Fluids

Radioactivity is the principal element in the geochemistry of helium. Unlike other rare gases, helium is formed as an end product of radioactive decay. It is continually being produced in the earth's crust by the disintegration of uranium, thorium, and other elements including polonium-210, radium-226, and samarium-150, all of which emit alpha particles (see Table I). Since uranium and thorium are by far the most abundant radionuclides, other elements which are radioactive are not further considered. Moreover, as the ratio of thorium to uranium is well documented in the literature [25, 27, 33, 43], we need to concern ourselves only with uranium.

Helium-generating potential, as defined here, is the potential or ability of certain elements or materials to generate helium; i.e., the helium-generating potential of any substance is the measure of its alpha-emission activity. For example, the ability of uranium and thorium to generate helium is 1.16×10^{-7} ml of helium per gram of uranium per year, and for thorium, 2.43×10^{-8} ml of helium per gram per year [60]. Alpha activity is difficult to measure because of its absorption by exceedingly thin layers of material, measured in a few microns in a solid. For this reason, gamma-ray activity is generally an acceptable substitute because it can be related to the alpha activity.

5.1. Procedure

Most of the samples checked for specific radiation activity (SRA) were chosen from the American Rock Collection (ARC) of Ward's Natural Science Establishment, Inc. This collection was assembled from the more common rocks found on the North American continent and a general classification of the rocks was utilized. Furthermore, each rock type in this collection; e.g., granite, is a sample which has average petrographic characteristics of all the granites.

Representative samples of crushed rocks of uniform particle size were weighed to an approximate 50.0 gm, to obtain the specific radiation activity of each sample.

5.2. *Equipment*

The data reported below were analyzed on a 400-channel pulse height analyzer (Model 402 of Technical Measurement Corporation), connected to an X-Y recorder (Model 2D2), with data print-out on an IBM typewriter.

Samples were placed on a sodium iodide (NaI) thallium-activated crystal mounted in a lead shield. As the gamma rays given off by the sample impinged on the crystal, a flash of light entered a photomultiplier tube. Each flash of light from the crystal struck a photosensitive surface on the inside face of the multiplier tube causing it to emit electrons and change it to an electric signal.

These signals were counted and stored in 200 channels of the 400-channel pulse height analyzer. The other 200 channels were used to store 40 min of background reading made prior to the running of the sample. This 40-min background count was substracted from the total gross count to obtain the net count emitted by the sample.

The results, plotted by the X-Y recorder with counts per channel vs. channel number, are shown in Fig. 2. Channels are numbered from left to

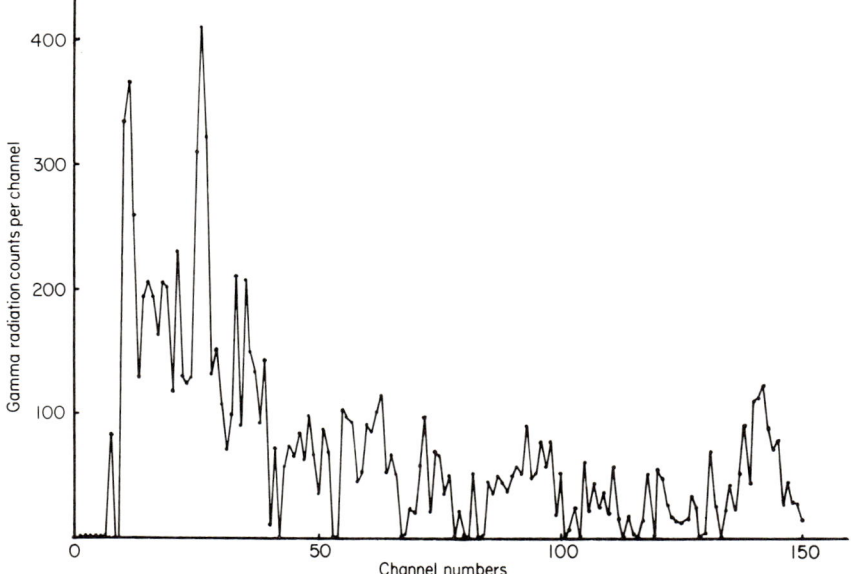

Fig. 2. The net radiation diagram of alkali granite, ARC 13, from Norfolk County, Massachusetts (Mississippian). The diagram illustrates the total net gamma-ray counts received in each of the channels.

ARC Number 13, Alkali Granite, Norfolk County, Massachusetts
(Mississippian) - Weight = 50.0 gm, 40 minute count.

```
00000 00000 99998 00000 99946 00001 00083 99970 99965 00335
00366 00259 00133 00194 00206 00194 00144 00206 00203 00119
00233 00130 00125 00129 00310 00411 00321 00134 00153 00108
00074 00100 00211 00090 00208 00151 00134 00093 00143 00010
00072 99951 00058 00074 00067 00084 00063 00099 00067 00036
00088 00070 99990 99995 00103 00097 00093 00045 00054 00091
00085 00102 00115 00052 00067 00051 99984 00002 00025 00020
00059 00099 00021 00071 00067 00035 00050 99989 00022 00004
99999 00053 99996 00002 00046 00036 00049 00043 00038 00051
00058 00052 00110 00048 00052 00079 00057 00079 00019 00053
00002 00006 00025 99990 00061 00022 00044 00024 00037 00020
00058 00015 99978 00019 00003 00001 00014 00052 00002 00056
00049 00027 00018 00017 00015 00017 00037 00025 00000 00004
00070 00025 99996 00030 00043 00023 00053 00095 00045 00112
00113 00124 00088 00073 00079 00027 00045 00029 00027 00015
00006 00000 99969 00004 00011 00028 99997 99991 00010 00016
00009 00006 00015 00015 00019 00004 00007 99992 00015 99999
99999 00007 99986 99998 00001 99986 99999 00010 00005 99990
00007 00017 00006 00014 00008 99998 99991 99996 00005 99999
00020 00010 99992 99994 00008 00003 00005 00002 00004 00006
```

	Left Cut	Right Cut	Spe. Rad. Act.
Peak on channel 33	74	90	0.0735
Peak on channel 35	90	93	0.1095
Peak on channel 39	93	72	0.1590
Peak on channel 44	58	67	0.0057
Total Specific Radiation Activity			0.34770

FIG. 3. Results of radiation counts recorded in each channel during the 40-min time period after subtraction of the background radiation.

right in sequence, as shown in Fig. 3. The greater the channel number, the higher the energy of the gamma-ray particle measured in MeV. Fig. 2 illustrates the total net gamma-ray count emitted during the 40-min time interval in each channel of the pulse height analyzer. Typed results of net counts received in each channel of the analyzer during the 40-min time are numerically tabulated (Fig. 3). Numbers listed in Fig. 3 correspond to the counts per channel as shown on the graph of Fig. 2.

5.3. Theory of Specific Radiation Activity

Mathematically, a body may be described as being homogeneous if the value of some measured parameter varies randomly from place to place. Random, as defined here, is the type of frequency distribution exhibited by randomly varying lithologic properties which are dependent on the nature of that property, i.e., specific radiation activity (SRA).

The fundamental theory of radioactive decay is expressed by the number of atoms of radioactive matter which have disintegrated after a unit time. These in turn are proportional to the number of atoms which were not influenced by the disintegration process during the unit time.

The number of atoms which have disintegrated after a unit time is proportional to the gamma-ray counts. Also, the number of atoms which have disintegrated after a unit time is proportional to the percent of uranium in the sample. Therefore the gamma-ray counts should be a measure of the percent uranium in the sample.

Two samples of a known U_3O_8 value were run on the pulse height analyzer to determine their respective specific radiation activity. Calibration of a scintillation counter with known SRA values due to uranium indicated that in the region between channels 33 and 44 the least interference was involved. Channels between 33 and 44 were chosen to determine the specific radiation activity due to uranium.

5.4. Helium-Generating Potential of Igneous Rocks

The chief naturally occurring radioactive elements are uranium and thorium, both of which are widely distributed in rocks and in mineral deposits. The helium-generating potential of igneous rocks, or any other rocks, may be calculated if their uranium and thorium concentrations are known.

As previously stated, only uranium will be considered here because the above described equipment does not readily discriminate thorium from other radioactive elements such as potassium. However, the general ratio of thorium to uranium of three to one is a fact well established in the literature [1,3,15]. For this reason, the uranium content suffices for determining helium-generating potential.

Table III shows the SRA for 19 igneous rocks, representing most major rock types. Fifteen of these came from the ARC, and the remaining four from private sources.

Knowledge of the distribution of uranium and thorium in igneous rocks, which make up most of the earth's crust, is fundamental to an understanding of the helium generating potential of these rocks. The distribution of uranium and thorium is not yet well documented [25], because many problems of sampling, analysis, and interpretation remain. Although most of the fresh igneous rocks have a uranium content lower than 7 ppm, some contain as much as 200 ppm uranium [1]. A high content of uranium and thorium usually can be correlated with other composition. Some igneous rocks, especially those at or near the acid end of the series, are usually significantly more radioactive than other rock types.

TABLE III. Specific radiation activity and uranium percent for igneous rocks

ARC	Locality (County, state)	Age	Lithology	SRA	Percent U($\times 10^{-5}$)
1	Lake, Oregon	Miocene	Obsidian	0.20175	5.9798
2	Klamath, Oregon	Pleistocene	Scoria	0.17800	5.2759
6	Chaffee, Colorado	Tertiary	Rhyolite porphyry	0.55050	16.3166
11	Washington, Vermont	Devonian	Biotite granite	0.17650	5.2314
12	Merrimack, New Hampshire	Late Paleozoic	Muscovite-Biotite granite	0.68500	20.3032
13	Norfolk, Massachusetts	Mississippian	Alkali granite	0.34770	10.3057
14	Essex, Massachusetts	Mississippian	Hornblende granite	0.29200	8.6548
18	Teller, Colorado	Post-Oligocene	Syenite	0.14725	4.3644
26	Stoarns, Minnesota	Precambrian	Granodiorite	0.48650	14.4197
29	Siskiyou, California	Late Tertiary	Hornblende andesite	0.13900	4.1199
31	Essex, Massachusetts	Early Paleozoic	Diorite	0.34450	10.2109
33	Chaffee, Colorado	Eocene	Vesicular basalt	0.17850	5.2907
35	Somerset, New Jersey	Triassic	Basalt	0.15200	4.5052
37	Hampshire, Massachusetts	Triassic	Diabase	0.14225	4.2162
41	Essex, Massachusetts	Early Paleozoic	Hornblende gabbro	0.23475	6.9580
	Flagstaff, Arizona	Recent	San Francisco peaks lava flow	0.71911	21.3142
	Apache, Arizona	Pliocene	Syenite from Dineh bi Keyah oil field	0.32435	9.6136
	Hilo, Hawaii	Recent	Kilauea iki lava flow of 1960	0.27725	8.2175
	LaPaz, Bolivia	Late Tertiary	Basalt from Andes Mts.	0.17800	5.2759

The variations in the radioactivity of igneous rocks are systematic and are relatable to the chemical, mineralogical, petrographic, and structural features of the rocks. However, Heinrich [34] has demonstrated that the change in the uranium content of successive members of a magma series is less systematic than the change in the major elements. This variability may stem from (1) diverse paths open to trace elements during magmatic differentiation, (2) differences inherent in sampling techniques of igneous rocks, and (3) differences in postcrystallization histories of the rocks. Therefore, one can conclude that the uranium and thorium content of igneous rocks is a function not only of the initial concentration of these elements, but also the postcrystallization histories of rocks.

The differences between the values taken from the literature [1,27] and those shown in Table III seem to be due to calibration of the gamma-ray spectrometer, interference from other radioactive sources such as radium and potassium, and the uranium used as a standard.

Nearly all igneous rocks contain trace amounts of uranium [21]. Much of the radioactivity in igneous rocks is concentrated in mildly radioactive common accessories such as zircon, sphene, and apatite. Highly radioactive minerals (monazite, allanite, uraninite, and thorite) though much scarcer than the common accessories, may·be of widespread yet spotty distribution in igneous rocks. These accessory minerals in granitic rocks are normally more erratically distributed throughout the rock than are the essential minerals. In general, rocks which are rich in magnesium, iron, and calcium, and poor in silica and alkalies, tend to be poor in radioactive elements; i.e., their helium generating potential is lower. Consequently, the helium generating potential of igneous rocks decreases from acidic to intermediate to basic types, respectively.

Unlike the major rock-forming minerals, the amounts of thorium and uranium contained in the solid phases of the mineral assemblage forming at any particular stage, depend not only upon concentration of those elements in the magma, but also upon the rate of precipitation of suitable host minerals, such as zircon, sphene, and apatite [48].

A detailed petrographic analysis of igneous rocks of the ARC group indicates that the sources of radioactivity due to uranium and thorium were from: (1) moderately radioactive accessory minerals, (2) weakly radioactive essential minerals, (3) interstitial material along grain boundaries and molecular structural defects of minerals, (4) fluid inclusions in minerals, and (5) intergranular fluids. A maximum concentration of uranium and thorium is found in the youngest members of a series, regardless of the composition of the original magma. Most of the igneous rocks greatly enriched in uranium are late-stage differentiates of granite or syenite. These rocks may approach pegmatitic composition, although not always the texture.

5.5. Geochemical Controls of Uranium and Thorium Fractionation

During most of the magmatic cycle, both uranium and thorium are in a tetravalent state, and the crystallization paths of both elements are parallel because of close similarity in ionic radii. During the intervals of crystallization when the water content of the melt is very low, magmas saturated with either constituent probably do not exist. Therefore, no precipitation of discrete uranium or thorium minerals takes place. Instead, their ionic radii and charge dictate that part of the U^{4+} and Th^{4+} which are present will enter appropriate host minerals. The U^{4+} and Th^{4+} affinity is probably for zirconium in zircon, and for calcium in apatite and sphene.

Uranium in unaltered igneous rocks occurs as U^{4+}. The uranium and thorium ions are concentrated in late magmatic fractions and in accessory minerals largely because their relatively large ionic radii hinder their entrance into the structure of the common essential silicates. Other factors governing the distribution of uranium and thorium are their low initial concentrations in magmas and their high valence.

Nearly all compounds of uranium and of thorium are known to be relatively insoluble in aqueous solutions in the laboratory [25]. Possibly at some time during the " granite " stage, the buildup of water may so reduce the solubility of both uranium and thorium that sporadic precipitation of actual uranium and thorium minerals may occur. As differentiation proceeds to the highly hydrous pegmatite stage, more and more uranium forms discrete minerals and less and less enters common accessory minerals. Also, the low concentration of the two elements prevents the precipitation of phases in which they are principal constituents. They may also be concentrated in residual fluids to crystallize finally in the pegmatitic stages.

5.6. Helium-Generating Potential of Sedimentary Rocks

Gamma-ray spectrometry yielded the uranium content of 24 samples of sedimentary rocks. The samples represented various geologic ages. Of the 24 samples, 13 came from the ARC and the remaining 11 were from various other sources. Table IV illustrates the results of the analyses of these sedimentary rocks.

Almost all the uranium contained in the earth is believed to be present in the upper lithosphere, preferentially concentrated in acidic igneous rocks, and in part as discrete uranium minerals. Some uranium exists in extremely thin intergranular films formed by the solidification of the last residual traces of magmatic fluid [34].

Weathering of all types of rocks releases uranium which may be deposited later in many kinds of sediments and in rocks. Uraninite and pitchblende, which occur as primary constituents of some pegmatites, are readily altered

Table IV. Specific radiation activity and uranium percent for sedimentary rocks

ARC	Locality (County, state)	Age	Lithology	SRA	Percent U ($\times 10^{-5}$)
51	Livingston, New York	Devonian	Calcareous shale	0.15675	4.6460
52	Monroe, New York	Silurian	Argillaceous shale	0.21125	6.2614
53	Greene, New York	Devonian	Arenaceous shale	0.23350	6.9209
54	Greene, Pennsylvania	Pennsylvanian	Carbonaceous shale	0.24438	7.2434
58	Cuyahoga, Ohio	Mississippian	Gray sandstone	0.18100	5.3648
60	Middlesex, Connecticut	Upper Triassic	Micaceous sandstone	0.20925	6.2021
61	Ulster, New York	Devonian	Fine grain sandstone	0.16880	5.0032
62	Hampshire, Massachusetts	Triassic	Feldspathic sandstone	0.14600	4.3274
63	Wyoming, New York	Devonian	Argillaceous sandstone	0.16625	4.9276
69	Oneida, New York	Ordovician	Argillaceous limestone	0.05900	1.7487
71	Erie, New York	Devonian	Cherty limestone	0.11800	3.4975
74	Monroe, New York	Silurian	Dolomitic limestone	0.05775	1.7117
75	Kent, Michigan	Mississippian	Gypsum	0.21400	6.3429
	Canon City, Colorado	Cretaceous	Uraniferous Dakota sandstone	8.65650	256.5761
	Major, Oklahoma	Permian	Gypsum	0.06300	1.8673
	Murray, Oklahoma	Ordovician	Dolomite	0.03600	1.0670
	Murray, Oklahoma	Ordovician	Asphaltic sandstone	0.33701	9.9889
	Murray, Oklahoma	Ordovician	Sylvan shale	0.10975	3.2530
	Carter, Oklahoma	Ordovician	Viola limestone	0.18799	5.5720
	Comanche, Oklahoma	Cambrian	Glauconitic sandstone	0.19550	5.7946
	Comanche, Oklahoma	Permian	Anhydrite	0.03750	1.1115
	Comanche, Oklahoma	Permian	Halite	0.08000	2.3712
	Comanche, Oklahoma	Ordovician	Simpson sandstone	0.00000	0.0000
	Rogers, Oklahoma	Pennsylvanian	Coal	0.33500	9.9293

during weathering processes to hydrated oxides, phosphates, and silicates; and some uranium is leached from them probably as soluble uranyl complexes.

Some uranium- and thorium-bearing minerals, such as monazite, are not readily altered in place but are readily reduced by abrasion during transportation with other clastic sediments. Some other uranium-bearing minerals are minor primary constituents of igneous rocks, and are resistant to mechanical disintegration and to chemical decomposition. These minerals, of which zircon is the most abundant, accumulate in placers and in the heavy mineral fractions of clastic sediments. Most secondary uranium minerals are susceptible to alteration and leaching, and are quickly reduced by abrasion during transportation with clastic sediments.

Uranium, which is dissolved in surface and ground water during weathering of rocks, may be redeposited nearby or may be carried into the drainage system and ultimately into the oceans. It is partially removed from aqueous solutions by precipitation as insoluble compounds, by adsorption on several kinds of sediments, and by substitution for calcium and possibly other elements deposited in chemical sediments.

The adsorption appears to be minimal when deposition occurs in waters containing either free oxygen or carbonate ions, and will increase markedly as the content of oxygen and carbonate decrease. Uranium is readily dissolved by the processes of weathering, and significant amounts of uranium are adsorbed by gel precipitates, by clay minerals, and by organic sediments. For this reason, no highly concentrated deposits are to be expected by direct deposition from weathering processes, rather, they are generally dispersed in sedimentary formations.

The bulk of the uranium deposits in the United States are found in the sedimentary rocks of Jurassic age on the Colorado Plateau, mostly in clastic sediments of terrestrial origin. Most of this uranium mineralization occurs as pore fillings and impregnations as cement, and as replacement of organic material.

5.7. Types of Uranium Concentration in Sedimentary Rocks

Uranium will be deposited epigenetically in all types of sediments and rocks. It may be introduced by hydrothermal solutions or by ground water, and will be deposited where evaporation or the chemical environment causes precipitation. Probably, some uranium will be removed from aqueous solutions by adsorption in clays and other sediments.

The nature of uranium deposits in sediments and in sedimentary rocks falls into three categories: (1) indigenous (native) to clastic sediments, (2) deposited syngenetically with the sediments, and (3) deposited epigenetically after sedimentation ceases. It can be postulated that probably all clastic sediments contain some indigenous uranium. Syngenetic deposition of uranium occurs in many clays, evaporites, and carbonaceous sediments. Epigenetic deposition of uranium is found in all types of sediments and sedimentary rocks. Some sediments may contain uranium deposits resulting from two, or possibly all, of these categories.

The uranium that is indigenous to clastic sediments will be contained in grains and fragments of discrete uranium minerals; partly in substituted form in other minerals; and possibly in part, in intergranular films between the crystalline constituents of clastic particles of igneous rocks. Most of the indigenous uranium content of clastic sediments exists as heavy minerals which consist of chemically undecomposed residues of weathering, such as sand and gravel-size particles.

Uranium may be removed from solutions by direct precipitation as insoluble compounds, or by adsorption in a variety of sediment types. Compounds of uranium may be precipitated from solutions saturated by evaporation, and they may be precipitated from solutions which have become strongly reducing. Uranium in the form of carnotite in coal is believed to be a result of the reducing environment produced by the organic content of the coal.

5.8. Helium-Generating Potential of Metamorphic Rocks

The helium generating potential for seven metamorphic rocks is shown in Table V. Practically no literature is available on the distribution of uranium and thorium in metamorphic rocks. Hence it is not possible to compare and contrast the results shown here with other results. However, the generalizations made on the helium generating potential for the two previous rock groups probably applies to the metamorphic rocks.

Variations in the helium generating potential of metamorphic rocks are a function of the uranium and thorium concentration of the parent rock which has undergone metamorphism. Hence a complete history of each rock would be essential to evaluate properly the values shown here.

Intense heat radiated from an intruding magma causes contact metamorphism. Regional metamorphism affecting broad areas is the result of both heat and pressure upon deeply buried rocks. In both types of metamorphism, fluids in the rock will augment the chemical changes.

Water is the principal fluid, but in addition such chemical elements as chlorine, fluorine, boron, and others may emanate from intrusive masses and react with the surrounding rocks. Many minerals are stable only within limited ranges of pressure, temperature, and chemical equilibrium. Upon metamorphism, changes including crystallization and chemical recombination can be expected to occur.

TABLE V. Specific radiation activity and uranium percent for metamorphic rocks

ARC	Locality (County, state)	Age	Lithology	SRA	Percent U ($\times 10^{-5}$)
79	Worchester, Massachusetts	Precambrian	Biotite gneiss	0.11450	3.3937
81	Clinton, New York	Precambrian	Hornblende gneiss	0.27050	8.0175
83	Manhattan, New York	Precambrian	Mica schist	0.31150	9.2328
86	Mitchell, North Carolina	Precambrian	Hornblende schist	0.02200	0.6520
91	Washington, New York	Cambrian	Red slate	0.09850	2.9195
93	Sauk, Wisconsin	Precambrian	Quartzite	0.11950	3.5419
96	Pickens, Georgia	Cambrian	Pink marble	0.09250	2.7417

5.9. Helium-Generating Potential of Fluids

Commonly, as much as 40% of the uranium in most fresh-appearing igneous rocks is readily leachable [25]. The exact distribution of this radioactivity is largely unknown. Leachable radioactivity may occur (1) in altered structure of primary silicates, (2) as interstitial materials derived from late magmatic or hydrothermal solutions, (3) in partly soluble radioactive accessories, and (4) as adsorbed ions in disseminated weathering products such as iron oxide, and also at grain boundaries.

The distribution of uranium, however, may be modified by the leaching action of surface and ground waters, resulting in some cases in a diminished uranium content. This uranium diminution is pointed out by Heinrich [34] who believed that in particular cases a diminished phosphate content may accompany somewhat increased uranium content. Moreover, it is believed that connate water and petroleum can partially leach uranium and thorium in sedimentary rocks. This leaching is believed to have taken place subsequent to migration of the petroleum and brine solutions into structural and stratigraphic-type traps. Following migration and entrapment these fluids have the potential to generate helium because they have become alpha-particle emitters.

The existence of uranium and thorium in percolating ground waters and thermal springs is well documented [1]. The petrographic study of some uranium-bearing ores makes this generalization possible: the inverse relation between grain size of the apatite nodules and their uranium content is constant for a given mining district. Any downward movement of uranium-bearing solutions would enrich the finer grain fractions.

In order to determine whether the fluids such as brines, ground water (fresh), and crude petroleum and natural gas have the ability of transporting uranium and generating helium, representative samples of such fluids were examined by gamma-ray spectrometry. The equipment shown in Fig. 4 consisted of a hollow cylindrical lucite tube core holder, a reservoir tank, and a manometer. A hypothetical unconsolidated core consisting of a homogeneous mixture of Simpson sand (Ordovician age from central Oklahoma), and uraniferous Dakota sandstone (Cretaceous age from Canon City, Colorado) was used. The Simpson sand was chosen because it is essentially free of radioactivity. Different types of fluids were injected into these synthetic cores to study the extent of dissolution and leaching of the uranium mineral from the unconsolidated sandstone.

Fluids such as tap water and crude oils of different specific gravity (rated as API gravities) were injected into separate cores for each run. The injected fluids were tested for radioactive content before injection into the core, and after they had flowed through the core. After dissolving or leaching some of

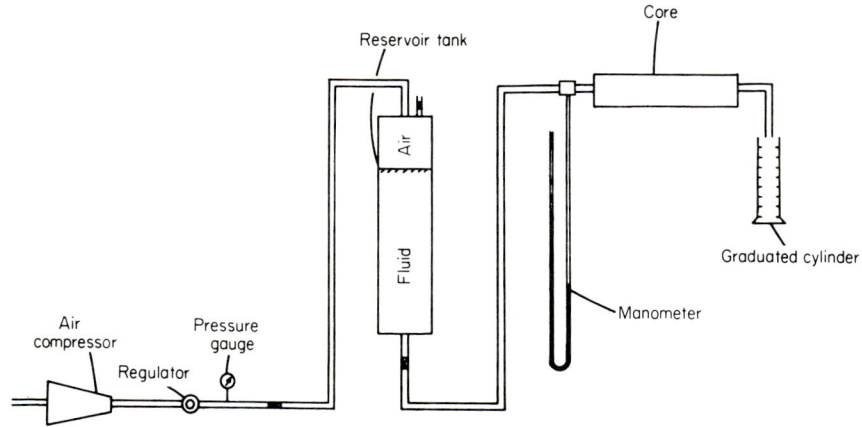

Fig. 4. Schematic drawing of the apparatus used in leaching uranium out of an unconsolidated sand mixed with radioactive material.

the uranium and thorium, each of the liquids became an alpha emitter. Therefore they demonstrated the ability to generate helium while they are trapped in a structure within the rocks of the earth. The radioactive filtrate was that fluid which was collected at the outlet of the core, and the radioactive sand residue was the material in the core after the fluid had been passed through it.

The pressure drop across the core was just sufficient to establish the flow. After collecting 1000 ml of injected fluid at the outlet of the core, 200 ml were analyzed for dissolved uranium. Table VI shows the results.

These results lead to the conclusion that all three categories of rocks and many fluids have the ability to generate helium. Obviously the magnitude of the generating potential of every rock or each fluid is a direct function of its alpha emitters. Uranium concentrations much higher, and much lower than those presented in Tables III–VI do exist. Such high concentrations however, are exceptional and local in extent.

Comparison of the data presented in Tables III–V shows that igneous rocks are generally richest in uranium content. Nonetheless almost all of the uranium mined commercially in the United States is from sedimentary rocks. The Morrison formation (Jurassic age) of western Colorado is the major uranium-bearing formation. It is a well-bedded sandstone carrying complex oxides of uranium, thorium, and vanadium. Sediments and sedimentary rocks are thus shown to be potential helium producers. Also, fluids such as brines and crude oil moving through porous and permeable rocks that are enriched in uranium and thorium yield fluids rich in alpha-emitting elements.

TABLE VI. Specific radiation activity and uranium percent for fluid and core fillings

Description	SRA	Percent U ($\times 10^{-5}$)
Tap water, University of Oklahoma, from Garber Sand (Permian)	0.11570	3.4293
24.7 API° crude oil, from West Texas	0.00000	0.0000
37.5 API° crude oil, from West Texas	0.00353	0.0104
Radioactive filtrate, tap water after passing through the core	0.18125	5.3722
Radioactive filtrate, 24.7 API° crude oil after passing through the core	1.95818	58.0399
Radioactive filtrate, 37.5 API° crude oil after passing through the core	0.37970	11.2542
Radioactive sand residue—after flow of tap water	35.96760	1066.0688
Radioactive sand residue—after flow of 24.7 API° crude oil	34.33749	1017.7529
Radioactive sand residue—after flow of 37.5 API° crude oil	36.44626	1080.2562

Some crude oils are appreciably radioactive [34], and abnormal concentrations of radioactive elements and their decay products including uranium and thorium have been found in some oil fields of the United States. Although only one of the crude oils used in this study showed traces of radioactivity, crude oil has been shown to leach uranium compounds from the host rock.

The uranium and thorium in petroleum may be present as colloidal particles or as dissolved metallic–organic complexes. If they are present in colloidal form, the uranium content of the source rock would have to be relatively large. If they occur in solution, it is expected that they are concentrated in a surface-active fraction coating the pore walls of the reservoir rock. The latter assumption is more likely to be true because large concentrations of uranium and thorium in the reservoir rocks are improbable.

Heinrich [34] estimated that the average uranium content in petroleum is 100 gr per ton. The highly variable radioactivity of petroleum and the fact that crude oils from nonuraniferous provinces contain only negligible amounts of uranium [54] suggest an epigenetic origin for the uranium. Also, a study of compositional analyses of many natural gases from uraniferous provinces such as the Colorado Plateau confirms the fact that high helium content of these gases is a result of high uranium concentration of the reservoir rock.

The highest helium concentrations in the natural gases coincide with the presence of uraniferous asphaltites. This fact is concluded from relatively high specific radiation activity of the asphaltic sandstone (Ordovician in Oklahoma) shown in Table IV.

The association of gamma-ray anomalies, uraniferous asphaltites, radium-bearing brines [64], and uranium-bearing petroleum, suggests that the major part of the helium is radiogenic, derived from petroleum source rocks and rocks through which the fluids have moved.

The association of uranium with asphaltic materials can be explained in three general ways:

(1) The uranium was transported in petroleum which was subsequently oxidized; the uranium may have been reconcentrated subsequently. But the organic material and the uranium are essentially contemporaneous.

(2) The asphaltite represents material derived from petroleum that was present in the rocks prior to uranium mineralization, and subsequently served as a precipitant for the uranium. A variation of this idea states that hydrothermal solutions depositing uranium also converted petroleum to asphaltite [66].

(3) The uranium was deposited first as uraninite (UO_2), and subsequently served to polymerize petroleum or natural gas to liquid phases that partly dissolve and replace preexisting uraninite.

6. Distribution and Occurrence of Helium

Knowledge of the distribution of uranium and thorium in specific regions of the outer crust of the earth is fundamental to the understanding of the potential helium generators or alpha emitters, and the earth's heat flow. The abundance of the radioactive elements and the heat flow data may have a close relation to helium generating potential of crustal rocks in any specific region.

The rate of production of helium is 1.16×10^{-7} ml/gm of uranium per year, and it is 2.43×10^{-8} ml/gr of thorium per year. Uranium concentration in granites is estimated to be 3.2×10^{-3}–6.4×10^{-4} % [1,15,25]. Rankama and Sahama [60] stated that the helium content of igneous rocks is 0.003 gm/ton, and that the helium content of these rocks is never more than 10^{-4} ml/gm.

6.1. Extraterrestrial Helium

As introduced in Section 4, helium is present in the hotter stars, as well as in the sun's atmosphere although other inert gases are absent. A high cosmic abundance of helium may be explained by the energy-producing thermonuclear processes which take place in the interior of the stars [48], which constantly change hydrogen to cosmogenic or extraterrestrial helium.

6.2. Terrestrial Helium

Helium and its natural isotope ^{3}He are present in minerals, rocks, natural gases, sea water, thermal springs, in fluids in the earth, and in the atmosphere, in highly variable proportions. Keesom [38] has suggested that some of the ^{4}He may be primordial, however there is general agreement that most of the helium gas in the crustal rocks is a product of alpha decay.

The helium content of common rocks is quite low compared to that in minerals such as beryl and spodumene. A metric ton of granite containing 2 ppm uranium and 10 ppm thorium, will produce $0.22 + 0.29$ ml, respectively, or a total of 0.51 ml of helium at standard temperature and pressure per million years. Most of that helium will be trapped in the crystals of the rock, but some of it may be released by recrystallization, and may migrate into whatever porous formation or passageways are available. Some of the helium could flow through fractures which may have formed after the consolidation of the host rock.

The scarcity of helium in terrestrial materials presents a problem. It is generally accepted that the earth's present atmosphere is of secondary origin [49]. In the process of planetary accretion, the volatile components, including the rare gases, are likely to have been largely lost from the primitive earth.

The present atmosphere has been developed through processes of degassing, weathering, and oxygen accretion brought about by photosynthesis of plants [8a]. There seems to be little doubt that present-day terrestrial helium is the result of nuclear disintegrations during the lifetime of the earth.

6.3. Crustal Erosion

Rock weathering will release the gases occluded in the crust to the atmosphere. An early estimate of the amount of weathered rock, based on the sodium content of the oceans [16], gave a value of 8.33×10^{23} gm, and this figure was substantiated by Goldschmidt [33]. Mayne [49] has calculated the minimum amount of helium thus released in 10^9 yr to be about 2×10^{17} gm, or 1.5×10^5 atoms of helium cm^{-1} sec^{-1} for the earth's surface.

This cannot be the main source of supply of the helium for the atmosphere. Rocks may, however, lose a fraction of their gas by erosion, at least if it is radiogenic. Nevertheless it is plausible that all the rare gases in the atmosphere came from continuous degassing of the earth, not crustal erosion. In fact, there is additional loss of gas through leakage even after cooling of the earth.

6.4. Helium in Crustal Rocks

The rate at which helium is generated in the crust is determined by the uranium and thorium content of the crustal rocks. Fleischer [27] summarized all of the available data on the radium content of granitic and basaltic rocks, and from this information it has been deduced (assuming a thorium to uranium ratio of 3.5), that helium production in these rocks is approximately 3.5×10^{-3} ml gm^{-1} and 0.5×10^{-3} ml gm^{-1} in 10^9 years, respectively, for each rock type. The granitic mass of the crust is 15×10^{24} gm and the mass of basalt to a depth of 5 km underlying the continents is 7.5×10^{24} gm as estimated by Mayne [49]. Table III and other sources [1,2,15,33] confirm higher uranium content for granitic rocks than for the basaltic rocks. This leads to the conclusion that the overall helium generation in the crust is coming from the surface granitic layer.

The following abundance values given by Rankama and Sahama [60] are considered most reliable. For thorium, the average concentration for igneous rocks in general is 11.5 gm per ton. Other thorium concentrations in igneous rocks are (1) for acidic rocks, 13.0 gm per ton; (2) for intermediate rocks, 9.97 gm per ton; and (3) for basic igneous rocks, 5.0 gm per ton. The uranium concentrations in terms of grams per ton are 3.96, 2.61, and 0.96 for acidic, intermediate, and basic igneous rocks, respectively.

Thorium to uranium ratios in igneous rocks are 3.2, 4.0, 4.5 for acidic, intermediate, and basic rocks, respectively. These values illustrate the potential of helium generating rocks and also the probable amount and distribution of helium in these rocks. Finally, the sedimentary rocks have the potential to generate helium according to their local concentration within the sedimentary section. Goldschmidt [33] gave the following uranium and thorium concentrations for the sediments: arenaceous sediments, 5.4; argillaceous, 12.0; shales, 10.1; and limestones, 1.1; all in units of grams per ton thorium concentration. The uranium content of sedimentary rocks is 1.1 gr per ton as an average value for all the marine sediments.

Helium has a geological occurrence and distribution that is unique among the elements. On the average, it is continuously increasing in amount in the earth's crust, being formed at the expense of the elements that are alpha emitters. On the other hand, the crust is continuously losing helium at a rate that is less than the amount being formed by the alpha emitters. Mayne [49] has concluded that helium is generally increasing in the earth's crust. Therefore the amount of helium in the crustal rocks is related to the age of these rocks and their content of uranium and thorium.

Almost all commercially significant deposits of helium in the United States are located in southwestern Kansas, the Oklahoma and Texas panhandles, and New Mexico. The helium content of natural gases in this area varies widely, from a trace to as much as 10 %, depending on the age, geometry, and the depositional history of the reservoir rock. The common assumption is that helium contents are low in the reservoir rocks because the helium atoms diffuse through the molecular lattice structure of the rock. It is possible, however, to suggest factors or a combination of factors which may cause a variation of helium in a natural gas:

(1) The uranium and thorium concentrations in the host rock which generates the helium.

(2) The geologic age of the source rock. It is believed that uranium and thorium are more abundant in the relatively younger rocks. As the rock increases in age, the helium content, or the helium generated from that particular rock also increases. As the helium is generated, uranium and thorium concentrations will decrease.

(3) Selective leakage and retention of helium which is a function of mineral bulk density, crystal structure, grain size, and fracture characteristics. These passageways may permit transportation of the original helium *in situ* to suitable traps and therefore may bring about a variable helium ratio.

(4) Conditions such as localized high concentrations of alpha emitters in sedimentary rocks as well as in the basement rocks, could be a major factor in the high helium content of some natural gases. Also, it should be pointed out that an increase in temperature can cause helium migration to cooler environments. Magmatic intrusions, and regional or contact metamorphism are sources for such heating.

6.5. Helium in the Hydrosphere

The general presence of helium in sea water and mineral springs in minute quantities indicates only minimal solubility of this inert gas [12,13]. Helium in waters of different oceans is assumed to have come from an influx of the gas through the bottom sediments. An examination of all available data leads one to attribute the helium to the decay of uranium, thorium, and their daughter isotopes in the basaltic layer of the earth's crust underlying the oceans. The degassing of the earth may well have contributed to the amount of helium present in ocean waters. Solubility of helium in sea water decreases with a decrease in temperature and an increased salinity.

6.6. Helium in the Atmosphere

The abundance of helium in the atmosphere is based on its release from the rocks of the crust of the earth and the rate of its escape from the earth's gravitational field. Many authors [8,17,20,26] have assumed that the atmosphere is now in equilibrium with respect to helium concentration. Therefore, the rate of escape must be equal to the rate of crustal liberation.

On the basis of spectroscopic evidence, an abundance figure for helium in the universe has been estimated as 3.5×10^7 atoms per 10,000 atoms of silicon (Fig. 1). The comparable figure for the abundance of helium in the earth is only 3.5×10^{-7} [68], so there is a considerable deficiency. Because of its specific gravity there has been a steady loss of helium from the earth's surface to the atmosphere and thence to space. This loss by escape of this small atom from the atmosphere of the earth is ascribed to a gravitational field that is inadequate to retain helium. Badhwar [7] has suggested a value of 5.2 ppm for the helium content of the atmosphere.

7. Migration of Helium

The distribution of this noble gas in all parts of the universe is a clear evidence of its migratory nature. Helium atoms are so light and small that they continue to escape from their host rocks, and they do migrate through the rocks of the earth's crust. The factors influencing the rate of escape are the geometry of the rock, the physical characteristics of the environment, the temperature, and the pressures. For example, migration of helium occurs in localities where helium-rich natural gases exist with no apparent source rock containing alpha emitters. A high helium content in such a case may be explained by the fact that at one stage of the heating of the earth, helium escaped to a cooler environment and was trapped by overlying impermeable barriers.

7.1. Migration of Alpha Emitters

Measurements made on the leaching of uranium and thorium from rocks (Table VI), indicate that crude oil and other natural fluids have the ability to transfer the alpha emitters *in situ* and carry them away. The presence of minor gamma-ray activity in the normal tap water analyzed by us, and waters of different origin analyzed by other authors [1,12,59,67] confirms the fact that radioactive materials are leachable and can be partially transported away from their original site. Modes of migration are either in solution or in colloidal suspensions.

7.2. Migration of Helium at High Pressures

Amyx, Bass and Whiting [4] showed in an extension of Klinkenberg's original work [40] that porous media exhibit relatively high apparent permeabilities when subjected to low pressure gas flow. This deviation, however, as defined by Klinkenberg's relationship is

$$(7.1) \qquad K_a = K[1 + (b/P_m)]$$

where K_a = apparent permeability, K = true permeability, b = constant, and P_m = mean pressure = $\frac{1}{2}(p_1 + p_2)$. This is true only in the range of pressures below 517 cm of mercury. Above this pressure, all gases studied

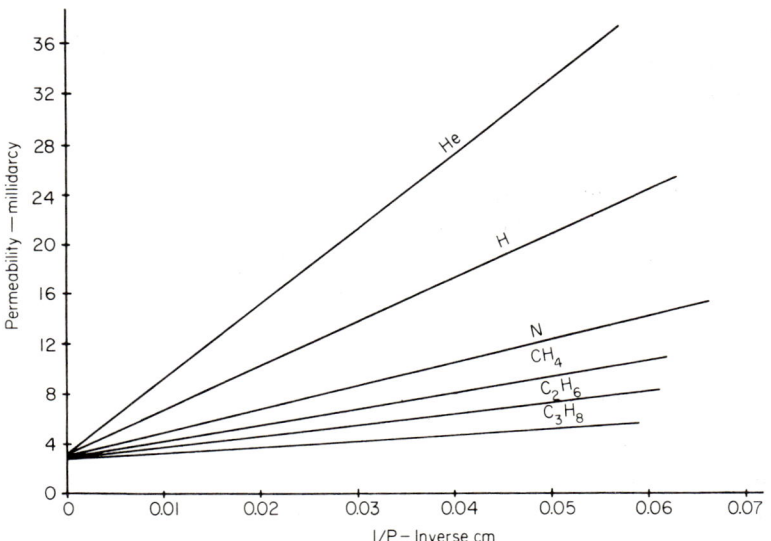

FIG. 5. Gas apparent permeabilities versus inverse of the mean pressure.

by Klinkenberg and Amyx, Bass, and Whiting flowed similarly through the permeable media. At this pressure of 517 cm of mercury, each medium yielded an essentially similar value for the absolute permeability. Figure 5 shows the apparent permeabilities as a function of mean flowing pressures for various gases.

Helium in pure state will never occur and has never occurred at pressures of 517 cm of mercury magnitude, because its generation is an extremely slow process. After production, this gas usually will migrate away from the site of its generation. Consequently, it can not be entrapped in place at a rate which permits gradual pressure buildup of 517 cm of mercury in magnitude. Moreover, if one admits such an occurrence, i.e., helium accumulated to pressures around 517 cm of mercury, it flows almost like any other gas, as indicated in Fig. 5.

7.3. Migration of Helium at Low Pressures

Migration of helium at low pressures, i.e., below 50 mm of mercury, may occur only if the rate of helium generation is exceedingly high and sufficient pore space is available. The pore space may be near or at some distance from the site of generation.

Such a geometrical situation could occur in nature, especially in sedimentary rocks where local concentrations of uranium and thorium are unusually high, and associated porosities may also be high. The helium generated there could accumulate in pore spaces of sandstones and in leached-out limestone caverns. With a slight buildup of pressure caused by accumulation of the helium itself at a rate exceeding its escape, the entrapped helium could migrate when pressures required to establish flow are reached.

To study the characteristics of this migration of helium at a low pressure of 0.262 atm (200 mm mercury), 71 core plugs were used. Most of these core plug samples had dimensions of 3.90 cm in length, and a diameter of 2.52 cm. A broad spectrum of rock types was obtained, mostly from the University of Oklahoma Core Library. Exact depths and lithologies of each sample were checked for accuracy using electrical log diagrams. Of the 71 cores, 62 were selected from the Morrow formation (Pennsylvanian) and Chester limestone (Mississippian). The remaining nine were igneous rocks—two from a well in Apache County, Arizona, and seven were from outcrops at a number of locations.

These 71 samples were analyzed to determine their transmissibility of helium at a relatively low pressure of 0.262 atmosphere. Cores from the Morrow were used in the study because this formation consists of varied compositions of limestone, shale, and sandstone, and because of the varied physical characteristics. Helium and air flow rates at a pressure differential of 0.262 atmosphere for each core were measured by soap bubble meter and stop watch at room temperature of 26.5°C. Figure 6 is a schematic drawing of the apparatus used for flow rate measurements. Porosities were determined by a digital voltmeter porosimeter at Core Laboratories, Oklahoma City, Oklahoma. The data obtained are shown in Table VII.

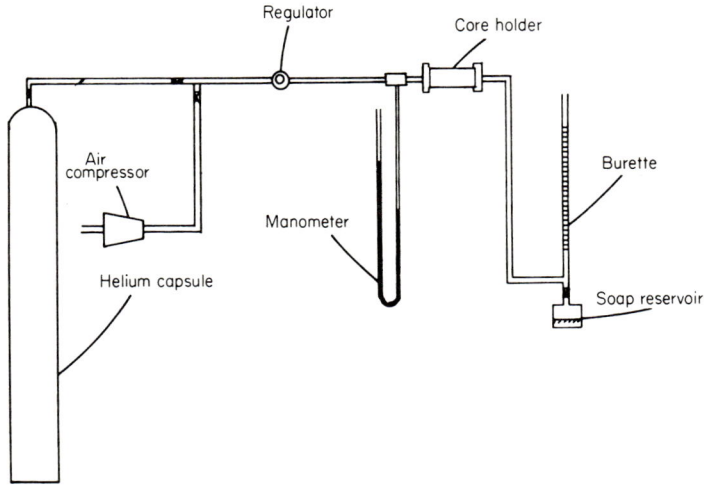

FIG. 6. Schematic drawing of the apparatus used in determining air and helium flow rates.

This table reveals that, in general, sedimentary rocks, especially sandstones, have the ability to transmit efficiently the helium stored or entrapped in them at low pressures. Crystalline sedimentary rocks and nonfractured igneous rocks are poor conductors of helium at corresponding low pressures. Figure 7 showing the helium flow rate as related to air flow rate for all the rocks studied, indicates a good correlation between these two parameters. However, there is no correlation between the air and helium flow rates and the porosities of the core plugs.

7.4. Diffusional Migration of Helium

Helium occurs widely in nature in very low concentrations. It is found in natural gases, entrapped in crystal structures, in ground waters, in deeper brines, and in the earth's atmosphere. Apparently the source of helium is a continuous process, being generated at a constant rate by the radioactive decay of larger atoms. Part of it is lost to the atmosphere, and a similar amount is lost from the atmosphere to space. Little work has been done on the mode of migration of helium within the earth to the atmosphere. It was therefore

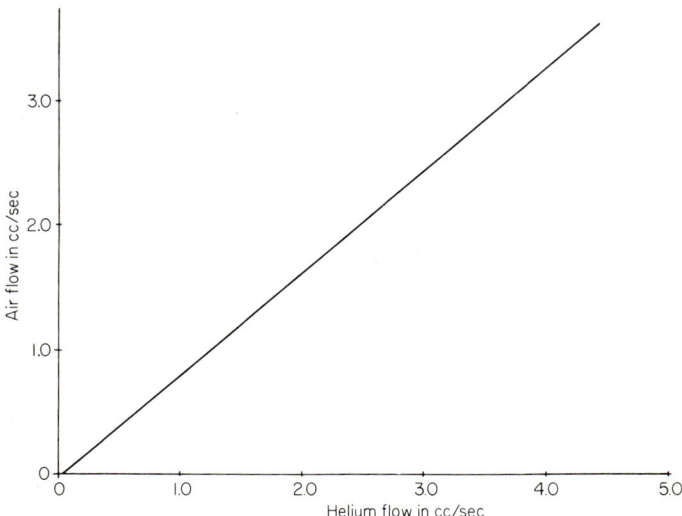

FIG. 7. Helium flow versus air flow for the cores of Table VII at a pressure differential of 0.262 atm.

interesting to examine how migration through sedimentary rocks takes place. For simplification in this phase of the investigation, the helium generated by the sedimentary rocks is neglected at the outset.

7.5. Subsurface Conditions

The dominant rocks in the earth's crust fall into two contrasted groups: a group of light colored rocks, or the sial; and a group of dark colored, heavier, basaltic rocks or sima.

For our purpose here, however, it is convenient to divide the crustal rocks into two classes different from the above named: Precambrian rocks, or "basement rocks," and the sedimentary rocks ranging in age from Paleozoic through the Tertiary and Quaternary. The above classification is one of age rather than geological nature.

Although the sedimentary rocks vary widely in character, the entire stratigraphic column may be regarded as a porous medium in which the pore size distribution varies throughout its horizontal and vertical dimensions. Basement rocks are denser than the sediments although they may be fractured and exhibit some porosity. The migration of helium from its primary sources can then be considered to take place throughout a porous–permeable medium.

7.6. Migration of Helium

It is clear that the major portion of the movement of helium will take place within the pore structure in view of the disparity between the diffusivities of helium in solids and in liquids. Because diffusion is the only means of mass transfer through the rock matrix, a comparison of diffusivity in a solid and in a liquid is indicated.

Diffusivity of helium through a solid, namely silica, is of the order of 10^{-10} cm^2 sec^{-1}, and through a liquid such as water, this factor is 10^{-5} cm^2 sec^{-1} [25]. A comparison of these diffusivity values points out that movement through the rock matrix represents only a minor portion of the overall movement of the helium. However, this portion of the helium movement is important as a controlling factor in the supply of helium to the pore structure. For example, a rock subjected to tensions resulting in microfracturing has a pore structure which presents a large surface area for the escape of helium from a rock matrix. Consequently in such regions of tension the concentration of helium in the pore spaces may be expected to be higher than that in a similar but nonfractured rock.

Within the pore spaces helium exists in two ways: as discrete bubbles alone or in association with other gases, and in aqueous or hydrocarbon solution. Consequently there are three possible modes of helium migration: migration as bubbles, migration in solution by fluid flow, and migration by molecular diffusion.

It is unlikely that helium moves as bubbles because in the first place capillary action will prevent such a movement, and secondly, the geometry of the pore structure would be expected to trap the bubbles. For these reasons, this mode of migration can be dismissed.

The possibility of migration of dissolved helium associated with fluids needs more consideration. It is known that in some layers of sediments there is a lateral flow of ground water. This movement in the shallow sediments may have a velocity as high as hundreds of meters per year. The proportion of the helium in the pore space associated with the fluid in that pore space, and therefore the proportion of the helium which can migrate in this manner, is dependent on the partial pressure of the helium.

Total pressure in a given pore is determined by the pressure of the overburden and, therefore, the partial pressure of the helium depends on the quantities of other natural gases present. Since helium is the least soluble of all the components of the natural gas, it is plausible that very little of the helium present is in solution. Moreover, the degree and extent of the flow of underground fluids are not well known, and hence it is not possible to generalize on this mode of migration.

Diffusion as a means of helium migration has been suggested by Emerson [23] to account for the variation in helium concentration with depth as associated with natural gases. The cause of diffusion is the activational energy of each helium atom resulting in random motion. If there is a higher concentration of helium atoms in a given region compared with another region, the random motion of the atoms will result in a mass transfer of the helium to the region of lower concentration. This rate of transfer is dependent upon the concentration gradient between the regions and the diffusion coefficient or diffusivity characteristic of the medium. It is therefore evident that molecular diffusion of helium does take place within the pore structure.

7.7. Diffusivity Process in Sediments

The diffusion process may be defined as the tendency toward equilibrium. It is a process which tends to establish uniform concentration. Physically, diffusivity at a point in a porous medium is defined as the ratio between the rate of helium flux per unit area of the medium perpendicular to the flow, and helium concentration gradient at that point. Therefore the diffusivity can be said to be determined by two factors: the transfer medium and the geometry of the pore structure in which the transfer medium exists.

The variation of the diffusivity with transfer medium depends on the chemical nature of the medium, its temperature, pressure, and also the concentration of helium in the medium. The diffusivity of a porous medium may be illustrated quantitatively by the following example: in a uniformly granular system, if the granules are spherical, the diffusivity of the system is independent of the size of the spheres. However, if the granules are disk-shaped, then the diffusivity varies widely in relation to the dimensions and thickness of the disks.

7.8. Simplification of Diffusivity Factors for Analysis

The mechanism of migration of helium through sedimentary rocks is a complex problem. Some simplification is required before mathematical techniques can be applied. Up to this point, the problem has been reduced to the local migration of helium from basement rocks as a source, through overlying uniform layers of sediments in the stratigraphic column, all of which is considered to be a porous medium. The mechanism of migration which is considered the most probable is diffusion, and as a consequence of the local uniformity, lateral effects can be ignored and the problem considered to be one-dimensional.

Every layer could be characterized by a gross or average diffusivity and the complete sediments are assumed to exist at the time of the origin of the diffusion process which evidently is contemporaneous with the start of sedimentation. This then is an assumption to reduce the extremely complex problem to a simpler form. The above simplifications are called for by uncertainties and lack of knowledge of so many factors. It should also be recalled that helium generated from the sediments is ignored, and the initial concentration of helium in the sediments is zero.

7.9. Mathematical Models

From the molecular theory of gases [35] one can derive the diffusion equation of gases as follows: Consider a slab of rock which has the thickness dx as shown in Fig. 8. If q is the net flux or net diffusion through the slab,

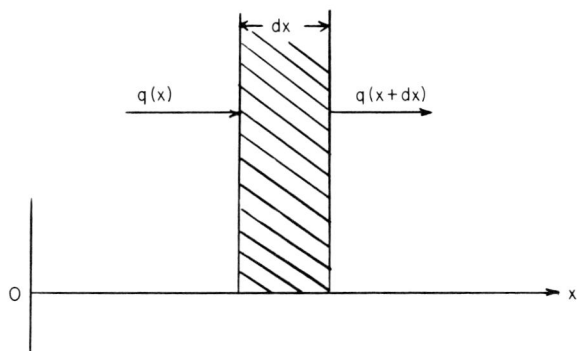

FIG. 8. Helium flux through a sedimentary slab of thickness dx.

which is a model of a single formation or of a number of formations with uniform diffusivity t, the time, and c is the concentration of helium, one could write

$$(7.2) \qquad (\partial c/\partial t)\, dx = q(x) - q(x+dx)$$

Applying Taylor expansion on the term $q(x+dx)$, yields

$$(7.3) \qquad q(x+dx) \cong q(x) + (\partial q/\partial x)\, dx$$

substitution of $q(x+dx)$ term from Eq. (7.3) into Eq. (7.2) results in

$$(7.4) \qquad (\partial c/\partial t)\, dx = q(x) - [q(x) + (\partial q/\partial x)\, dx]$$

Equation (7.4) can be simplified to

$$(7.5) \qquad (\partial c/\partial t)\, dx = -(\partial q/\partial x)\, dx$$

But the net flux or net diffusion q is defined as

$$(7.6) \qquad q \equiv -D(\partial c/\partial x)$$

Differentiating Eq. (7.6) with respect to x

$$(7.7) \qquad \partial q/\partial x = -D(\partial^2 c/\partial x^2)$$

and substituting Eq. (7.6) into Eq. (7.5) will yield

$$(7.8) \qquad (\partial c/\partial t)\, dx = D(\partial^2 c/\partial x^2)\, dx$$

which simplifies to the model proposed here, a general unidimensional molecular diffusion equation of the form

(7.9) $$D(\partial^2 c/\partial x^2) = \partial c/\partial t$$

Applying a Laplace transformation [35] on the variable t gives

(7.10) $$\partial^2 \bar{c}/\partial x^2 = (p/D)\bar{c} = R^2 \bar{c}$$

where $R^2 = p/D$ using the assumption made in the beginning that initial concentration of helium in the sediments is zero. Here D is the diffusion coefficient, c the concentration of helium and $\bar{c}$ its Laplace transform, x is the distance from the basement, t is time, and p is the Laplace transform parameter.

Equation (7.10) will have solutions of the form

$$\bar{c} = A \cosh Rx + B \sinh Rx$$

where A and B are constants to be determined by the boundary conditions of the problem. These boundary conditions will take one of the following forms:

(a) $c = 0$ for $t \geq 0$ at some $x = x_0$
(b) $c = C_0 \exp(-\beta t)$ for $t \geq 0$ at some $x = x_0$
(c) $\partial c/\partial x = 0$ for $t \geq 0$ at some $x = x_0$

where c is the concentration of helium which is a function of time and distance, x is the distance measured from Precambrian basement rocks, t is the time measured from beginning of sedimentation, and C_0 and β determined by the abundances and decay constants of the source elements.

The extent of sedimentary layers, their physical characteristics such as grain shape, homogeneity, rock type, porosity, permeability, and diffusivities are quite different, hence no attempt is made to consider this diffusional problem further. Moreover, the form of the boundary conditions in specific regions are essentially unknown. Also, some local conditions such as temperature and pressure gradients which directly affect the diffusion rates are highly variable factors.

In addition to that generated in basement rocks, if one considers the helium generated by the sediments Eq. (7.10) becomes

(7.11) $$D(\partial^2 c/\partial x^2) + C_p = \partial c/\partial t$$

where $c = f_1(x, t)$, $C_p = f_2(x, t)$, and c is the concentration at any point due to the helium generated from igneous rocks and C_p is the helium derived from sedimentary rocks. Here the effects of temperature and pressure were neglected in the analysis of diffusional migration.

8. Accumulation or Entrapment of Helium

As a result of radioactive disintegration, each alpha particle, after the loss of its original energy, will behave as an atom of helium gas, being principally occluded in the interstices of the rock matrix. There will be movement of helium from the rock particle surface in the course of geologic time due to energy from diastropic movements and from rock metamorphism, and the mobile gas will migrate upward whenever channels of permeability are afforded by the stratigraphic circumstances.

The concept that helium in the petroleum gases is derived from relatively shallow igneous rocks containing high proportions of uranium and thorium minerals is not always borne out. In many areas helium and nitrogen gases do occur in reservoirs relatively close to basement rocks, such as along the granite ridge of the buried Nemaha Mountains in central Oklahoma, but in other areas such as northwestern Oklahoma, the gas from shallow zones contain a larger percentage of helium than do those from deeper strata.

This decrease of helium content with depth may be a result of local migration, but in general, there is no universally clear correlation between helium content and local radioactivity. Pierce [55] suggested that the unusually large volumes of helium in the Panhandle gases of Texas and Oklahoma are due to the occurrence of residual deposits of radioactive petroleum deposits.

Helium will have a tendency to escape from deeply buried rocks at some stage of increase in temperature. If this transient helium were accumulated and trapped by impermeable barriers in a cooler environment, there would be a helium-rich gas such as is present in the panhandles of Texas and Oklahoma.

8.1. Ability to Hold Helium

From consideration of helium flow rates in Table VII, it is apparent that the ability to hold helium varies for different rock types. Data presented in Table VII show that shales and so-called cap rocks are least permeable to helium. Also, crystalline limestones, anhydrites, and nonfractured igneous rocks exhibit an ability to retain helium in a structural or stratigraphic trap. Such rock types are given the name of "helium holders."

Paleozoic rocks that are relatively close to basement rocks may contain some helium that was derived from the basement rocks. The helium was diffused vertically and locally trapped in these impermeable beds. These rocks are highly compacted, and act as a seal to prevent the flow of helium from structural or stratigraphic traps. Fluids such as oil, or oil and water may also serve as an imperfect seal. Because accumulations of high concentrations of helium require long periods of time, it follows that higher helium contents may be associated more often with the natural gases from the older geological formations.

TABLE VII. Air and helium flow through selected core plugs

Code number	Core cut drilled depth[a]	Location (county in Oklahoma)	Lithology	Air flow (ml/sec)	Helium flow (ml/sec)	Percent porosity
			Selected igneous and metamorphic rocks			
3	H	Dineh bi Keyah Oil Field, Apache County, Arizona	Syenite	0.000	0.000	11.5
44	V	Same	Syenite	0.000	0.000	9.8
29			Bytownite gabbro	0.000	0.000	0.8
55			Gabbro	0.000	0.000	1.1
65			Syenite	0.000	0.000	0.6
45			Biotite granite	0.000	0.000	1.2
46	ARC 13	Norfolk, Massachusetts	Alkali granite	0.000	0.000	0.7
10			Pyritic schist	0.091	0.102	5.5
36			Gneiss	0.000	0.000	1.6
		Selected cores from Morrow formation, N.W. Oklahoma. Counties are named.				
34	H 4852	Beaver	Gray calcareous shale	0.000	0.000	1.5
57	V 4852	Beaver	Gray calcareous shale	0.000	0.000	1.1
22	H 10330	Blaine	Gray calcareous shale	0.009	0.009	6.8
47	H 4425	Texas	Gray calcareous shale	0.000	0.000	3.0
6	H 9147	Woodward	Dark gray calc. shale	0.000	0.000	3.7
25	H 9244	Woodward	Dark gray calc. shale	1.063	1.200	1.9
38	H 7777	Woodward	Gray shale	0.107	0.172	10.4
48	V 9147	Woodward	Calc. sandy shale	0.000	0.000	1.6
50	H 9219	Woodward	Gray shale	0.000	0.000	3.8
67	V 9219	Woodward	Gray shale	0.000	0.000	2.4
28	H 7794	Beaver	Calcareous sandstone	0.000	0.000	3.6

GEOCHEMISTRY AND GEOLOGY OF HELIUM

31	V 7794	Beaver	Calcareous sandstone	0.000	0.000	4.4
70	V 7867	Beaver	Calcareous sandstone	0.000	0.000	1.7
4	V 7010	Harper	Light gray sandstone	2.500	2.520	15.4
5	H 7010	Harper	Light gray sandstone	2.500	2.520	14.7
11	H 7000	Harper	Light gray sandstone	0.184	0.189	15.4
14	V 7000	Harper	Light gray sandstone	0.117	0.166	18.8
19	H 6959	Harper	Calcareous shaly sand	0.055	0.072	12.3
21	H 5670	Harper	Arkosic sandstone	0.040	0.260	7.4
23	V 6959	Harper	Slightly calc. sand	0.020	0.030	11.6
24	H 6946	Harper	Light green sandstone	0.023	0.028	13.9
32	H 5667	Harper	Light gray dense sand	0.000	0.000	7.1
39	H 7010	Harper	Gray shaly sand	0.000	0.000	0.3
52	V 5667	Harper	Arkosic sandstone	0.000	0.000	7.0
53	V 7010	Harper	Gray shaly sand	0.000	0.000	1.6
56	V 6946	Harper	Light gray dense sand	0.000	0.000	14.7
69	V 5670	Harper	Coarse light gray sand	0.000	0.000	7.5
1	V 6304	Texas	White coarse sand	1.000	1.163	17.1
7	V 4436	Texas	Light gray dense sand	0.000	0.000	9.1
8	V 4444	Texas	White coarse sand	1.662	1.785	21.6
9	H 6314	Texas	White coarse sand	1.315	1.350	16.3
12	H 4422	Texas	White coarse sand	4.170	5.000	20.2
13	V 6314	Texas	White coarse sand	1.315	1.570	16.6
15	H 4722	Texas	Light gray dense sand	0.006	0.008	9.6
16	H 4444	Texas	White coarse sand	0.526	0.625	21.8
17	V 4722	Texas	White dense sand	0.052	0.065	6.9
20	V 6332	Texas	White coarse sand	1.665	2.000	18.6
26	H 6304	Texas	White coarse sand	0.735	0.860	17.5
35	H 4436	Texas	Light gray dense sand	0.000	0.000	8.1
51	V 4423	Texas	White coarse sand	4.150	5.000	19.2

(Continued)

V, vertically; H, horizontally.

TABLE VII.—*Continued*

Code number	Core cut drilled depth[a]	Location (county in Oklahoma)	Lithology	Air flow (ml/sec)	Helium flow (ml/sec)	Percent porosity
		Selected cores from Morrow formation, N. W. Oklahoma. Counties are named.				
54	H 4431	Texas	White sandstone	0.000	0.000	9.6
61	V 4431	Texas	White sandstone	0.000	0.000	7.6
2	V 7806	Woodward	Light green shaly sand	0.011	0.016	19.8
18	H 7806	Woodward	Fine dense sandstone	0.023	0.030	17.9
33	H 7835	Woodward	Fine dense sandstone	0.000	0.000	3.4
40	H 8668	Woodward	Fine dense sandstone	0.000	0.000	3.1
41	H 9199	Woodward	Light gray calcareous sand	0.000	0.000	0.6
43	V 7835	Woodward	Fine dense sand	0.000	0.000	4.0
49	V 8667	Woodward	Fine dense sand	0.000	0.000	3.9
64	H 9214	Woodward	Gray calcareous sand	0.000	0.000	6.7
66	H 9217	Woodward	Gray dense sand	0.051	0.054	2.8
71	V 9217	Woodward	Gray dense sand	0.000	0.000	4.2
42	H 7867	Beaver	Gray shaly limestone	0.000	0.000	1.6
37	H 6977	Harper	Gray sandy limestone	0.000	0.000	3.2
68	V 6977	Harper	Gray sandy limestone	0.000	0.000	3.8
27	H 8633	Woodward	Dense crystalline limestone	0.000	0.000	0.8
58	V 8635	Woodward	Dense crystalline limestone	0.000	0.000	0.2
59	V 9199	Woodward	Gray crystalline limestone	0.000	0.000	1.2
62	V 9214	Woodward	Gray crystalline limestone	0.000	0.000	6.5
		Selected cores from Chester Limestone, N.W. Oklahoma				
30	H 8721	Woodward	Light dense limestone	0.000	0.000	3.1
60	V 8721	Woodward	Light dense limestone	0.000	0.000	1.2
63	V 8712	Woodward	Light dense limestone	0.000	0.000	3.7

[a] V, vertically; H, horizontally.

8.2. Effect of Radiation

From the studies by Faul [25] and Funkhouser and Naughton [30] on helium loss and retentivity, one can assume that the rate of helium escape is in proportion to the damage of molecular structure and the amount of helium present. Most of the radiogenic helium present is produced on grain boundaries and is consequently diffusible. The emission of the alpha particles which subsequently produce helium atoms causes structural and mechanical damage to the molecular lattice of the alpha emitter. One can conclude that the radiogenic gases such as helium and argon may be trapped in these lattice defects caused by radiation damage.

8.3. Conditions Favoring Helium Accumulation

Rogers [62] in his classic work, emphasized the fact that helium was most likely to be found at shallow depths. However, Silurian gases in Vinton County, Ohio, contain a greater quantity of helium than do gases from Mississippian beds at the same locality [11]. At Pinta Dome field, Arizona, the deeper zone, 315 m, contains gas with 8.09 % helium content, but the shallow zone, at 230 m, contains 6.98 %.

Recent data from Keyes field in Oklahoma, the Hatcher field in Colorado, and other localities, seem to contradict the above conclusions. A chart prepared by Pierce [54] clearly indicates that the helium content of natural gases increases with the age of the gas-producing reservoir. Although it is not feasible to develop a generalized relation between maximum helium generated from given rock types and the geologic conditions necessary for the highest helium concentration, the following conclusions seem warranted:

(1) Helium accumulations may be found in the strata of any geologic age, but they appear to be most concentrated in Paleozoic rocks where presumably the disintegration of radioactive elements has been progressing for longer periods of time than in the younger beds.

(2) Helium accumulation need not be restricted to shallow depths. However, because the diffusive qualities of helium increase with increasing pressures and temperatures, it may be that the efficiency of the seal over the helium-containing reservoir must increase in proportion to depth to some critical point. At this point even the most efficient sedimentary rock seal is not able to prevent the diffusion of helium.

8.4. Retentivity and Diffusivity of Rocks

Some common-rock forming minerals do not retain all the helium generated in their structures during their geological history. According to Goldschmidt [33], the retentivity of feldspar is low, whereas minerals with molecular close-packed structures such as magnetite, pyroxenes, and amphiboles, appear to have the ability to retain helium.

The size of the helium atom is large compared with the vacant spaces in the molecular lattice of minerals which retain helium better than others. If in the lattice the volume of the space unoccupied by ions is large, the diffusion of helium in such lattices takes place more rapidly than in those with a closely packed structure. These selective retention and loss factors are significant causes of variable helium content in minerals and hence in rocks.

"Excess" helium is that which can not be explained by accumulation during the decay of the known radioactive elements. Excess helium and other rare gases such as argon, in minerals such as beryl [19] are probably a consequence of the presence of helium in magmatic gases in the immediate environment of crystallization and may form inclusions. This excess amount could not have been derived from the selective retention of helium in beryl.

9. Natural Gas Analyses

Most of the chemical analyses of natural gases in the United States are performed and published by the U.S. Bureau of Mines in Information Circulars, Report of Investigations, and Bulletins. In addition, a number of authors [27,31,71] have analyzed natural gases and have reported their compositions in mole percent of components, i.e., methane, ethane, propane, nitrogen, helium, and argon.

This chapter will review the relationship between the abundances of helium and argon in natural gases of different composition, environment, and age. Composition of helium-rich gases is variable and in some gas fields a correlation has been suggested between the helium and nitrogen content.

The helium, nitrogen, and argon content of 40 natural gases from Zartman's work [71] are presented in Table VIII. Nitrogen values vary from 0.1 to 42.5%. The origin of nitrogen in these natural gases has been explained in terms of (1) the introduction of atmospheric nitrogen, (2) the release of nitrogen by bacterial decomposition of nitrogeneous compounds, (3) the release of nitrogen by the inorganic chemical breakdown of organic compounds, and (4) the liberation of inorganic (possibly radiogenic) nitrogen from igneous rocks.

9.1. Helium–Argon Ratio

If essentially all the helium which is contained in natural gases is produced by radioactive disintegration of uranium and thorium, and if the radiogenic factor of argon is produced by the electron capture decay of potassium, a number of theoretical calculations can reveal the values to be expected for the radiogenic helium and argon abundances, and their ratio. The abundances of helium and argon observed in natural gases can be obtained by assessing the variation of helium and argon production with time, with the natural distribution of uranium, thorium, and potassium in the rocks, and the process of gas migration and accumulation into gas reservoirs.

A number of gas transport mechanisms could also produce a fractionation of helium and argon. If during the time of gas accumulation the helium and argon were transported in solution in connate water, fresh water, or in petroleum, the helium to argon ratio would be influenced by the relative solubilities of these gases in the transporting medium. The solubilities of argon and helium in crude oil are unknown in this important transmitting medium, but an elaboration on helium in crude oil solutions follows. Argon is not considered in this next section because our focus is on helium and only traces of argon have been reported from the gases which were studied.

9.2. Helium–Crude Oil Solution

An experiment to study the mechanism by which a gas is dissolved by a liquid, especially the solution of helium in crude oil, was carried out in a PVT cell using crude oil of 37.5° API gravity, and pressures as high as 310 atmospheres at a room temperature of 25.5 °C. Understanding involves the concept of vapor pressure of a solution. A kinetic-molecular description of the manner in which the equilibrium vapor pressure is established has been given by Barrow [9], in which equilibrium was described as the balancing of the rate of evaporation with the rate of condensation.

The vapor pressure is a function of the kinetic energy of the molecules and the attractive forces which exist between them. In our case the system is an oil reservoir, and is so thermal that any consideration of the kinetic energy of the molecules can be omitted and attention can be focused on the intermolecular forces of attraction.

Four types of contributions to intermolecular attractions can be recognized in ordinary liquids [9]. These are dipole–dipole attraction, the induction effect, London dispersion forces, and hydrogen bonding. The first two require the molecule to exhibit a dipole moment, while the last force of attraction requires the molecule to contain hydrogen with a partial positive charge, or an electron-rich atom as the result of bonding.

As helium neither exhibits a dipole moment nor is electron-rich, these three forces are absent. The London dispersion forces result from instantaneous dipole moments and are extremely weak in the case of helium. Consequently, the intermolecular forces of attraction on helium are very weak and it has little tendency to condense at reservoir temperatures. As a result, helium vapor pressures are very high, in fact, they may be abnormally high. Crude oil, on the other hand, contains many dipole moments and much hydrogen bonding which causes its vapor pressure to be much less than that of helium.

TABLE VIII. Natural gas analyses from different localities

No.	State	County	Depth (ft) to producing horizon	Geologic system	N$_2$ (%)	He (ppm)	A (ppm)	Producing zone's lithology
1	California	Glenn	3400	Cretaceous	1.6	47.5	57.4	ss
2	California	Sacramento	4100	Eocene	1.57	41.2	52.2	ss
3	California	Sacramento	5400	Eocene	0.86	37	67	ss
4	California	San Joaquin	2340	Eocene	30.97	96.6	140	ss
5	California	San Joaquin	2500	Eocene	3.76	101	88.2	ss
6	California	Sutter	5400	Cretaceous	9.15	37.6	53.1	ss
7	California	Yolo	2500	Eocene	2.65	85	125	ss
8	Colorado	Moffat	8075	Paleocene	4.03	101	32	ss
9	Colorado	Moffat	4885	Paleocene	2.40	63	39	ss
10	Colorado	Moffat	2851	Eocene	2.30	140	118	ss
11	Colorado	Wallace	5006	Pennsylvanian	36.6	26000	1400	ss
12	Louisiana	Webster Parish	8554	Jurassic	1.0	152	26.5	ss
13	Louisiana	Webster Parish	9174	Jurassic	0.87	151	42.0	ss
14	Michigan	Hillsdale	3755	Ordovician	22.2	1350	360	lm
15	Mississippi	Forest	7303	Cretaceous	1.48	359	81.9	ss
16	New Mexico	Harding	2075	Permian	0.12	44.5	28.0	sh + ss
17	New Mexico	Harding	2074	Permian	0.15	46.6	29.2	sh + ss
18	New Mexico	Lea	5330	Permian	1.98	348	57.3	ss

#	State	County	Depth	Age				Rock
19	New Mexico	Lea	6420	Permian	2.6	480	66.8	ss
20	New Mexico	Lea	3900	Permian	3.4	69.4	117	lm
21	New Mexico	Rio Arriba	2630	Cretaceous	0.94	172	46.1	ss
22	New Mexico	San Juan	6366	Pennsylvanian	42.5	62200	5630	lm
23	New York	Erie	3011	Silurian	5.6	1640	77.1	ss
24	Oklahoma	Cimarron	4628	Pennsylvanian	24.9	22600	1080	ss
25	Pennsylvania	Cameron	6627	Devonian	0.5	203	6.8	ss
26	Pennsylvania	Mercer	590	Mississippian	1.9	1575	35.8	ss
27	Pennsylvania	Venango	1902	Devonian	0.64	805	14.9	ss
28	Texas	Coke	5200	Pennsylvanian	6.5	1720	376	ss + lm
26	Texas	Crocket	8200	Pennsylvanian	2.5	593	117	lm(reef)
30	Texas	Hartley	3200	Permian	16.6	9370	877	dolo
31	Texas	Moore	2560	Permian	8.95	4180	470	dolo
32	Texas	Moore	3525	Permian	9.70	4480	482	dolo
33	Texas	Moore	3000	Permian	9.6	4170	461	dolo
34	Texas	Moore	3000	Permian	15.2	7000	710	dolo
35	Utah	Carbon	3100	Permian	0.60	232	79.5	ss
36	Utah	Carbon	3100	Permian	0.55	187	65.4	ss
37	Wyoming	Sweetwater	5400	Cretaceous	2.08	75	27	ss
38	Wyoming	Sweetwater	2375	Eocene	1.22	158	73.4	ss
39	Wyoming	Sweetwater	2605	Cretaceous	1.80	757	66.8	ss
40	Wyoming	Uinta	2768	Cretaceous	3.09	152	17.5	ss

Next it is important to consider the theory and results of the dependence of the relative liquid composition on the component vapor pressures. Ideal gases are found to conform to an ideal relation expressed by Dalton's law: "the total pressure of a mixture of ideal gases is equal to the sum of the pressures of the components."

Since liquids exist only because of molecular interactions, no such ideal liquid solutions can be expected in the same sense as an ideal gas solution. Some solutions behave in a simple enough manner, however, to warrant the use of the term "ideal solution." Such solutions obey Raoult's law which states: "the solution of gas within a liquid is directly proportional to the pressure exerted by the gas above the liquid." This linear relation can be expressed as

$$P_a = x_a P_b^\circ \tag{9.1}$$

where P_a and P_b are the vapor pressure of a and b above a solution of mole fraction x_a and x_b, and the vapor pressures of the pure components are P_a° and P_b°.

Since the partial pressure of a gas is proportional to the number of moles of the gas per unit volume, the mole fractions of the vapor can be written as

$$x_{a \text{ vap}} = \frac{P_a}{P_a + P_b} \quad \text{and} \quad x_{b \text{ vap}} = \frac{P_b}{P_a + P_b} \tag{9.2}$$

or

$$x_{a \text{ vap}} = \frac{x_a P_a^\circ}{P_a + P_b} \quad \text{and} \quad x_{b \text{ vap}} = \frac{x_b P_b^\circ}{P_a + P_b} \tag{9.3}$$

The ratio of the mole fractions of the vapor components in the liquid is therefore given as

$$\frac{x_{a \text{ liq}}}{x_{b \text{ liq}}} = \frac{x_{a \text{ vap}}}{x_{b \text{ vap}}} \frac{P_b^\circ}{P_a^\circ} \tag{9.4}$$

This expression can be used to calculate the composition of an ideal solution in equilibrium with a vapor of any composition. The qualitative result which should be noted is that the liquid will be relatively richer if P_a° is less than P_b°; i.e., if a is less volatile than b.

From the expression in Eq. (9.4) and the knowledge that the vapor pressure of pure helium is much greater than that of crude oil, it is obvious that very little helium is in the liquid state, or

$$\frac{x_{\text{He liq}}}{x_{\text{oil liq}}} = \frac{x_{\text{He vap}}}{x_{\text{oil vap}}} \frac{P_{\text{oil}}^\circ}{P_{\text{He}}^\circ} \tag{9.5}$$

Using the realistic vapor pressures of 50 cm of mercury for crude oil and infinite pressure for helium (since helium is above its critical temperature at reservoir conditions) one could see that the ratio of helium to crude oil is infinitely greater in the vapor phase than in the liquid phase:

$$(9.6) \qquad \frac{\infty}{50} \frac{(x_{He})}{(x_{oil})_{liq}} = \frac{(x_{He})}{(x_{oil})_{vap}}$$

Consequently, no helium will go into solution with crude oil in the reservoir.

9.3. Noble Gas Abundances in Natural Gases

The helium content of the gas samples examined varies between 37 and 62,200 ppm (1 ppm $= 10^{-4}$ %) and the radiogenic argon content varies between 6.8 and 5630 ppm. Because any attempt to explain the occurrence of the rare gases in natural gases must account for the absolute amounts and concentrations as well as for the ratio of radiogenic helium to argon, factors affecting these abundances need scrutiny.

Whereas the helium to argon ratio is only slightly time dependent when considered over times comparable to the age of the earth, the actual production of these gases is strongly time-dependent. The radiogenic helium and argon content of a natural gas reservoir is not necessarily proportional to the age of the source rock, but is, rather, a complicated function of the accumulation history of the gas.

It is possible that much of the radiogenic gases are incorporated into the natural gases by a "sweeping-up" effect during the time of migration from source to the reservoir rock [71]. In such an event, the helium and argon content of the rocks at the time that they were traversed by the accumulating gases would be an important factor. The sweeping effect of liquids such as connate water and petroleum could also partially dissolve the uranium, thorium, and potassium of the rocks through which fluids migrated.

Following the entrapment of these fluids, due to their insoluble nature, helium and probably argon, escape out of the solution and will enter into the gas cap which lies directly above the entrapped petroleum. However, if there were little helium or argon to escape before gas migration, the length of time between the consolidation of the rock and the petroleum accumulation would determine the concentration of helium and argon in the pore spaces. Studies of a number of oil and gas fields [44] have indicated that this time between source rock deposition and petroleum migration may vary up to hundreds of millions of years.

9.4. Helium Distribution in Natural Gases

More than 4000 natural gas analyses reported by the U.S. Bureau of Mines include data on helium. These gases come from a wide variety of geological formations. The large number of analyses permits a significant frequency curve. Zartman et al. [71] used 3000 analyses of the samples containing less than 0.8 % helium. Because an excessive number of samples that were analyzed were from known helium-producing areas, all analyses reporting over 0.8 % helium were discarded in their study. Results of Zartman's study [71] indicate the following statistical parameters:

Mean   = 0.2170 % σ = standard deviation = 1.59
Median = 0.0610 % μ = mean of original population = 2.80
Mode   = 0.0049 %

Figure 9 taken from Zartman [71] illustrates the frequency curve and

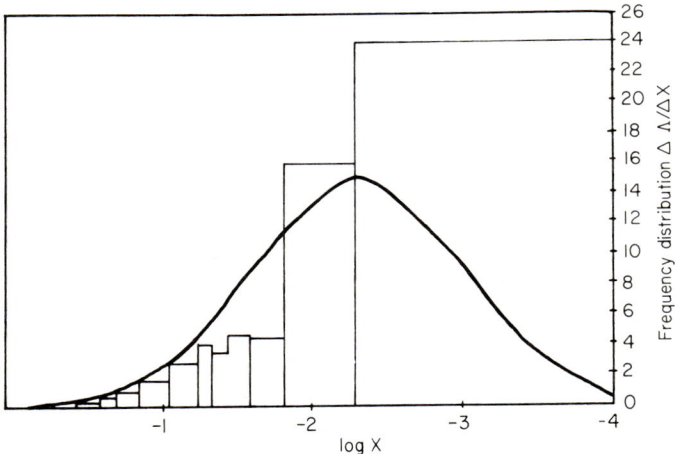

FIG. 9. Frequency curve and histogram showing the distribution of helium in natural gas.

uses these parameters for the frequency function:

(9.7) $$d\Lambda(x)/dx = (1/2\pi\sigma x) \exp -[(\ln x - \mu)^2/(2\sigma)^{1/2}]$$

This curve may be compared with the histogram which is unimodal in character. From this observation, one can conclude that the high helium content does not, in a statistical sense, represent low-probability events on the tail of a continuous probability curve. It follows that the helium, argon, and other radiogenic gases in natural gases have been obtained from rather average rock types.

The variation in abundance of radiogenic helium and argon in the gases studied has probably resulted from leakage and entrapment factors, solubility, porosity and age of the source rocks. Major differences in the abundances of helium may be caused by large variations in the abundances of uranium, thorium, and alpha-emitting fluids in the source rocks.

10. Conclusions

The following conclusions have been drawn from the study of generation, migration, distribution, and entrapment of helium:

Uranium concentration in all the major rock types and many fluids has been determined by using a 400-channel pulse height analyzer. By knowing the quantity of uranium in any sample, the amount of helium generated by that sample could be estimated.

Although elements like radium and polonium are also alpha emitters, as well as uranium and thorium, their abundances are relatively much smaller than the uranium and thorium. Inasmuch as uranium to thorium ratios are documented in the literature, only uranium was considered.

Thorium to uranium ratio of three to one is as related earlier. Considering the generation rates of helium from these two elements, one can conclude thorium has almost an equal potential to generate helium as does uranium.

Uranium content of the sedimentary rocks examined, shown in Table IV, indicates that these rocks are potential helium generators, obtaining their radioactivity mainly from the cementing materials and a possible asphaltic content. In igneous and metamorphic rocks, radioactivity due to uranium and thorium stems from the accessory minerals.

Alpha emitters such as uranium, thorium, and radium, as well as helium, have the ability to migrate.

The radioactivity of igneous rocks increases from basic to intermediate and acidic types, respectively. Sedimentary rocks have radioactive contents which increase from limestones, the least radioactive, through sandstones, shales, and to clays which are the most radioactive.

Helium accumulations are generally found in the beds of any geologic age with no restriction on depth or concentration of alpha emitters.

The wide variation in helium content of many natural gases is due to the combination of many factors which govern its production, entrapment, and escape. These variables are: the local concentration of alpha emitters, selective retentivity, diffusivity, geometry of migration passages, and age of the host rock. Moreover, variation in abundances of radiogenic helium and argon are due to factors of leakage, porosity, and solubility.

Natural gases that are rich in helium are generally rich in nitrogen and argon. The ratio of helium to argon in the analyses of the gases studied suggests that these two gases exhibit a radiogenic origin. However, ratios and correlations are not as well demonstrated for nitrogen as for argon.

There is no universal and clear correlation between helium content and local radioactivity. It is believed, nevertheless, that the helium in most of the natural gas deposits has been derived from the alpha emitters within or surrounding the host rocks.

Examination of the samples from Dineh bi Keyah oil field, Apache County, Arizona, indicates a relatively high specific radiation activity. Its impermeability to helium, and its high uranium content, may be correlated with the high helium content of 6.2 % of the natural gas produced from that field.

Based on the flow rate studies of different rock types, shales and caprocks are the best helium holders or seals, to retain helium.

The most probable mode of migration of helium is molecular diffusion from the rock matrix. Also, fluids such as ground water, connate water, and petroleum have the ability to carry helium as they migrate through rocks. Joints and fractures, up to a size of a major fault, are possible transmitting channels of helium that was generated from deeper, basement rocks to overlying or adjacent rocks.

Although solubilities of uranium and thorium are low in petroleum and brines, a favorable solubility factor could produce a highly radioactive fluid that could in turn yield extremely high helium contents.

Subsequent migration of petroleum and brines into structural and stratigraphic traps may cause the transportation, or the leaching of certain radioactive minerals, in these fluids. Upon entrapment of such fluids, helium will evolve from the solution because of its insoluble nature, and will be trapped in overlying gas deposits.

References

1. Adams, J. A. S., and Lowder, W. M. (Eds.). (1964). "The Natural Radiation Environment," 1069 pp. Univ. of Chicago Press, Chicago, Illinois.
2. Ahrens, L. H., Press, F., Runcorn, S. K., and Urey, H. C. (eds.). (1965). "Physics and Chemistry of the Earth," Vol. 6, 510 pp. Pergamon Press, New York.
3. Aller, L. H. (1961). "The Abundance of the Elements," 283 pp. Wiley (Interscience), New York.
4. Amyx, J. W., Bass, D. M., and Whiting, R. L. (1960). "Petroleum Reservoir Engineering," pp. 498–500. McGraw-Hill, New York.
5. Anderson, C. C., and Hinson, H. H. (1951). Helium-bearing natural gases of the U.S. Analysis and Analytical Methods. *U.S. Bureau Mines Bull.* **486**.
6. Axford, W. I. (1968). The polar wind and the terrestrial helium budget. *J. Geophys. Res.* **73**, No. 21, 6855–6859.
7. Badhwar, G. D., Deney, C. L., and Kaplon, M. F. (1969). Differential energy spectrum of proton, helium nuclei, and electron, *J. Geophys. Res.* **74**, No. 3, 744–754.
8. Baranov, V. I., *et al.* (1967). The age of the earth's crust and dynamics of radiogenic gas supply to the atmosphere. *Geochem. Intern.* **4**, No. 6, 1121–1129.
8a. Berkner, L. V., and Marshall, L. C. (1967). The rise of oxygen in the earth's atmosphere with notes on the Martian Atmosphere. *Advan. Geophys.* **12**, 309–331.

9. Barrow, G. M. (1966). "Physical Chemistry," 843 pp. McGraw-Hill, New York.
10. Barsukov, O. A., et al. (1965). "Radioactive Investigation of Oil and Gas Wells." Macmillan, New York.
11. Beebe, B. W. (Ed.). (1968). Natural gases of North America. *Amer. Ass. Petr. Geol., Memoir* **9**, 2.
12. Bieri, R. H., et al. (1967). Geophysical implication of the excess helium found in Pacific Ocean. *J. Geophys. Res.* **72**, No. 10, 2497–2511.
13. Bieri, R. H., et al. (1968). Noble gas contents of marine waters. *Earth Planet. Sci. Letter* **4**, No. 5, 329–340.
14. Carman, P. C. (1956). "Flow of Gases through Porous Media," 182 pp. Academic Press, New York.
15. Cherdyntsev, V. V. (1961). "Abundance of the Chemical Elements," 304 pp. Univ. of Chicago Press, Chicago, Illinois.
16. Clarke, F. W. (1924). The data of geochemistry (fifth edition). *Geol. Surv. Bull.* **770**, 841.
17. Cook, G. A. (1961). "Argon, Helium and the Rare Gases," Vols. 1 and 2, 818 pp. Wiley (Interscience) New York.
18. Craft, B. C., and Hawkins, M. F. (1959). "Applied Petroleum Reservoir Engineering," 437 pp. Prentice-Hall, Englewood Cliffs, New Jersey.
19. Damon, P. E., and Kulp, J. L. (1958). Inert gases and the evolution of the atmosphere. *Geochim. Cosmochim. Acta* **13**, 280–292.
20. Damon, P. E. (1957). Terrestrial helium. *Geochim. Cosmochim. Acta* **11**, 200–203.
21. Day, F. H. (1963). "The Chemical Elements in Nature," 372 pp. Harrap, London.
22. Dobbin, C. E. (1953). Geology of natural gases rich in helium, nitrogen, carbon dioxide, and hydrogen sulfide, in Geology of natural gas symposium. *Amer. Ass. Petr. Geol.* pp. 1053–1072.
23. Emerson, D. E., Stroud, L., and Meyer, T. O. (1966). The isotopic abundances of neon from helium bearing natural gas. *Geochim. Cosmochim. Acta* **30**, 847–854.
24. Evans, R. D. (1955). "The Atomic Nucleus," 972 pp. McGraw-Hill, New York.
25. Faul, H. (Ed.). (1954). "Nuclear Geology," 414 pp. Wiley, New York.
26. Ferguson, E. E., et al. (1965). A new speculation on terrestrial helium loss. *Planet. Space Sci.* **13**, 925–928.
27. Fleischer, M. (Ed.). (1962). The data of geochemistry (sixth edition). *U.S. Geol. Surv. Prof. Paper*. **440**.
28. Fowler, W. A. (1964). The origin of the elements. *Proc. Natl. Acad. Sci. U.S.* **52**, 524–548.
29. Fox, W. F. (1966). Relation of natural gas analyses to geology and reservoir parameters in Beaver County, Oklahoma. M. Engr. Thesis (unpublished). Univ. of Oklahoma, Norman, Oklahoma.
30. Funkhouser, J. C., and Naughton, J. J. (1968). Radiogenic helium and argon in ultramafic inclusions from Hawaii. *J. Geophys. Res.* **73**, No. 14, 4601–4607.
31. Gerling, E. K. (1967). Argon isotopes and helium in natural hydrocarbon gases. *Geochem. Intern.* **4**, No. 3, 498–506.
32. Goldman, D. T. (1965). "Nuclides and Isotopes. Chart of the Nuclides," 8th ed., Educational relations. General Electric Co., N.Y.
33. Goldschmidt, V. M. (1954). "Geochemistry," 730 pp. Oxford Univ. Press (Clarendon), London and New York.
34. Heinrich, E. W. (1958). "Mineralogy and Geology of Radioactive Raw Materials," 654 pp. McGraw-Hill, New York.
35. Hirschfelder, J. O., Curtiss, C. F., and Bird, R. B. (1964). "Molecular Theory of Gases and Liquids," 1249 pp. Wiley, New York.

36. Howard, R. A. (1963). "Nuclear Physics," 578 pp. Wadsworth, Belmont, California.
37. Junge, C. E. (1963). "Air Chemistry and Radioactivity," 382 pp. Academic Press, New York.
38. Keesom, W. H. (1942). "Helium," 494 pp. Elsevier, Amsterdam.
39. Kesebir, M. (1968). Relation of natural gas analyses to geology and reservoir parameters in Chesterian series in Beaver County, Oklahoma. M. Engr. Thesis (unpublished). Univ. of Oklahoma, Norman, Oklahoma.
40. Klinkenberg, L. J. (1941). The permeability of porous media to liquids and gases. *Amer. Petr. Inst., Drilling Production Practice* pp. 200–207.
41. Krumbein, W. C., and Graybill, F. A. (1965). "An Introduction to Statistical Models in Geology," 475 pp. McGraw-Hill, New York.
42. Lambert, I. B., and Heier, K. S. (1967). The vertical distribution of uranium, thorium, and potassium in the continental crust. *Geochim. Cosmochim. Acta* **31**, 377–390.
43. Lambert, I. B., and Heier, K. S. (1968). Estimates of the crustal abundances of thorium, uranium, and potassium. *Chem. Geol.* **3**, No. 4, 233–238.
44. Leverson, A. I. (1967). "Geology of Petroleum," 2nd Ed., 363 pp. Freeman, San Francisco, California.
45. Lifshits, E. M., and Andronikashvili, E. L. (1959). "A Supplement to Helium," 167 pp. Consultants Bureau, New York.
46. Lord, H. C. (1968). Hydrogen and helium ion implantation into olivine and enstatite. *J. Geophys. Res.* **73**, No. 16, 5271–5280.
47. MacDonald, G. J. F. (1964). Dependence of the surface heat flow on the radioactivity of the earth. *J. Geophys. Res.* **69**, 2933–2946.
48. Mason, B. (1967). "Principles of Geochemistry," 329 pp. Wiley, New York.
49. Mayne, K. I. (1956). Terrestrial helium. *Geochim. Cosmochim. Acta* **9**, 174–182.
50. Miller, R. D., and Norrell, G. P. (1965). Analysis of natural gases of the U.S., 1963. *U.S. Bureau Mines Infor. Circular* **8241**.
51. Miyake, Y. (1965). "Elements of Geochemistry," pp. 160–165. Maruzen, Tokyo.
52. Munnerlyn, R. D., and Miller, R. D. (1963). Helium-bearing natural gases of the U.S. *U.S. Bureau Mines Bull.* **617**.
53. Murray, E. G., and Adams, J. A. S. (1958). Thorium, uranium, and potassium in some sandstones. *Geochim. Cosmochim. Acta* **13**, 260–269.
54. Pierce, A. P., *et al.* (1955). Radioactive elements and their daughter products in panhandle and other oil and gas fields in the U.S. *U.S. Geol. Surv. Prof. Paper* **300**, 527–531.
55. Pierce, A. P., *et al.* (1964). Uranium and helium in the Panhandle gas field Texas, and adjacent areas. *U.S. Geol. Surv. Prof. Paper* **454-G**.
56. Puri, P. K., and Aditya, P. K. (1968). Flux of primary helium nuclei near Hyderabad during 1965. *J. Geophys. Res.* **73**, No. 13, 4393–4395.
57. Ragland, P. C., Billings, K., and Adams, J. A. S. (1967). Chemical fractionation and its relationship to the distribution of thorium and uranium in a zoned granite batholith. *Geochim. Cosmochim. Acta* **31**, 17–33.
58. Rankama, K. (1954). "Isotope Geology," 535 pp. McGraw-Hill, New York.
59. Rankama, K. (1963). "Progress in Isotope Geology," 705 pp. Wiley (Interscience), New York.
60. Rankama, K., and Sahama, T. G. (1950). "Geochemistry," 912 pp. University of Chicago Press, Chicago, Illinois.
61. Reid, R. C., and Sherwood, T. K. (1967). "The Properties of Gases and Liquids." McGraw-Hill, New York.
62. Rogers, G. S. (1921). Helium-bearing natural gas. *U.S. Geol. Surv. Prof. Paper* **121**.

63. Roy, R. G., et al. (1968). Heat flow in the United States. *J. Geophys. Res.* **73**, No. 16, 5207–5221.
64. Scott, M. R. (1968). Thorium and uranium concentrations and isotope ratios in river sediments. *Earth Planet. Sci. Letter* **4**, No. 3, 245–252.
65. Smales, A. A., and Wager, L. R. (eds.). (1960). "Methods in Geochemistry," 464 pp. Wiley (Interscience), New York.
66. Smith, F. G. (1963). "Physical Geochemistry," 634 pp. Addison-Wesley, Reading, Massachusetts.
67. Suess, H. E. (1965). *Prog. Oceanography* **3**, 409.
68. Tiratsoo, E. N. (1967). "Natural Gas, a Study," 386 pp. Plenum Press, New York.
69. Whalen, B. A., and McDiarmip, I. B. (1968). Direct measurement of auroral alpha particles. *J. Geophys. Res.* **73**, No. 7, 2307–2313.
70. Wheeler, H. P. (1956). Helium, in mineral facts and problems. *U.S. Bureau Mines Bull.* **556**, 347–358.
71. Zartman, R. E., et al. (1961). Helium, argon, and carbon in some natural gases. *J. Geophys. Res.* **66**, No. 1, 277-306.

EVALUATION OF PRECIPITATION RECORDS IN WEATHER MODIFICATION EXPERIMENTS

Floyd A. Huff

Illinois State Water Survey, Urbana, Illinois

	Page
1. Introduction	60
1.1. Purpose and Scope	60
1.2. Sources of Analytical Data	61
1.3. Definition of Terms	63
1.4. Data Stratifications	63
2. Effects of Natural Precipitation Variability in Evaluating Cloud Seeding Experiments	64
2.1. Data and Analytical Procedures	64
2.2. Results of Analyses	65
2.3. General Conclusions	69
3. Use of Precipitation Climatology in Assessing Potential Benefits of Weather Modification	70
3.1. Natural Distribution Characteristics	70
3.2. Evaluation of Potential Cloud Seeding Benefits	73
4. Space and Time Variability Relationships	78
4.1. Mesoscale Spatial Variability	78
4.2. Variation of Point Precipitation with Distance	82
4.3. Time Distribution Models of Storm Rainfall	88
4.4. Sequential Variability and Lag Correlations	91
4.5. Diurnal Distribution of Storm Precipitation	91
5. Other Relevant Climatological Studies	92
5.1. Climatological Distribution of Areal Mean Precipitation	92
5.2. Application of Published Climatic Data on Point Precipitation	95
5.3. Frequency Distribution of Various Storm Factors	97
6. Precipitation Measurement Requirements	97
6.1. Spatial Correlation of Point Precipitation	98
6.2. Correlation of Storm Mean Precipitation between Areas	102
6.3. Sampling Errors in Measurement of Mean Precipitation	103
6.4. Detection of Storm Precipitation	107
6.5. Areal Extent of Storm Precipitation on Fixed Sampling Areas	110
7. Statistical Evaluation Techniques	111
7.1. Use of Areal Precipitation Measurements in Evaluating Seeding Experiments	111
7.2. Use of Area-Depth Curves in Evaluating Precipitation Modification Experiments	114
7.3. Use of Rain Cell Measurements in Precipitation Modification Experiments	115
7.4. Estimating Natural Distribution of Storm Precipitation in Target Area from Control Area Data	117
8. Downwind Seeding Effects	119
8.1. General Conclusions on Downwind Effects	120
8.2. Analyses of Dense Raingage Network Data	120

9. Design of a Precipitation Modification Experiment....................... 123
   9.1. Selection of Sites... 124
   9.2. Sampling Design and Verification Methods............................. 126
   9.3. Precipitation Measurement Requirements 127
   9.4. Time of Year... 129
   9.5. Time of Day.. 130
   9.6. Weather Types 130
   9.7. Length of Experiment 131
References... 133

1. Introduction

One of the key problems encountered in precipitation modification experiments has been the evaluation of results. Evaluation of modest increases or decreases in surface precipitation resulting from cloud seeding operations is exceedingly difficult because of the great natural variation of precipitation in space and time. This problem has been compounded by the lack of adequate statistical data to define quantitatively the natural variability characteristics in most regions of the United States. McDonald [1] pointed out that precipitation variability above all else has stood in the way of efforts to secure an immediate answer to the question of the efficacy of cloud seeding. In the ensuing years, numerous conflicting evaluations resulting from both commercial cloud seeding operations and scientifically oriented weather modification experiments have continued to appear in the literature. It is likely that part of this controversy which still remains with us arises from inadequate precipitation measurements during seeding experiments or a lack of understanding of natural precipitation variability, or both, which can readily lead to erroneous conclusions in short-term experiments, as demonstrated by Huff [2].

1.1. Purpose and Scope

The intent of this review paper is to present an appraisal of the problems involved in weather modification experiments that result from the space and time distribution characteristics of natural precipitation and the measurement thereof. An effort will be made to place the natural realities in their proper light in order to provide a basis for the rational evaluation of efforts to augment precipitation. In so doing, major emphasis will be placed upon the use of comprehensive studies of precipitation climatology in assessing the magnitude of the verification problem, estimating the potential output from cloud seeding operations, defining in quantitative terms the space and time variability of precipitation, and determining precipitation measurement requirements. An appraisal of several statistical evaluation techniques will be made through use of natural precipitation data. Finally, the application of knowledge accumulated from various precipitation studies in the design of a specific precipitation modification experiment will be illustrated.

This paper is based primarily upon findings in a recently completed research project in Illinois, which is located in the continental humid climate of the midwestern United States [3,4]. Only the most salient findings in various studies are summarized, but references are provided for those interested in more detailed accounts. These studies were made possible by the availability of a large sample of precipitation data from dense raingage networks in Illinois, a state which is representative in climate, topography, and food production of the midwestern agricultural region. The findings should be useful also as first approximations for other regions of the country. Although aimed at weather modifications, much of the information presented herein should be useful to hydrologists, climatologists, and other users of precipitation data.

In most cases, the Illinois studies involved sampling areas of 50–550 miles^2 for which data from the dense raingage networks were available to define accurately the time and space properties of precipitation. Cloud seeding capability on areas of this size must eventually be determined for efficient agricultural application. Furthermore, water supply augmentation from seeding would frequently involve treatment of relatively small basins. For example, Stall [5] made a study of low flows of Illinois streams for use in impounding reservoir design. The drainage areas for the 164 basins used in his study had a median area of 170 miles^2 and only 25% of the basins had areas exceeding 600 miles^2. Light to moderate droughts in which potential benefits from cloud seeding could be very substantial sometimes encompass only a few hundred square miles. Also, in a scientifically oriented experiment in which statistical analyses of surface precipitation are the primary means of verification, climatic and topographic homogeneity can not usually be maintained over large areas. This complicates the verification problem.

1.2. Sources of Analytical Data

Location of the five dense raingage networks supplying basic data in various studies are shown in Fig. 1, along with the sampling patterns in the two most important networks. The East Central Illinois Network was the primary network used in the studies. It consisted of 49 recording gages arranged in a nearly uniform grid pattern in a 400 miles^2 rural area of relatively flat terrain in which elevations ranged from 650 to 910 ft MSL. The network was operated from 1955 through 1966 with no significant changes in gage locations and provided a sample of 1344 storms during this period. The Little Egypt Network of 49 gages in 550 miles^2 was operated during 1958–1966. It ranked second in importance in the studies and provided a

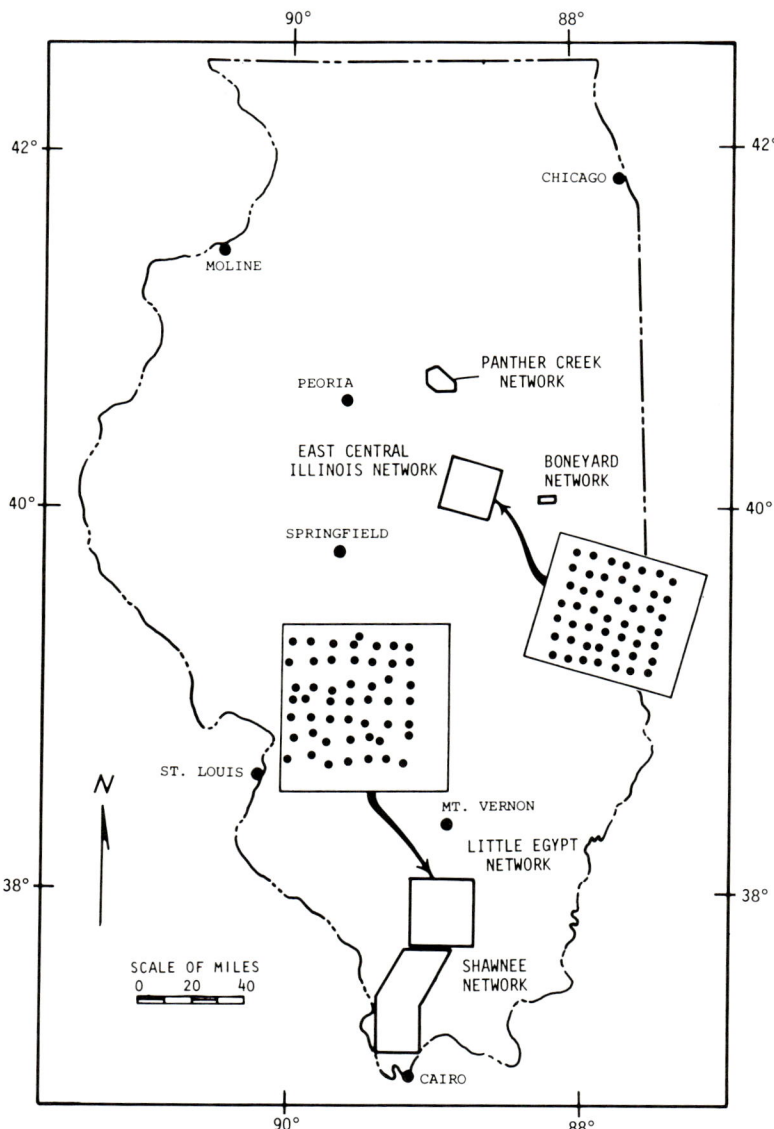

FIG. 1. Raingage network locations.

1200-storm sample. The Little Egypt Network consisted of 25 recording and 24 nonrecording gages during most of its history and was located in rural flatlands, with total relief ranging from 350 to 600 ft MSL.

The Boneyard Network, an urban network of 10–11 recording gages on 10 miles^2, provided a sample of 1233 storms in the 1955–1966 period used in the research. The Panther Creek Network in north central Illinois enveloped an area of 100 miles^2 in rural, relatively flat terrain. During the 1954–1961 period used in certain studies it consisted of nine recording gages and yielded a sample of over 700 storms. The Shawnee Network of 44 recording gages was installed in 1964 in an area of 700 miles^2 extending southward from the Little Egypt Network. This network straddles a relatively narrow east–west oriented hill area in which the hills rise abruptly from the flatlands, although the total relief ranges only from 300 to 1025 ft. It provided a sample of 215 storms.

1.3. Definition of Terms

Several terms used frequently throughout this review are defined as follows. A storm is a precipitation period separated from preceding and succeeding precipitation on the sampling area by six hours or more. This definition has been found most suitable for separating storms resulting from different synoptic causes on the sampling networks.

Winter refers to the three-month period, December–February. Spring includes March, April, and May. Summer incorporates June, July, and August. September through November constitutes Fall. The growing season or warm season includes the five-month period, May–September. The cold season or water supply replenishment season encompasses the seven months from October through April.

1.4. Data Stratifications

Frequently, cloud seeding is not conducted on all types of storms, and seeding selectivity is likely to increase as knowledge advances in weather modification. Statistical verification of seeding results could be facilitated by data stratification if seeding effects vary substantially with the meteorological characteristics of storms, as indicated by recent work [6,7]. Consequently, an evaluation of the general effects of data stratification on the sampling requirements for verification was made as part of the Illinois studies. Usually, the data were grouped according to season, synoptic storm type, precipitation type, storm duration, and average or total precipitation.

Four major precipitation types were used in the studies. These included thunderstorms (TRW), rain showers (RW), steady rain (R), and snow (S).

Other types were too few for separate treatment. When more than one precipitation type prevailed during a storm, the storm was assigned to the type responsible for the major portion of the precipitation.

Seven basic synoptic weather types were used. These included the four frontal types (cold, warm, static, and occluded), prefrontal squall lines, air mass instability showers, and low center passages. In most cases, storm durations were divided into five groups consisting of those storms lasting 3 hr or less, 3.1–6.0 hr, 6.1–12.0 hr, 12.1–24.0 hr, and over 24 hr.

2. Effects of Natural Precipitation Variability in Evaluating Cloud Seeding Experiments

Attention will be focused first on demonstrating how natural variability can affect verification efforts in precipitation modification experiments because this is a major problem confronting the investigator. This was done in the Illinois studies by performing hypothetical seeding experiments on network data using two basic statistical sampling designs frequently employed in cloud seeding experiments. These designs were tested on various stratifications of the precipitation data to determine how the interference level of natural variability is affected by such divisions which may be necessary or desirable in seeding experiments.

2.1. Data and Analytical Procedures

The two statistical designs were: the *random experimental* which involves randomization of days over a single target area into seeded and nonseeded days with nonseeded as the control; and, the *cross-over target-control* which requires random interchange of target and control areas among seeding days. The random experimental design has been used in two well-documented United States experiments [8,9] and the cross-over design has been employed in the Israeli experiments [10] and others. Major advantages of the cross-over design are that it tends to minimize the effect of any natural rainfall differences that may exist between target and control areas and, statistically, it requires less verification time than the random experimental design. However, the random experimental method minimizes the contamination problem between target and control and eliminates the possibility of other control area effects that may be initiated by the target seeding.

The data sample consisted of all storms which produced measureable precipitation at any network gage, and the average network precipitation was the basic measure of storm precipitation used. Most of the analyses utilized data from the East Central Illinois and Little Egypt Networks. A randomization program on the IBM-7094 computer was used to separate the natural precipitation data into sets of hypothetically seeded and nonseeded storms with both statistical designs. The randomizing was repeated 100 times on each storm group to provide a probability distribution of differences between the two groups. The differences, of course, are the result of natural variability and provide a measure of the background interference that may be associated with evaluation of actual cloud seeding experiments. Several of the numerous analyses performed in this study will be described here to illustrate the magnitude of the variability effect. A full account can be found in [2,4].

The network data were separated into air mass storms and all storms combined in most analyses. The air mass category was selected for special attention because the two major United States experiments were concerned with this storm type [8,9] and some investigators have suggested that nonfrontal convective rainfall offers the greatest opportunity for augmentation by seeding.

2.2. Results of Analyses

In one phase of the study, the random experimental design was applied to summer air mass storms on the Little Egypt Network for the five-year period, 1960–1964, which coincides with the Missouri cloud seeding experiments by Braham [9]. For this period, 116 qualifying storms were sampled with average network rainfall ranging up to nearly two inches. Figure 2 shows the frequency distribution curve derived from the computer randomization program. This curve indicates that a rainfall increase of 50%, due

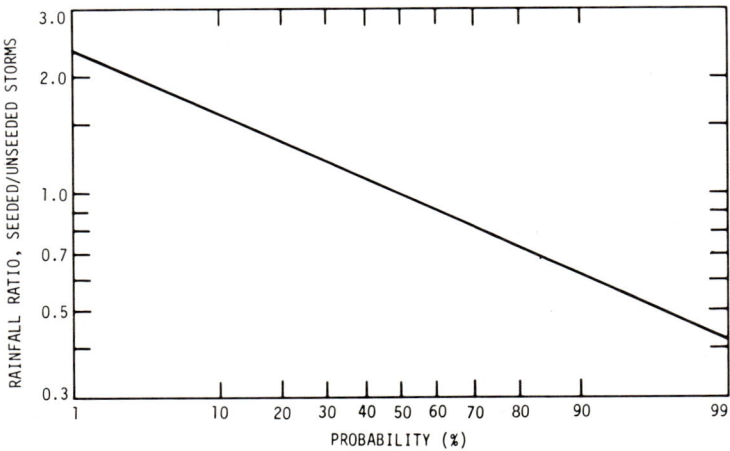

Fig. 2. Distribution of differences between hypothetically seeded and unseeded air mass storms on Little Egypt Network.

entirely to natural rainfall variability, can be expected in the seeded sample in approximately 14% of the experiments carried out under similar circumstances. Similarly, employing this type of randomized experiment, a 20% increase, resulting strictly from natural rainfall variability, will occur in 30% of the five-year experiments involving seeding of summer air mass storms in the Midwest or similar climatic areas. Figure 2 clearly illustrates the difficulty confronting the experimenter who is endeavoring to verify possible increases of a few percent in convective rainfall resulting from cloud seeding, such as the 10 to 20% increases which has been reported in the literature [11].

Figure 3 is an example of the persistence that may prevail for several years in areas of apparently homogeneous precipitation climate. A SW–NE profile

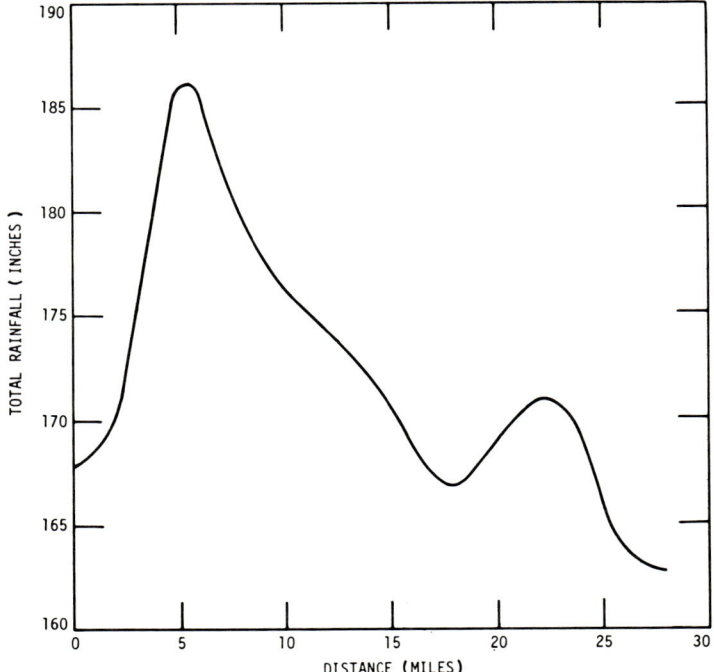

FIG. 3. SW–NE profile of May–September total rainfall on East Central Illinois Network, 1955–1964.

on the East Central Illinois Network is shown for the total rainfall during May–September (growing season) over a ten-year period, 1955–1964. Distance in miles is plotted against total rainfall. Within the 28-mile distance, the ten-year total rainfall showed a range of 14% with point amounts varying from

163 to 186 in. The implication is that the experimenter must be careful not to misinterpret spatial persistence in natural rainfall as seeding-induced changes, or as residual effects resulting from earlier seeding operations. This problem is most acute with a fixed target-control sampling scheme.

Comparison of the effects of natural precipitation variability in random experimental and cross-over target-control types of experiments indicates that the variability interference is appreciably less with the target-control experiments, as predicted by statistical theory, provided that the target and control are close together. As the distance increases, the superiority of the target-control method decreases. In the Illinois study, the superiority was relatively small when target and control were separated by 125 miles, the distance between the East Central Illinois and Little Egypt Networks.

However, when tests of the natural variability effect on cross-over target-control experiments were made on adjacent target and control areas (SW and NE halves on East Central Illinois Network), the statistical superiority of this method over the random experimental was very substantial. This is illustrated in Table I where the two sampling methods were applied to

TABLE I. Comparison of background interference from natural variability between random experimental and cross-over designs, based on 1960–1964 summer storms with means of 0.11 to 1.00 in.

Probability (%)	Seeded to unseeded ratios	
	Random experimental	Cross-over
5	1.23	1.11
10	1.17	1.09
20	1.10	1.06
30	1.05	1.03
50	0.98	1.00
70	0.91	0.96
90	0.83	0.91
95	0.78	0.88
Number of storms	72	72

summer storms with means of 0.11 to 1.00 in. during the five-year period, 1960–1964. Probabilities are expressed in terms of the ratio of precipitation in hypothetically seeded storms or areas (target rainfall) to that in unseeded storms or areas (control rainfall). For example, Table I indicates a probability of 5% that the natural variability in summer rainfall over a five-year period will result in an apparent seeding-induced increase of 23% with the single-area type of experiment, even though there was absolutely no seeding effect.

This naturally occurring increase at the 5% level decreases to 11% with the cross-over target-control type of experiment.

The effect of length of sampling period in the cross-over target-control method of verification is illustrated in Table II. Here, probability distribu-

TABLE II. Time distribution of seeded to unseeded ratios in cross-over target-control experiments on East Central Illinois Network for storm mean precipitation ≥ 0.01 in.

Probability (%)	Ratios for given period			
	1960	1960–1961	1960–1962	1960–1964
5	1.07	1.06	1.05	1.04
10	1.05	1.04	1.04	1.03
20	1.03	1.03	1.02	1.02
30	1.02	1.01	1.01	1.01
50	1.00	1.00	1.00	1.00
70	0.98	0.98	0.99	0.99
80	0.97	0.97	0.98	0.98
90	0.96	0.96	0.97	0.97
95	0.94	0.95	0.96	0.96
Number of storms	105	216	316	513

tions are shown for consecutive periods of one, two, three, and five years, based upon the use of all storms in which the network mean precipitation equalled or exceeded 0.01 in. Under the conditions of this hypothetical experiment, it was found that the background level of interference was relatively low compared with the single-area type of verification. Furthermore, the natural variability effect decreased only slightly when the sampling period was increased from one to five years, or more realistically, from a 100-storm to a 500-storm experiment. Examination of the other individual years showed the highest variability effect in 1963 when the ratio had a range of 1.09–0.89 from the 5 to 95% levels.

With adjacent target and control areas, such as used in Tables I and II, the contamination problem is acute, and some separation would undoubtedly be necessary under operational conditions. Data from the two Illinois networks (East Central Illinois and Little Egypt) can be used to provide estimates of the minimum and maximum effects of natural precipitation variability in the cross-over type of experiment, as illustrated in Table III. Here, the magnitude of the natural variability effects between adjacent and 125-mile separated target and control areas has been shown for summer

TABLE III. Effect of background interference from natural variability with cross-over design applied to adjacent target and control compared with target-control separation of 125 mi, based on 1960–1964 summer storms with means ≥ 0.01 in.

	Seeded to unseeded ratios	
Probability (%)	Adjacent target-control	Separated target-control
5	1.09	1.44
10	1.07	1.34
20	1.04	1.22
30	1.02	1.15
50	1.00	1.02
70	0.98	0.92
90	0.94	0.79
95	0.90	0.73
Number of storms	140	76

storms. Only those storms which produced rainfall in both networks were used in the separated target-control computations.

Investigation of the effects of stratifying the precipitation data according to areal mean precipitation indicated some advantage in the single-area experiments and no significant advantage in the cross-over target-control tests. Storm duration stratification was found to have a small effect on verification of seeding effects with both techniques. Seasonal stratification of data had a moderate influence. Stratifications were made also according to precipitation type and synoptic storm type. With the random experimental sampling, both types of stratification increased the background interference from natural variability in all groups from that obtained with all storms combined. In the cross-over target-control sampling tests, the effect of the stratifications was pronounced only in the case of air mass storms. With this synoptic storm type, the natural variability factor is much larger than with frontal storms or low-pressure centers, both of which showed variability effects similar to those obtained with all storms combined.

2.3. General Conclusions

Overall, the Illinois study offered strong evidence that natural precipitation variability can lead to fallacious interpretation of cloud seeding results in short-term experiments, unless the experimenter is aware of the potential pitfalls and evaluates his data accordingly. The background interference from natural variability is a substantially greater problem with the random experimental type of statistical sampling design than with the cross-over method. Accordingly, verification time at a given level of statistical reliability is shorter with the cross-over design. The natural variability problem is relatively great when cloud seeding is concentrated on nonfrontal storms of short duration in the warm season. The air mass instability shower usually meets the above criteria. Strictly from the statistical evaluation standpoint, data stratification is usually undesirable because of the resulting reduction in test sample sizes. However, from consideration of both varying meteorological characteristics between storms and seeding operational capabilities, stratification may be both logical and desirable in many experiments.

3. Use of Precipitation Climatology in Assessing Potential Benefits of Weather Modification

In the previous section, problems created by natural precipitation variability in the verification of cloud seeding experiments were discussed. Now, we will turn our attention to how precipitation climatology can serve as a valuable tool in assessing the potential benefits of cloud seeding under various assumptions of future capability and seeding technology. Definition of benefits to be derived from seeding-induced augmentation of precipitation is one of the most pertinent problems facing the weather modification field.

3.1. Natural Distribution Characteristics

In achieving the above objective, it is necessary first to express natural precipitation characteristics in the form of climatological distributions readily adaptable for assessing the potential benefits. At the same time, these climatological distributions can be a valuable asset in developing optimum seeding methods in any given climatic region, since they tell us how nature distributes its precipitation with respect to season, storm type, precipitation type, and other meteorological parameters. Types of analyses employed in the Illinois study [12] will be discussed briefly in the following paragraphs to illustrate some of the methods that can be used in climatological assessments.

Data from 1344 storms collected during a 12-year period on the East Central Illinois Network (Fig. 1) were used to determine the precipitation distribution characteristics on areas ranging from a point to 400 miles^2. Distributions were defined by relating magnitude of storm precipitation to both cumulative percent of total precipitation and cumulative percent of total number of storms per season and year. Stratifications of the data according to storm intensity and duration, precipitation type, synoptic weather type, and relative wetness or dryness of seasonal and annual precipitation were tested to evaluate their influences upon the distribution characteristics. Frequency distributions of both rainfall depth and number of storms were developed for various data classifications.

The two major beneficiaries from successful precipitation modification would be agriculture and water supply (municipal and industrial). In the midwest, the growing season from May through September encompasses the period of principal agricultural needs in years of deficient rainfall. Similarly, the major contribution to water supplies is during the period from October through April when evapotranspiration is at a minimum. Consequently, a major effort was concentrated on determining natural distributions for these two periods and using them in evaluation of potential seeding benefits.

Significant differences in the natural distributions employed in this study were found when the data were grouped by storm duration on a seasonal and annual basis. This is illustrated in Fig. 4 which shows the average annual distributions of total precipitation and number of storm occurrences grouped by duration. Figure 5 provides a measure of the distribution of total precipitation with respect to the frequency of storm occurrences.

The three basic types of curves shown in Figs. 4 and 5 provide quantitative estimates of the average distribution characteristics that may be used for guidance in weather modification, hydrological design, and other problems in which such climatological information is useful. The effects of storm duration are clearly evident from these curves, and Fig. 5 illustrates well the climatological trend for a large percentage of the total precipitation to occur in a small percentage of the storm occurrences.

Investigations of the relationship between the distributions of point and areal mean rainfall and between storm and daily precipitation were made through use of the network data and 57 long-term Weather Bureau climatic stations in Illinois. Results indicated that for areas up to several hundred square miles, daily point rainfall totals can be used to obtain a reasonable estimate of the precipitation distribution characteristics, and, consequently, aid in the evaluation of potential seeding benefits for climatic regions lacking dense network data. Figure 6 illustrates the use of daily point precipitation data from climatic stations to obtain estimates of the natural distribution characteristics. Here, average percentages of total annual precipitation from daily amounts in the range from 0.11 to 1.00 in., a range of storm intensities in which seeding-induced increases appear feasible, are shown for the state. Point distribution patterns can be defined further by using data from recording gage stations in state climatic networks. Distributions could then be derived for storms of various durations similar to those obtained in the Illinois network studies.

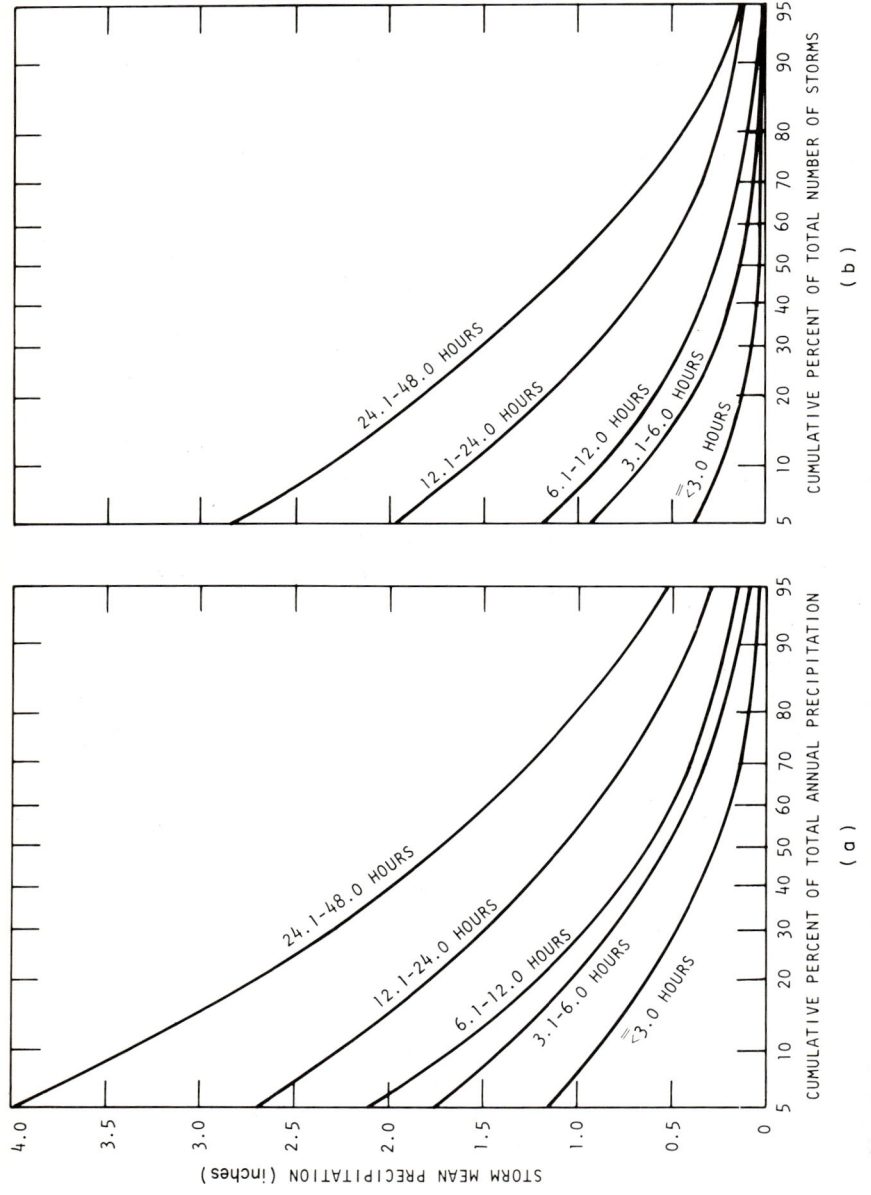

Fig. 4. Average annual distributions grouped by storm duration. (a) Distribution of total precipitation; (b) distribution of storm occurrences.

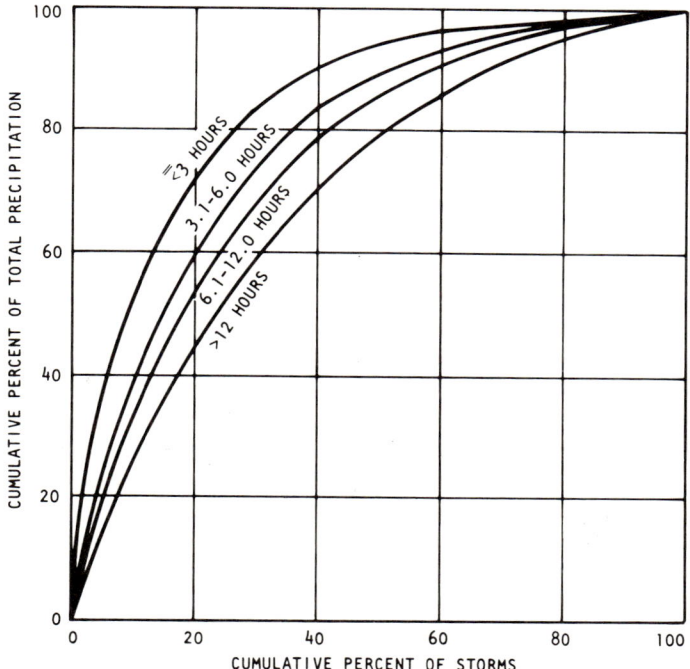

Fig. 5. Relation between distribution of total precipitation and storm occurences.

3.2. Evaluation of Potential Cloud Seeding Benefits

Natural precipitation distributions may be used with any hypothesis one wishes to apply with respect to his specific evaluation of weather modification capabilities and forecasting skill (if this is pertinent to his assessment). The following examples, in which upper cut-offs in rain intensity are used, assume a continuous seeding program on all rain-producing clouds. Presently, this would be exceedingly difficult (if not impossible), and the results therefore represent seeding benefits with optimum operational efficiency under the assumed capabilities. The primary purpose of the illustrations is to demonstrate how natural precipitation distributions may be applied as a useful tool in weather modification. The natural distributions will become increasingly more useful as scientific knowledge increases in the relatively crude science of weather modification.

Seeding increases in naturally heavy storms may not be feasible or desirable. Furthermore, relatively large percentage increases in very light storms would not produce significant added amounts for agricultural or water supply purposes, especially during dry periods when evaporation and infiltration

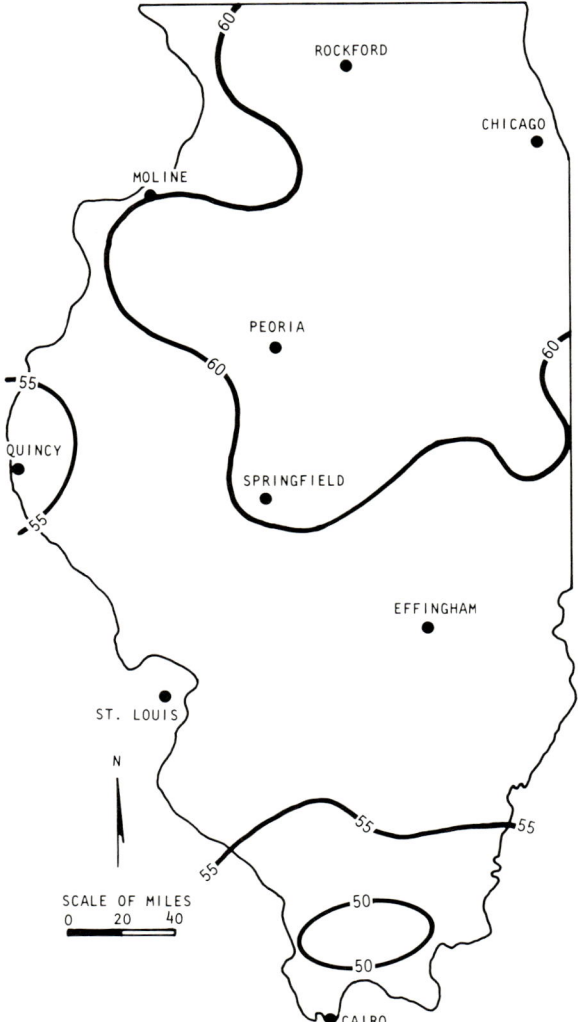

Fig. 6. Average annual percentage of total precipitation from daily amounts of 0.11 to 1.00 in.

are high and runoff is usually insignificant. As a first approximation to demonstrate use of climatological distributions such as illustrated in Fig. 4, it was assumed that seeding of midwestern storms which produce 0.10 in. or less naturally would not significantly help the agricultural or water supply needs, and that storms capable of producing over 1 in. of precipitation are too efficient naturally for substantial increases to result from seeding.

Nomograms were developed from the precipitation distributions to facilitate the calculation of potential benefits of cloud seeding during the agricultural and water-supply season under various assumed weather modification capabilities. One of these nomograms is illustrated in Fig. 7 which shows the

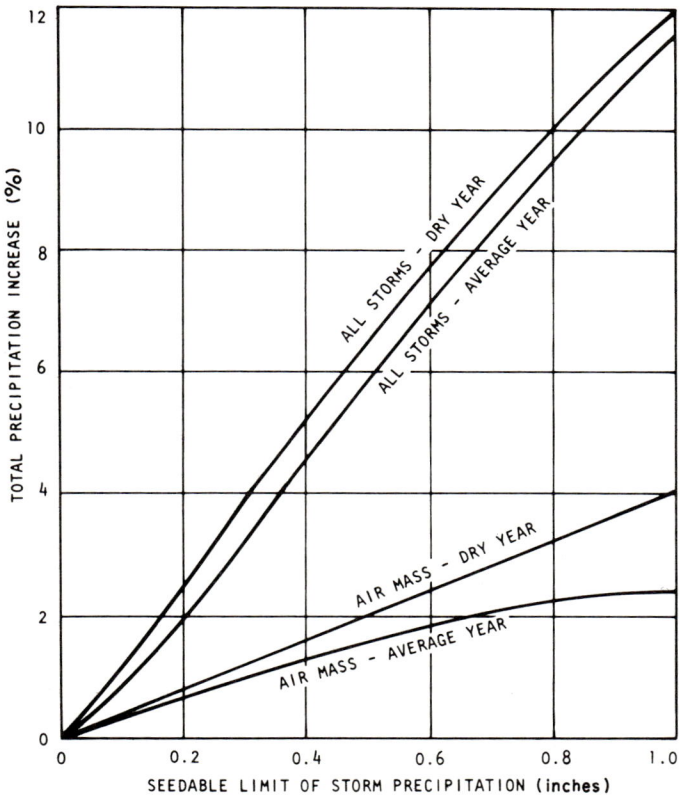

FIG. 7. Effects of 20% seeding-induced increases on total May–September rainfall.

effects of 20% increases from hypothetical seeding on total May–September rainfall on the 400 mile^2 area for (1) all storms combined under dry-year conditions, (2) all storms combined in an average year, (3) air mass storms in dry periods, and (4) air mass storms in an average year.

The use of the nomogram is illustrated in the following example. Assume a continuous seeding operation on the target and that cloud seeding is capable of producing an average increase of 20% in all storms in which the areal mean precipitation produced naturally is 0.5 in. or less, but that seeding storms of greater intensity (over 0.5 in.) will have no effect (positive or negative) upon the natural rainfall. The abscissa shows the upper limit of areal mean rainfall for a given assumption, and the ordinate shows the percentage increase in total seasonal rainfall resulting from the defined capability. Thus, in a typical dry year the curves indicate a realized rainfall increase of 6.4% compared to 5.9% in an average year. From a practical standpoint, this difference is insignificant, considering the sources of errors involved in the observation and analyses of rainfall data even under optimum conditions, which are closely met in these network studies. The foregoing example provides an estimate of maximum benefits under the assumed seeding capability. In practice, forecasting limitations and operational problems would make it extremely difficult to recognize and seed every storm. Thus, if one assumes only 80% of the natural storms are seeded, the realized increase in the example lowers from 6% to 5%.

Figure 7 can be used also to determine potential seeding benefits within specific ranges of storm mean rainfall. For example, assume that only storms in which the mean rainfall is greater than 0.10 and less than 1 in. are affected by seeding. Then simply subtracting the calculated increases at these two levels produces an estimate of the maximum benefit. Also, the effects of other assumed seeding increases, such as 10 or 30%, can be obtained by multiplying the nomogram answer by the ratio of the desired percentage to 20. Similarly, the nomogram can be used to calculate the potential effects of selective seeding. For example, one could determine the results of seeding only 20% of the storms in any mean rainfall range and for any assumed percentage increase in these seeded storms. If one believes all storm intensities are significantly affected by seeding, the 1-in. cutoff in Fig. 7 can be extended to include all storms.

If one considers certain storm types affected beneficially by seeding and others detrimentally, then quantitative precipitation forecasting skill would have a major effect on the benefits and detriments of seeding natural storms. Until this skill is better defined and seeding technology substantially advanced from its present status, one can only use the climatological distributions developed here for estimating weather modification potential under various assumed capabilities.

It was shown earlier that storm duration has a pronounced effect upon the mean rainfall distribution curves (see Fig. 4). Therefore, nomograms were constructed to permit consideration of this factor in estimating potential cloud seeding benefits in Illinois during the agricultural and water-supply seasons. A set of these curves is shown in Fig. 8 for average conditions and duration classifications which appeared most desirable from the various analyses performed in this study.

Figure 8 shows that the realized surface increases in seasonal precipitation are relatively small, especially in the October–April period, unless storms of

WEATHER MODIFICATION EVALUATION

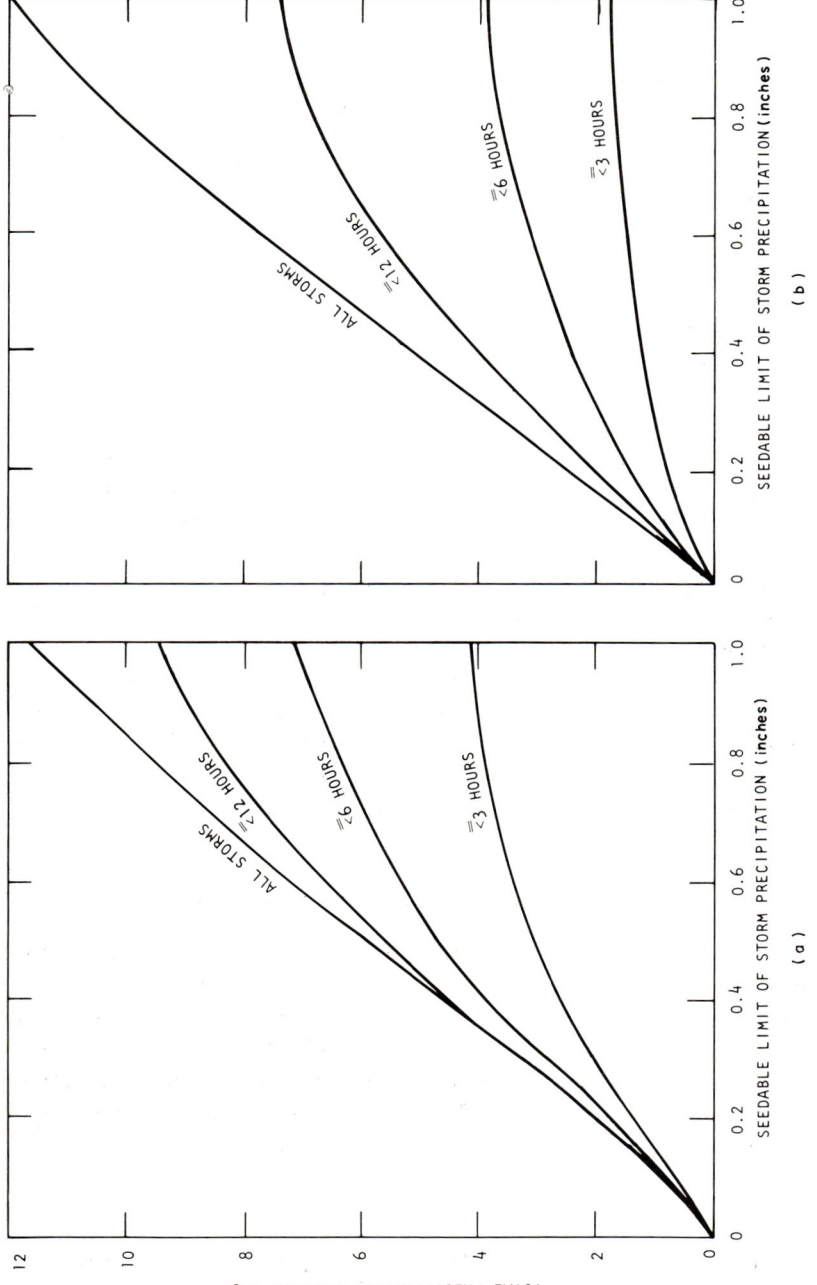

FIG. 8. Effects of 20% seeding-induced increases in storms of selected duration on total seasonal precipitation. (a) Growing season; (b) cold season.

relatively long duration can be successfully seeded. This results from the strong trend for precipitation to increase with increasing storm duration, and this trend is more pronounced in the colder part of the year when large-scale systems and steady types of precipitation are prevalent.

The 1966 Final Report of the Panel on Weather Modification and Climate of the National Academy of Sciences [11] stated that considerable statistical evidence had been found that increases of 10 to 20% can be induced by cloud seeding under favorable conditions. Figures 7 and 8 show that it would be difficult to achieve seasonal increases of this magnitude in the midwest by seeding on days with natural rainfall unless seeding intensifies heavy rainstorms (>1.0 in. mean), or very large percentage increases are produced in storms of light to moderate intensity. Soil erosion, reservoir silting, and flooding may result in more damage than benefits from stimulation of heavy storms, especially storms of short duration that deposit their water with high rates of rainfall.

Figure 7 indicates that seeding of warm-season, air mass storms cannot substantially increase the growing season rainfall by itself; therefore, both frontal and nonfrontal storms must be seeded under midwestern climatic conditions to achieve substantial aid for the farmer. Assuming a 20% seeding increase from all air mass storms which naturally produce areal means of 1 in. or less, the total May–September increase in a typical year would be 2 to 4% according to Fig. 7.

A general conclusion resulting from this study is that cloud seeding must produce relatively large rainfall increases in naturally occurring storms under favorable circumstances or initiate considerable rainfall from naturally nonprecipitating clouds, or both, if substantial contributions are to be made to the agricultural industry and municipal water supplies under midwestern climatic conditions.

4. Space and Time Variability Relationships

From discussions in the previous two sections, it is evident that natural variability should be considered in both the planning and verification of precipitation modification experiments. Failure to do so may lead to both poorly designed operational programs and invalid interpretation of experimental results. This section will be devoted to development of quantitative relationships to define more specifically the magnitude of the time and space variability with which the weather modifier is confronted.

4.1. Mesoscale Spatial Variability

Knowledge of the spatial variability of natural precipitation is especially pertinent in cloud seeding experiments involving target-control comparisons.

One method of defining it is through calculations of the relative variability in a given sampling area. Therefore, data from the dense raingage networks in Fig. 1 were analyzed to obtain quantitative measures of the spatial relative variability of precipitation on a mesoscale for storm, monthly, seasonal, and longer periods. The relative variability was obtained with the simple method recommended by Conrad and Pollak [13] in which the average deviation from the mean is divided by the sample mean and this number multiplied by 100 to obtain a percentage expression.

The spatial relative variability V of storm precipitation was found to be related exponentially to mean precipitation P. The relation was not significantly improved by adding other variables, such as storm duration and maximum storm precipitation. The relative variability tends to increase with increasing area and was substantially greater with unstable types of precipitation (rainshowers and thunderstorms) than with stable types (steady rain and snow). Grouped by synoptic storm type, the highest relative variability was obtained with air mass storms and the lowest with low center passages.

The areal effect is illustrated in Fig. 9 in which curves of V for thunder-

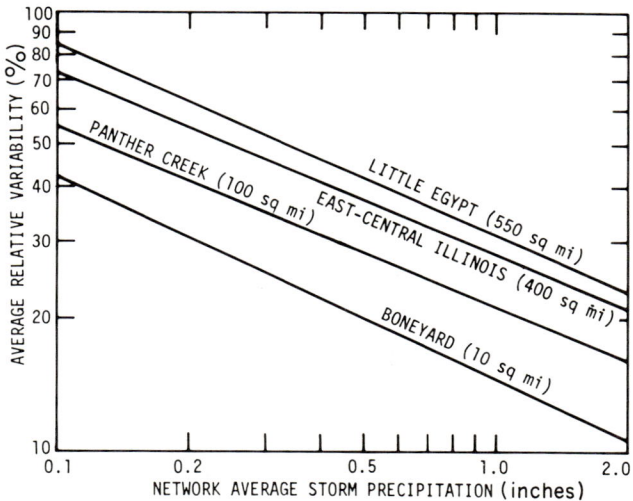

FIG. 9. Relation between storm relative variability and mean precipitation in thunderstorms.

storm rainfall (TRW) have been shown for four networks ranging in size from 10 to 550 mile^2. Because of the more uniform spatial distribution with the stable types of precipitation (R, S), the area factor is less important in determination of V, although a weak trend for it to increase with increasing area was found in the network analyses [14].

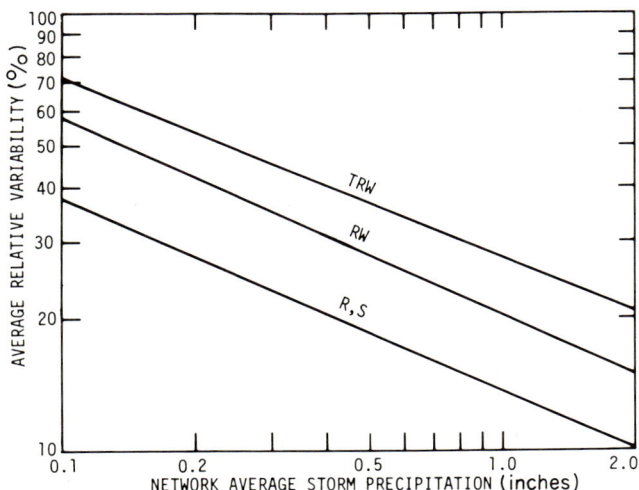

FIG. 10. Relation between storm relative variability and precipitation type on 400 miles² network.

The effect of precipitation type is indicated in Fig. 10 where it can be seen that V for TRW is approximately twice that for R and S on the 400-miles² network. The effect of seven basic synoptic weather types on V is illustrated in Fig. 11. Since high spatial variability intensifies the problem of rain modification evaluation, it is obvious from Fig. 11 that the seeding of air mass

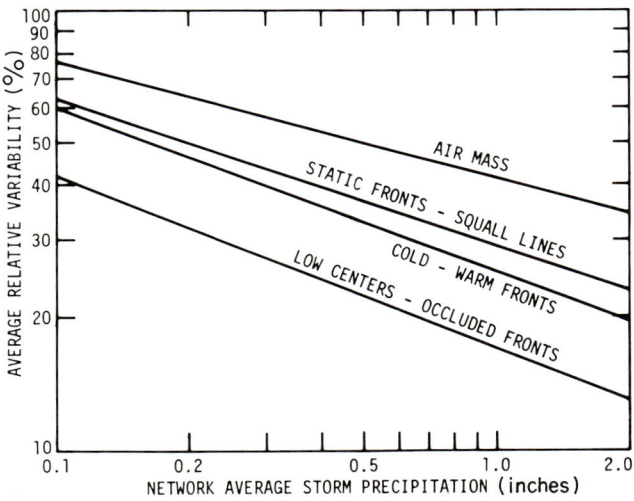

FIG. 11. Relation between storm relative variability and synoptic storm type on 400 miles² network.

instability showers, such as done on the much discussed Project Whitetop [15], presents an unusually difficult evaluation problem when surface rainfall is the primary verification tool.

Seasonal comparisons of RW and R on the East Central Illinois Network indicated that the average rainfall gradient, as measured by the relative variability, was approximately twice as large in summer as in winter for a given mean precipitation, and was about 60% greater in summer than in spring or fall. Other considerations being equal, it is apparent that precipitation modification effects could be evaluated statistically with a substantially shorter sampling period in the cold season, when major replenishment of water supplies takes place in the midwest, than in the crop-growing season when agriculture would benefit most by successful cloud seeding.

The relative variability, although a useful parameter to evaluate the general effects of various meteorological factors upon storm spatial variability, displays large differences between storms of similar precipitation volume, precipitation type, and synoptic storm type. Because of its interstorm instability, it is not a desirable parameter in itself for the evaluation of weather modification experiments. This is brought out further in Fig. 12 which shows

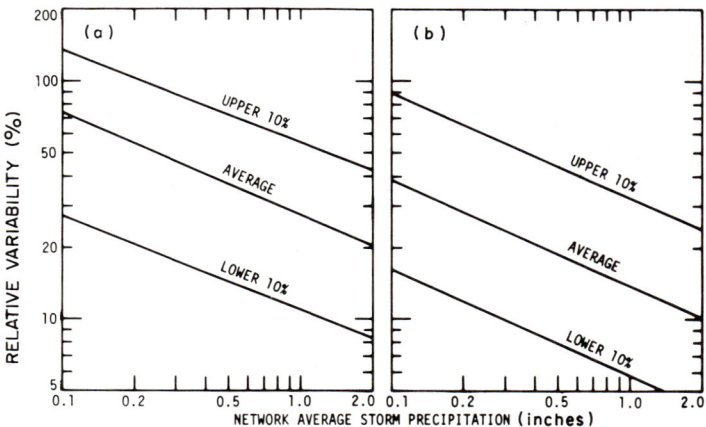

FIG. 12. Range of storm relative variability on 400 miles² network. (a) Thunderstorms; (b) steady rain.

the upper and lower 10% ranges about the average relative variability in thunderstorms and steady rainfall on the 400-miles² network (East Central Illinois). Relative variability varies by a factor of approximately five between the upper and lower limits.

For some purposes, knowledge of the relative variability over periods of a month, season, or years is of more interest than storm variability. One

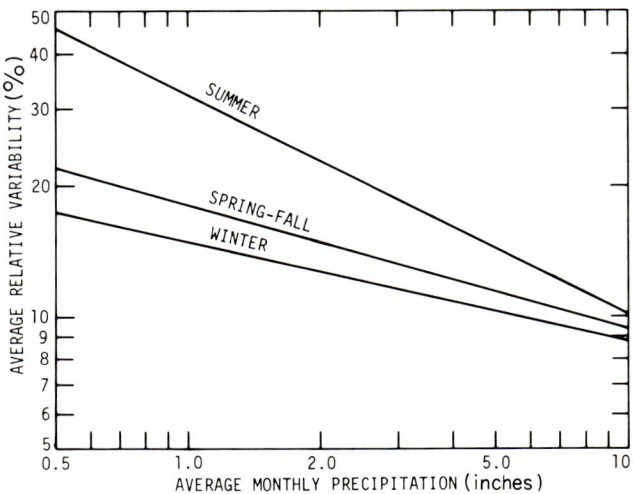

FIG. 13. Seasonal effects on monthly relative variability on 400 miles² network.

example is the establishment of sampling requirements for long-term weather modification experiments. As illustrated in Fig. 13, the monthly variability showed little difference between fall, winter, and spring months, but large differences between summer and winter. Determination of V was made also for extended periods of 3-60 months. Results indicated that V minimizes to an approximately constant value after time periods of 24 months are integrated into the calculation, and that it becomes minor when precipitation is summed for several years over an area of relatively homogeneous precipitation climate. For example, V decreased to a constant value of 4% on the 400-miles² network when the time-period integration exceeded 24 months.

4.2. Variation of Point Precipitation with Distance

Another method of evaluating spatial variability is through determination of the variation of point precipitation with distance. Several studies have been made in recent years to provide quantitative information on rainfall gradients in midwestern storms. Huff [16] presented relationships for rainfall gradients of total storm rainfall derived from approximately 200 warm-season storms. Huff, Shipp, and Schickedanz [17] investigated gradients of rainfall rate in warm season rainfall through use of 1-min rainfall amounts from a special network of raingages.

Recently, the earlier studies were expanded to determine gradients in monthly and seasonal precipitation [4]. This now makes available quantitative estimates of midwestern precipitation gradients for time intervals ranging from 1 min to several months. Although these studies certainly do not provide answers to all questions on rainfall gradients, they do add to our basic knowledge on the natural distribution of precipitation. Consequently, results are summarized briefly in the following paragraphs.

Figures 14 and 15 have been abstracted from [16] to illustrate quantitatively the magnitude of rainfall gradients in warm-season storms. Figure 14

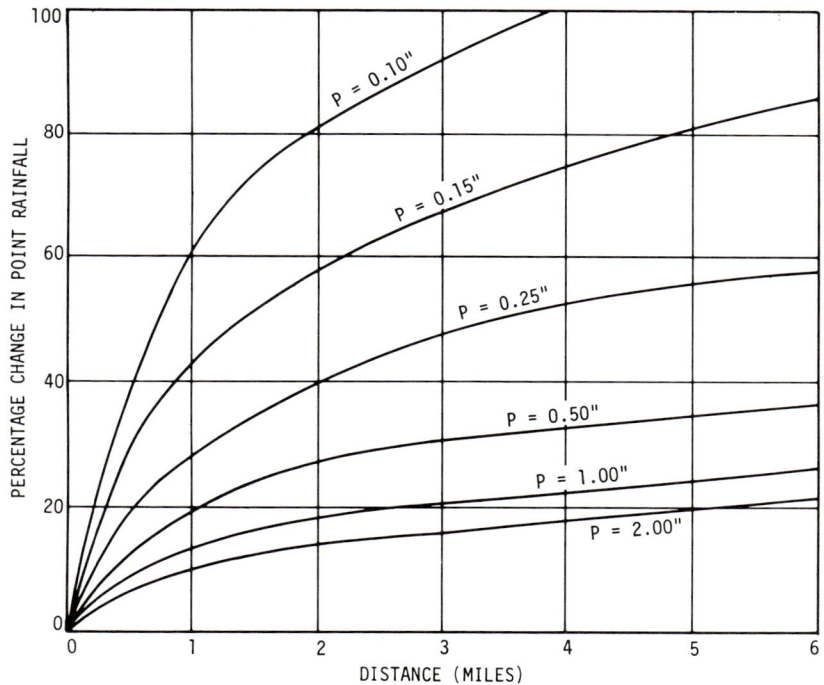

FIG. 14. Average rainfall gradients in warm season storms.

shows average percentage changes in point rainfall with increasing distance for starting point rainfalls of various amounts. The percentage changes remain nearly constant for amounts in excess of 2 in. The percentage curves reflect the known tendency for the relative variability to decrease with increasing rainfall volume.

The curves in Fig. 14 are for average values of rainfall gradient. Estimates of extreme gradients are equally desirable for some applications. Figure 15 illustrates the differences between average and extreme gradients, the extreme gradient being defined as one that will not be exceeded more than 5% of that time. In Fig. 15 a starting point rainfall of 1 in. has been assumed and the gradients calculated in the direction of lower rainfall from the

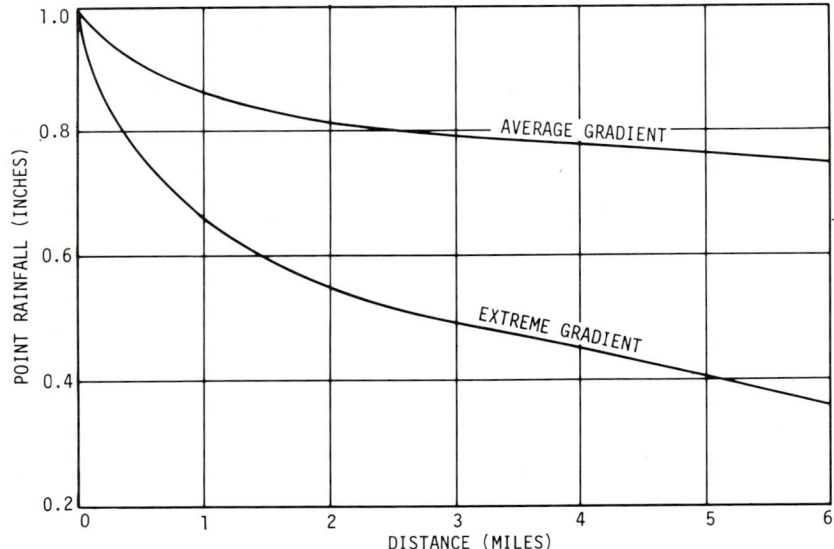

Fig. 15. Average and extreme rainfall gradients for 1-in. storm.

starting point. Thus, at 6 miles, the rainfall would have decreased from 1 in. to 0.75 in. under average conditions, but would have decreased to 0.36 in. or more in 5% of the storms of this magnitude. Empirically, the rainfall gradient in Illinois storms was found to be related to the cube root of the distance between sampling points and the square root of the rainfall [16]. Rainfall volume was found to be the most important independent variable in defining storm rainfall gradients and empirical prediction equations were not improved significantly by the addition of other rainfall parameters.

Table IV, taken from [17], shows the average gradient of 1-min point

TABLE IV. Average variation of point rainfall rates with distance

Starting point rate (in./hr)	Average difference (%) for given distance (mi)					
	1	2	4	6	8	10
0.1	64	74	81	87	90	93
0.2	61	69	78	82	85	87
0.5	56	63	71	75	77	79
1.0	52	58	65	69	72	74
2.0	49	54	60	64	67	69
5.0	45	49	55	58	60	62

rainfall with increasing distance from a given reference point for various rate intensities and distances. It provides a quantitative estimate of the magnitude of instantaneous rainfall rate gradients in convective cells, information which should be of interest to cloud physicists and others concerned with the mechanisms of precipitation development in clouds.

The major difference in monthly precipitation gradients between the growing season and cold season was found with relatively light monthly amounts of 2 in. or less. The growing season exhibited substantially larger gradients in these months. In general, however, total monthly rainfall did not exert a strong control on the variation of point rainfall with distance.

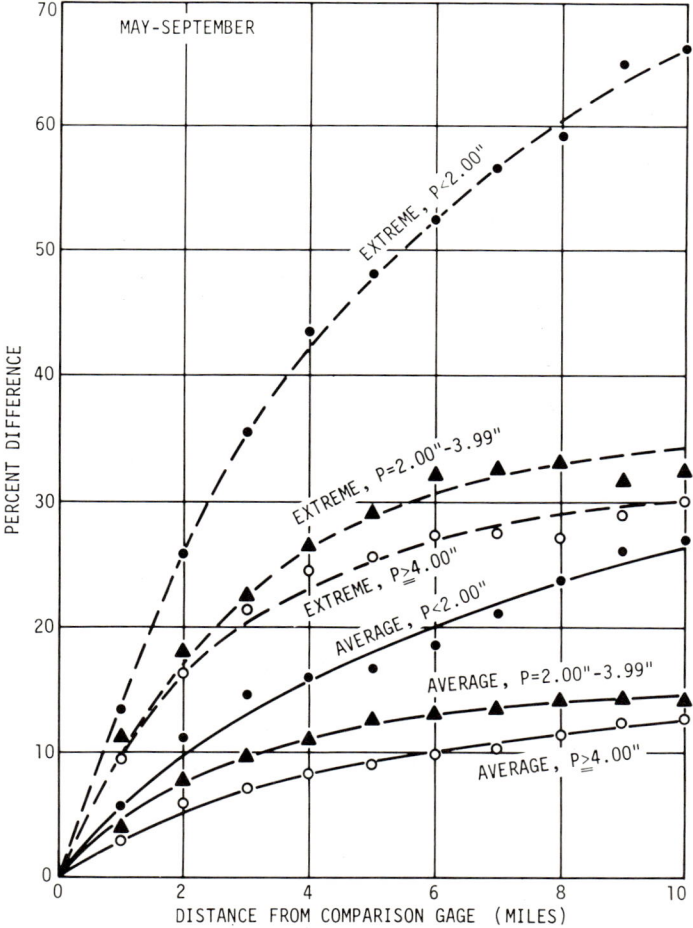

Fig. 16. Variation of monthly precipitation with distance in May–September period.

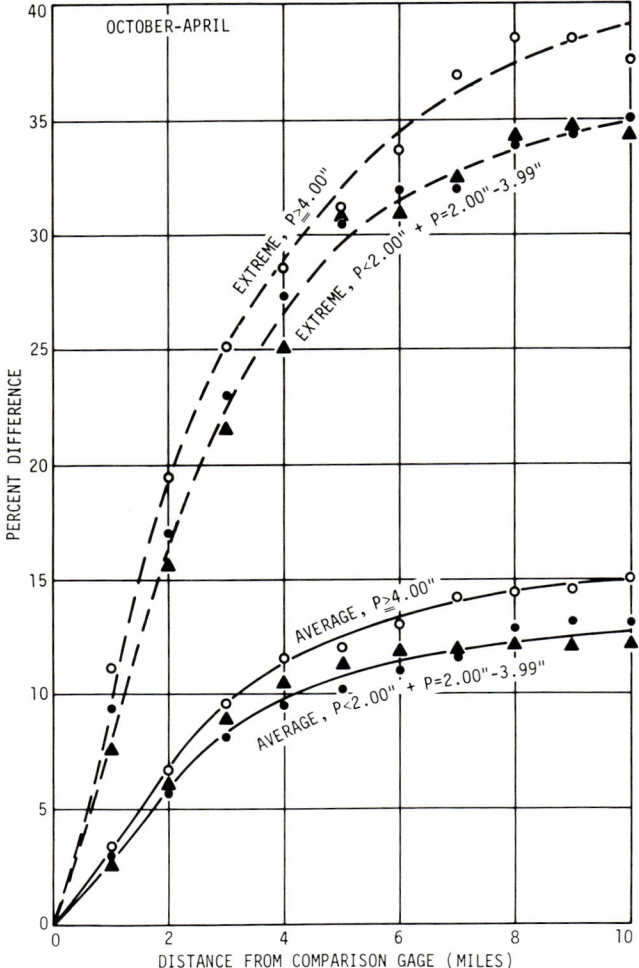

FIG. 17. Variation of monthly precipitation with distance in October–April period.

Instead, it was found that the monthly precipitation gradients were controlled to a large extent by the heavy storm patterns experienced during the month rather than the total volume of monthly precipitation.

Results of monthly analyses are summarized in Figs. 16 and 17 for the warm and cold seasons. The solid curves show the average difference from the comparison gage in percent with increasing distance. The extreme gradients (dashed lines) are defined by the upper 95% confidence band (2 standard deviations). Because differences were small and exhibited an erratic trend, the October–April curves for monthly precipitation of less than 2.00 in. and 2.00–3.99 in. have been combined.

The monthly analyses indicated that dense networks would be required to determine small precipitation changes with distance in specific months, as might be required in weather modification verification programs. For example, it was found that the average monthly difference during May–September reached 10% at a distance of 2 miles from a given point for monthly totals less than 2 in., and at 3 miles for amounts of 2–4 in.

The May–September total rainfall was used as a measure of the point rainfall gradient on a seasonal basis. The percentage variation in precipitation with distance decreased substantially from that in monthly precipitation, and the average remained essentially constant after reaching a value of 6% at 3 miles, as illustrated in Fig. 18.

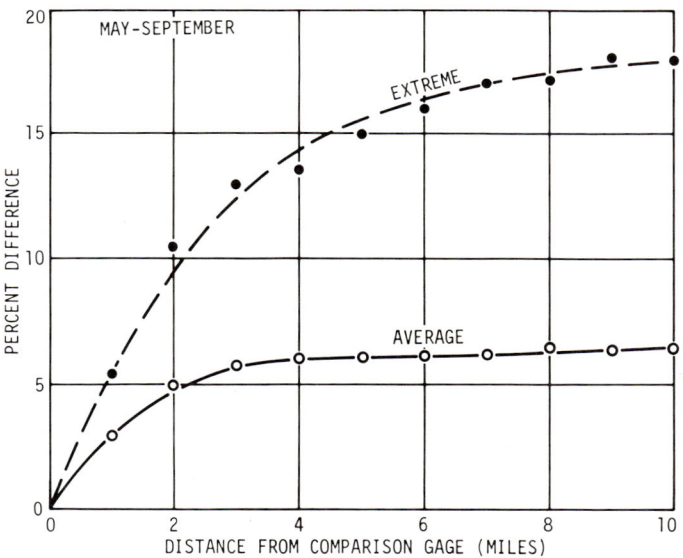

FIG. 18. Variation of seasonal precipitation with distance.

Extremely dense raingage networks would be required to measure accurately the spatial distribution of instantaneous rainfall rates. The operation, data collection, and data reduction would be extremely difficult with an adequate network of gages extending over hundreds or thousands of square miles as would be required in most weather modification projects. Unless radar can be adapted to the measurement of rainfall rates, the rate parameter is not very promising as an evaluator of weather modification effects through quantitative comparisons of rainfall intensity patterns or point intensities in seeded and nonseeded situations.

The problem lessens gradually as time integration increases, and the background interference from natural variations becomes relatively small with seasonal rainfall. Therefore, comparison of seasonal patterns and point rainfall differences on a seasonal basis should be useful in the verification of seeding-induced effects. The change in rainfall gradient with increasing time integration of precipitation is illustrated by the following example from the Illinois studies. The average percentage difference in point rainfall rates 2 miles apart with a moderate summer rate of 0.5 in./hr is 63%. With a moderate monthly rainfall of 3.50 in. in the May–September period, the average 2-mile difference is 8%, and this decreases to 5% for the five-month season, May–September.

4.3. Time Distribution Models of Storm Rainfall

If one hypothesizes, as have some investigators in weather modification, that cloud seeding will increase the duration and decrease the intensity of rainstorms, then the natural time distribution properties should be substantially modified in seeded storms. This modification constitutes a potential means of verifying the cloud seeding treatment, provided that the natural distribution has properties sufficiently consistent and stable so that modest changes can be recognized above the background noise level of the natural variability.

In an earlier study [18], time-distribution relations were developed from a sample of 260 storms on the East Central Illinois Network (Fig. 1). These were in the form of statistical models for storms of moderate to heavy intensity on areas ranging up to 400 miles^2. Rainfall distributions were grouped according to whether the heaviest rainfall occurred in the first, second, third, or fourth quarter of a storm. The models were presented in probability terms to provide quantitative information on interstorm variability and to provide average and extreme relations for various applications of the findings. It was found that the relations could be represented best by relating percent of storm rainfall to percent of total storm time and grouping the data according to the quartile in which rainfall was heaviest. The individual effects of mean rainfall, storm duration, and other storm factors were small and erratic in behavior when the foregoing analytical technique was used. Size of sampling area had a small but consistent effect upon the time distribution. This is illustrated in Table V in which differences between the average time distribution curve for areas of 50 to 400 miles^2 combined and specific area curves are shown for first-quartile storms at the 50% probability level.

The time distributions were expressed in probability terms because of the great variability in the characteristics of the distribution from storm to storm. Numerous factors contribute to the storm variance, but no single parameter dictates the characteristics of the distribution. Among the factors

TABLE V. Differences between average curve and specific areas for 50% probability level in first-quartile storms

Area (mi²)	Difference (%) for given cumulative percent of storm duration								
	10	20	30	40	50	60	70	80	90
Point	−9	−1	+5	+6	+6	+6	+5	+4	+3
10	−11	−3	+2	+1	+3	+2	+2	+2	+1
50	−2	+3	+3	+2	+2	+2	+1	0	0
100	−2	−3	0	0	0	0	0	0	0
200	−2	−3	−2	−1	−1	0	0	0	0
400	+6	+4	−1	−2	−2	−1	−1	0	0

are the stage of development of the storm, the size and complexity of the storm system, rainfall type, synoptic storm type, location of the sampling area with respect to the storm center, and the movement of the storm system across the sampling region. Probability distributions allow selection of a time distribution most appropriate for a particular application.

The statistical models discussed above were derived from data for relatively heavy storm intensities. However, as a result of other analyses performed on 1-min rainfall data [17], it was concluded that the results of the heavy-storm study are applicable, as a first approximation, for deriving the time-distribution characteristics of all warm-season storms.

Examination of the various statistical models derived with dense raingage network data indicated that their use as a primary verification tool in rainfall modification experiments is not promising. The interference level of natural variability between storms appears too great for the detection of modest changes in the time distribution that might result from cloud seeding.

The magnitude of interstorm variability in the time distribution characteristics of storm rainfall is quite apparent in Fig. 19 which shows the average statistical model derived from first-quartile storms on areas of 50 to 400 miles². This quartile accounted for approximately one-third of the total number of storms. The model is typical of midwestern, warm-season storms or cold-season storms in which unstable rain types are the major rain producers. Probability levels are shown from 10 to 90%, but the 50% level (median) has been stressed by a heavier line, since it is probably the most useful statistic from the standpoint of most users of rainfall data. The curves reflect the average rainfall distribution, and, therefore, do not exhibit the burst characteristics of a mass rainfall curve.

In Fig. 19, the 10% curve is typical of storms in which the rainfall is concentrated in an unusually short portion of a storm. It indicates a chance of

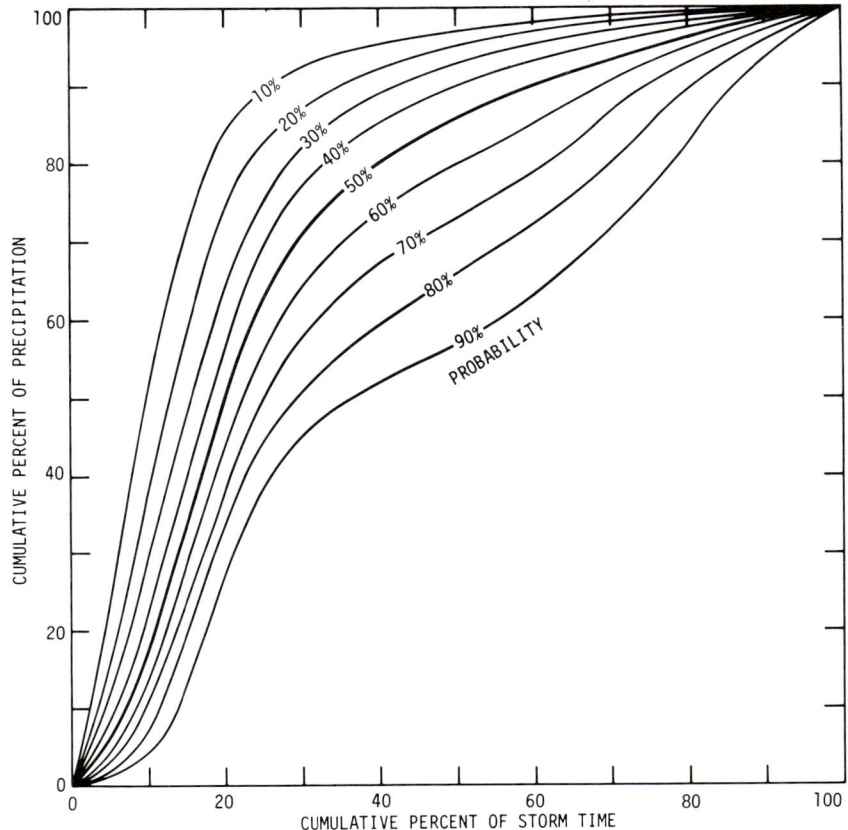

Fig. 19. Time distribution of first-quartile storms.

one in ten that a first-quartile storm will have at least 89% of its rainfall in the first quarter of the storm period and over 95% in the first one-half of the storm. The 50% curve shows 63 and 86% of the rainfall at 25 and 50% of the storm period. The 90% curve reflects an unusually uniform distribution for first-quartile storms. It may be interpreted as the distribution that will occur in 10% or less of the storms. Thus, this curve shows that in 10% of the storms, 39% or less of the rain will occur in the first quarter of the storm, and 57% in the first one-half of the storm.

The curves at various probability levels are indicative of certain storm types. For example, the 10% probability curve of first-quartile storms is the type of distribution associated with relatively short duration storms, such as the passage of an intense, prefrontal squall line in which light rain falls from the middle cloud deck system for substantial periods following the major rain bursts. Similarly, the distribution at the 90% level is likely to be associated with longer duration storms, in which the rain is more evenly distributed during the storm period, and is often dominated by a series of rainshowers or a combination of showers and steady rain.

4.4. Sequential Variability and Lag Correlations

In the Illinois studies, interstorm variability of areal mean precipitation was investigated through calculations of sequential variability and lag correlations between storms to provide additional information on time distribution properties. Storm data from the East Central Illinois Network (Fig. 1) for 1960–1964 were grouped by May–September and October–April periods for this purpose. The sequential variability [13] takes account of both the magnitude and sequence of the sample mean precipitation in characterizing the time variability, and, therefore, was considered superior to the standard or average deviation for evaluation purposes. Lag correlations for lags of 1 to 30 storms were used to investigate the presence of time trends.

In general, there was relatively large sequential variability and lack of correlation between storms in both seasons tested [4]. For example, in the five-year period 1960–1964 the lag correlations between consecutive storms was in the range of 0.0 to 0.3 in all years and less than 0.2 in four of the five years. Even within storms the lag correlation decreases rapidly [17]. In the referenced study, correlation coefficients of 1-min point and areal mean rainfall on a 50-gage network in 100 miles^2 were found to approach zero at an average lag of 15 min in warm season storms.

Overall, the analyses of sequential variability and lag correlations in a sample of over 500 storms provided additional support for the previous conclusion that time distribution parameters have limited applications as primary verification tools in rain modification experiments.

4.5. Diurnal Distribution of Storm Precipitation

Conceivably, the diurnal distribution of storm precipitation is another time distribution parameter that could be useful in the evaluation of weather modification experiments. Thus, if the diurnal distribution for a given region and season remained relatively stable from year to year and cloud seeding substantially altered the magnitude or time sequence of the maxima and minima, then the diurnal distribution could serve as a valuable verification tool. Cloud seeding is frequently performed in late forenoon and afternoon to attack developing cumulus in summer when agricultural requirements for water maximize. This seeding, if effective, would very likely affect the intensity and, perhaps, the time distribution of the afternoon maxima common in many regions of the country.

Consequently, a limited study of the natural diurnal distributions was undertaken through use of data from the East Central Illinois and Little Egypt Networks (Fig. 1) and from recording raingage stations of the U.S. Weather Bureau in the midwest [4]. The overall conclusion was that natural variability is a major problem in verifying seeding effects through use of diurnal rainfall distributions. However, it was also concluded that natural variability is no more of a problem than it is with other rainfall parameters that could be used in verification. Therefore, the statistical evaluation of seeding effects on diurnal distributions could serve as one of several tools in verification procedures. Evaluation of cloud seeding effects solely from measurements of point or areal mean precipitation, as frequently done in past experiments, is deplorable in that it wastes other useful precipitation measurements that could facilitate and improve the reliability of the verification.

Problems resulting from natural variability in diurnal distributions when comparing rainfall samples collected on seeded and nonseeded days will be illustrated and discussed in more detail in Section 8 which deals with downwind seeding effects from the Project Whitetop experiments.

5. Other Relevant Climatological Studies

In addition to the studies discussed in previous sections, it is most desirable to make a number of fundamental climatological studies that can provide useful background knowledge for designing precipitation modification experiments. In the following paragraphs, several of these studies will be illustrated through reference to the Illinois project. Although network data were employed in some of the studies, acceptable estimates of most climatological factors discussed here can be obtained from analyses of point precipitation data available in U.S. Weather Bureau publications.

5.1. Climatological Distribution of Areal Mean Precipitation

Network studies were made to determine the relationship between total precipitation, storm intensity, and frequency of storm occurrences on a seasonal and annual basis. Results showed that a large percentage of the total precipitation occurs in a small percentage of the storms, as illustrated in Table VI. On the average, approximately 50% of the total precipitation was found to occur in 10% of the storms and 95% in 50% of the storm occurrences. Also, the distributions appear to be quite insensitive to size of sampling area. Obviously, these climatic realities should be taken into consideration in the design of weather modification experiments.

The distribution of precipitation with respect to storm duration, synoptic weather type, and precipitation type can also provide useful information in

TABLE VI. Percentage distribution of total precipitation in storms on East Central Illinois Network, 1955–1966

Cumulative percent of total precipitation	Cumulative percent of total storm occurrences for given sampling period and area					
	May–September		October–April		Annual	
	50 miles^2	400 miles^2	50 miles^2	400 miles^2	50 miles^2	400 miles^2
5	0.5	0.4	0.4	0.4	0.5	0.5
10	1	1	1	1	1	1
20	3	3	3	3	3	3
30	5	4	5	4	5	5
40	8	7	7	7	7	7
50	11	10	10	10	10	10
60	15	13	13	14	14	14
70	20	19	19	18	19	19
80	28	27	25	25	27	27
90	41	37	41	40	39	40
95	51	50	52	53	52	51

TABLE VII. Average distribution of precipitation on East Central Illinois Network grouped by storm duration

Storm duration (hours)	Percent of total precipitation		Percent of storms	
	May–September	October–April	May–September	October–April
$\lesssim 3$	22	9	57	35
3.1– 6.0	21	11	20	19
6.1–12.0	20	21	12	22
12.1–24.0	23	40	8	19
24.1–48.0	13	16	3	5
>48.0	1	3	0+	0+

designing cloud-seeding experiments. As shown in Table VII, storm durations of three hours or less are most frequent in Illinois storms in both warm and cold seasons. A greater percentage of the total precipitation occurs in relatively short-duration storms during the growing season than in the October–April period. However, augmentation of precipitation from short-duration storms by 10 to 20% [12] would not increase the total precipitation substantially in either season.

Stratification of the precipitation distribution according to synoptic types provides the weather modifier with knowledge of the comparative importance of various types of storms as water producers. In the Illinois studies, this was done on both a seasonal and annual basis. Table VIII illustrates the con-

TABLE VIII. Average distribution of May–September rainfall on East Central Illinois Network grouped by synoptic storm type

Synoptic type	Percent of total rainfall	Percent of storms
Cold Front	39	36
Warm Front	14	11
Static Front	21	17
Occluded Front	2	2
Low Centers	7	8
Air Mass	17	26

tributions of rainfall from frontal and nonfrontal types in the growing season. The need to seed frontal type storms is apparent, since air mass storms contribute only 17% of the central Illinois rainfall.

Stratifications by precipitation type showed that thunderstorms predominate among warm-season rainfall types. Frequently, thunderstorms are the major producer of storm rainfall, but are intermingled with rainshowers during lighter rain periods. On the East Central Illinois Network, the combination of thunderstorms and rainshowers accounted for 88% of the May–September rainfall in the 1955–1966 period and were associated with 87% of the storm occurrences. In the cold season, the rainshower–thunderstorm combination were associated with 40% of the storm occurrences, but accounted for 55% of the total seasonal precipitation. Stable-type rains accounted for approximately 30% of the precipitation and storm occurrences in the October–April period. Snow or snow mixed with rain were recorded in nearly 30% of the storms, but accounted for only about 14% of the total seasonal precipitation.

Another climatic factor useful in planning precipitation modification experiments is the sequential distribution of wet days that occur naturally. This type of information is especially useful in those experiments where the residual effects of seeding on a target are of concern. Table IX shows the distribution of wet-day sequences on the East Central Illinois Network during a ten-year sampling period. Annually, approximately two-thirds of the storms did not last more than one day.

TABLE IX. Wet day sequences on East Central Illinois Network, 1955–1964

Number of consecutive days with precipitation	May–September		October–April		Annual	
	Total cases	Percent of total	Total cases	Percent of total	Total cases	Percent of total
1	173	60	257	72	430	67
2	70	24	62	18	132	20
3	24	8	21	6	45	7
4	14	5	12	3	26	4
5	5	2	3	1	8	1+
≥ 6	4	1	1	0+	5	1−

5.2. Application of Published Climatic Data on Point Precipitation

Some of the useful information that can be obtained readily from U.S. Weather Bureau publications of climatic data is illustrated in Table X. Based on the 50-year period, 1906–1955, the average percentage distribution of annual precipitation, stratified by daily amounts, is shown for stations representative of the northern, central, and southern parts of Illinois.

It is evident that, whereas daily precipitation falls most frequently into the light class, 0.01–0.10 in., only a very small amount of the total annual precipitation is accounted for on these days. Thus, if weather modification could

TABLE X. Average percentage distribution of annual precipitation grouped by daily amounts

Station	Percent of total precipitation from given daily amount (in.)			
	0.01–0.10	0.11–0.50	0.51–1.00	>1.00
Chicago	8	35	28	29
Peoria	5	32	29	34
St. Louis	5	29	28	38
Cairo	4	26	28	42
	Percent of days with given daily amount (in.)			
Chicago	48	36	11	5
Peoria	43	37	13	7
St. Louis	41	38	13	8
Cairo	39	36	15	10

either significantly increase the number of these occurrences or increase the rainfall amounts by 10 to 20%, or even 50%, the net result, on the average, would be only a small percentage increase in total precipitation. Climatologically, significant modification of precipitation on those days with amounts from 0.11 to 0.50 in. would appear to be most rewarding, since this group of storms accounts for over one-third of the total occurrences annually and nearly one-third of the total annual precipitation in Illinois. A substantial portion of the annual precipitation results from days with over one inch, but these days are relatively infrequent and the desirability of further increasing precipitation on these days of heavy storms is questionable, even if it were feasible. As shown in [4], published climatic data can be used also to determine climatological distributions similar to those in Table X for dry periods when cloud seeding would be most in demand.

An examination of the distribution of hourly precipitation, readily available from Weather Bureau records, can provide useful information for the weather modifier in determining potential benefits from cloud seeding and in planning seeding operations. Table XI shows the annual and seasonal average

TABLE XI. Annual and seasonal distribution of hourly precipitation in central Illinois

Period	Average number of hours per year	Percent of total hours with given hourly amounts (in.)				
		0.01–0.10	0.11–0.25	0.26–0.50	0.51–1.00	Over 1.00
Winter	135	90.7	7.6	1.2	0.3	0.1—
Spring	151	84.8	11.2	3.0	0.8	0.2
Summer	90	72.1	15.0	8.0	3.7	1.2
Fall	89	84.2	11.4	3.4	0.9	0.1
Annual	465	83.8	11.1	3.5	1.3	0.3

distributions of hourly precipitation as derived from Weather Bureau recording gages in central Illinois. From these data, it is apparent that precipitation modification undertaken in any season to increase the output from existing storms would be performed predominately on storms in which hourly amounts at a given point are 0.10 in. or less. An appreciable percentage of the hourly amounts exeed 0.25 in. only in summer when the central Illinois average is approximately 13%. The statistics in Table XI indicate that seeding operations would usually need to be carried on successfully for several consecutive hours on storm days to achieve substantial increases in total precipitation.

The diurnal distribution of total precipitation, frequency of precipitation occurrences, and average hourly rates are other climatological parameters that need to be considered in planning rain modification experiments and which can be readily derived for a given region from existing publications of hourly precipitation amounts from recording raingages of the U.S. Weather Bureau. In many areas, the natural distribution of snowfall is of major importance in the planning of cloud seeding experiments during the cold season. For example, intensification of naturally heavy snowstorms would normally be an overall disbenefit. Changnon [19] has indicated that severe winter storms which produce heavy snowfall in excess of 6 in. produce more damage in Illinois than any other form of severe weather. In areas where snowfall is a major form of precipitation, however, seeding of snowstorms would be required for substantial seeding-induced precipitation.

5.3. Frequency Distribution of Various Storm Factors

The frequency distribution of a number of storm factors not mentioned previously can also provide useful background information for those involved in weather modification, hydrologists concerned with structural design, and climatologists. For example, in the Illinois studies distributions were derived for the frequency of rain bursts (rain cell passages) stratified according to storm duration and quartile-type of storm, along with the time distribution and relative magnitude of maximum bursts in multiburst storms [18]. Other investigations were made of: the distribution of storm centers on areas of 50–400 miles^2 and the influence of storm mean rainfall and rainfall duration on the number of centers; the distributions of rain type, storm type, and storm duration in storms with different time distribution characteristics; and, the shape and orientation properties of storms.

6. Precipitation Measurement Requirements

The accurate measurement of precipitation is important not only in weather modification activities with which we are concerned here, but in numerous other fields such as hydrology, climatology, and agricultural research. Bergeron [20] has shown the importance of accurate rainfall measurements in studies of the mechanisms producing various types of precipitation. Dense networks operated in conjunction with his Project Pluvius in Sweden have provided an outstanding example of their use in revealing the substantial effects of small-scale changes in topographic features on the volume and areal distribution characteristics of precipitation.

As part of the precipitation evaluation studies discussed in this review paper, investigations were made of measurement requirements in storm, monthly, and seasonal precipitation. This was done through analyses of correlation patterns of point and areal mean precipitation, sampling errors in the measurement of areal mean precipitation, and rain-gaging requirements for detecting and measuring the areal extent of storms on sampling areas of various sizes.

6.1. Spatial Correlation of Point Precipitation

One approach to describing sampling requirements for precipitation measurement networks is through statistical correlation methods. Data from the Little Egypt and East Illinois Networks of Fig. 1 were used to investigate point correlation patterns on a storm, monthly, and seasonal basis. In addition, data from 3142 min in 29 storms during 1952–1953 on the Goose Creek Network of 100 miles^2 in central Illinois were used to obtain correlation patterns for 1-min rainfall amounts, the best available estimate of instantaneous rainfall rates. This special network, installed for radar-rainfall research, contained 50 recording gages with 12.6-in. orifices and 6-hr gears which permitted reading of 1-min amounts [17]. Spatial correlation patterns were established for the data grouped according to season, precipitation type, synoptic weather type, storm duration, and storm movement.

Correlation decay with distance, used to indicate sampling requirements for establishing rainfall patterns, was greatest in thunderstorms, rainshowers, and air mass storms. Conversely, minimum decay occurred with steady rain and the passage of low-pressure centers. Seasonally, the correlation decay with distance in storms was much greater in the growing season, May–September, than during the October–April period. For example, combining all storms and assuming a requirement for 90% explained variance, on the average, a gage spacing of 2 miles would be needed in the warm season compared with 6 miles in the cold season. However, if similar accuracy is required in air mass storms, the gage spacing must be decreased to 1 mile. If measurements are to be made only in low center passages, a spacing of 8–10 miles would be adequate for the above accuracy level. Examples of the average decay of correlation coefficient with distance in warm season storms on the dense raingage networks are shown in Figs. 20 and 21 for selected synoptic weather and rainfall types, and for 1-min rainfall rates.

Spatial correlation increased, on the average, with increasing duration in storms lasting up to approximately 12 hr, after which the trend reversed. Erratic trends were found when the storms were grouped according to network mean precipitation. As expected, general improvement in correlation occurred when the storms were grouped by wind direction and storm movement.

Directionally, slightly higher correlations were obtained among stations oriented in W–E and SW–NE directions compared with those lying in N–S

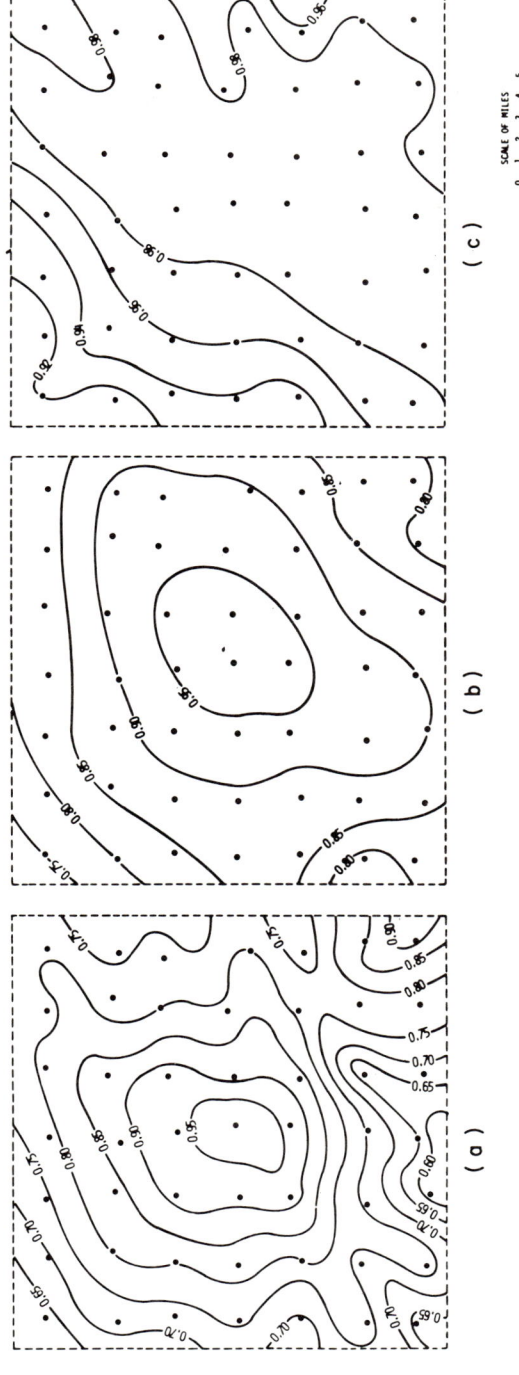

Fig. 20. Correlation patterns of synoptic types in warm season storms. (a) Air-mass storms; (b) frontal storms; (c) low centers.

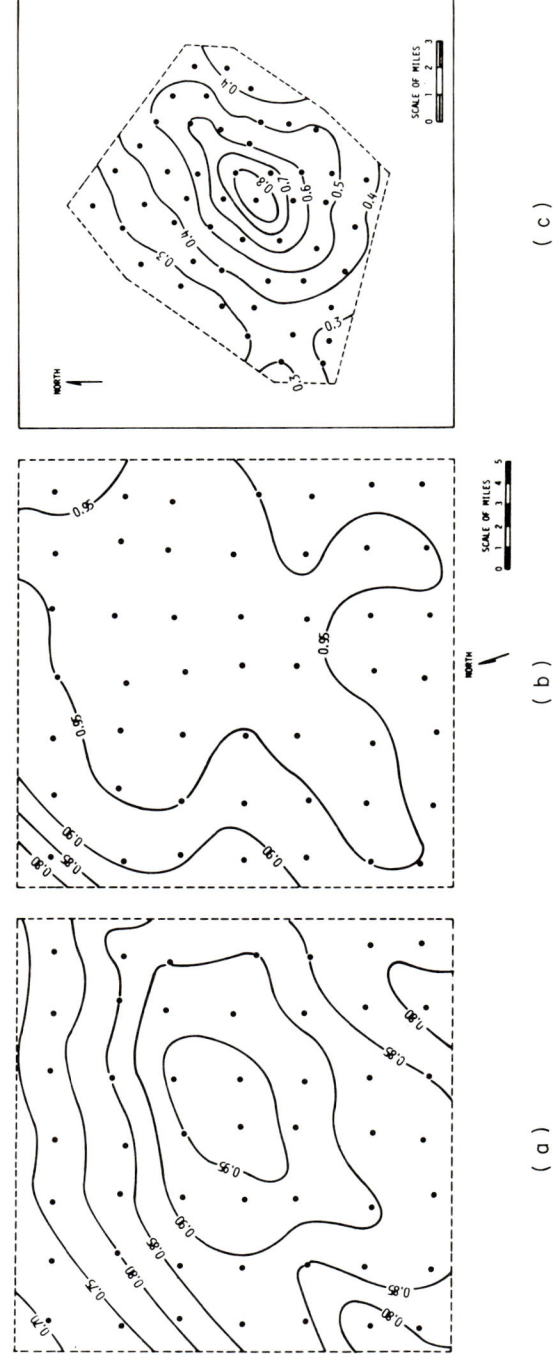

FIG. 21. Correlation patterns associated with rain types and rain rates in warm-season storms. (a) Thunderstorms and rainshowers; (b) steady rain; (c) one-minute rainfall rates.

and NW–SE directions across the networks, reflecting the more frequent movement of storms from the SW and W directions. Large differences were not found between total storm, monthly, and seasonal correlation relations, so that a total storm sampling network should satisfy sampling needs for all of these periods. However, sampling requirements are much greater when the measurements of rainfall rate are needed. For example, with a minimum acceptance of 75% explained variance between sampling points, it was found that a gage spacing of approximately 0.3 mile would be needed for 1-min average rates compared with 7.5 miles for total storm rainfall in warm season storms.

TABLE XII. Variation of average correlation coefficient with distance about central gage in East Central Illinois and Goose Creek Networks during May–September storms

Group	Average correlation coefficient for given distance (miles)					
	1	2	4	6	8	10
Precipitation types						
TRW-RW	0.98	0.96	0.92	0.88	0.85	0.82
R	0.99	0.98	0.96	0.94	0.93	0.92
Synoptic weather types						
Fronts	0.98	0.96	0.94	0.91	0.88	0.86
Low centers	1.00—	0.99	0.99	0.98	0.97	0.96
Air mass storms	0.97	0.94	0.87	0.79	0.76	0.74
Storm duration groups						
$\lesssim$ 3 hr	0.96	0.91	0.82	0.75	0.70	0.65
3.1–6.0 hr	0.97	0.95	0.90	0.86	0.81	0.76
6.1–12.0 hr	0.98	0.96	0.93	0.91	0.89	0.87
12.1–24.0 hr	0.97	0 95	0.82	0.72	0.69	0.66
All storms combined	0.98	0.96	0.93	0.89	0.86	0.84
1-min rain rate (Goose Creek)	0.77	0.60	0.40	0.31	—	—

The correlation decay with distance is illustrated further in Table XII, for several data stratifications in warm season storms. Table XIII provides information on gage spacing in various types of warm- and cold-season storms required to maintain selected levels of correlation and variance explained. The cutoff of 20 miles in this table was dictated by the network sizes. A detailed description of all phases of the correlation studies is provided in [21].

TABLE XIII. Average storm relation between raingage spacing and correlation coefficient

r	r^2 (%)	May–September storms							
		All storms	TRW, RW	R + RW	R	Lows	Fronts	Air mass	1-min rates
		Spacing (mi) for given correlation (r) and explained variance (r^2)							
0.95	90	2	2	3	10	10	2	1	0.2
0.90	81	5	4	7	>20	>20	5	3	0.4
0.85	72	8	7	15			9	5	0.6
0.80	64	12	10	>20			13	7	0.8
0.75	57	17	14				18	9	1.0
0.70	49	>20	>20				>20	13	1.2

r	r^2 (%)	October–April storms						
		All storms	TRW, RW	R, R + RW	Lows	Fronts	S	R + S
		Spacing (mi) for given correlation (r) and explained variance (r^2)						
0.95	90	6	5	13	8	4	1	2
0.90	81	18	10	>20	>20	13	2	8
0.85	72	>20	15			>20	4	13
0.80	64		20				6	>20
0.75	57						11	
0.70	49						20	

6.2. Correlation of Storm Mean Precipitation between Areas

To aid further in the clarification of sampling problems in weather modification experiments, analyses were made of the correlation of storm mean precipitation between contiguous and separated sampling areas. Data from the East Central and Boneyard Networks in the 1960–1964 period were used in this study.

The contiguous sampling areas were the southwestern and northeastern halves of the East Central Illinois Network. The centers of these contiguous areas are approximately 14 miles apart. Several methods of stratifying the precipitation data were investigated to evaluate seasonal and meteorological effects on the degree of correlation. The analyses were restricted to storms in which the areal mean equaled or exceeded 0.01 in. on the 400-miles² network.

Correlation coefficients were determined between the 400-miles² network and the Boneyard urban network of 10 miles² to obtain a measure of the change in correlation of storm mean precipitation with increasing distance between areas. The centers of these networks are approximately 30 miles apart in a W–E direction, and their closest boundaries are separated by approximately 16 miles.

Combining all storms with network means of 0.01 in. or more, the results indicated that correlation of mean storm precipitation between contiguous areas is somewhat better in the October–April period (0.99) than in the May–September period (0.94). Stratification of precipitation according to mean rainfall may be desirable in the growing season period for establishing verification procedures in weather modification. However, there appears to be no significant advantage in doing so in the colder part of the year, according to the various correlation analyses performed in this study.

The areal correlations provided some support for data grouping by synoptic weather and precipitation types. For example, air mass storms had a correlation coefficient of 0.88 in warm-season storms between contiguous areas compared with 0.94 for frontal storms and 0.99 for low-center passages. Steady rains (0.99) correlated somewhat better than the showery types (0.93).

As expected there was a substantial difference in correlation between the contiguous areas and the East Central Illinois–Boneyard combination. Combining all data for the five years, there was a decrease in explained variance of 23% when the distance between areal centers was increased from 14 to 30 miles. Table XIV illustrates how the correlation may vary between nearby

TABLE XIV. Correlation of storm mean precipitation between separated areas in May–September storms

Year	Correlation coefficient	Variance explained (%)
1960	0.82	67
1961	0.80	64
1962	0.81	66
1963	0.64	41
1964	0.96	93
Average	0.81	66
Median	0.82	66
1960–1964 combined	0.77	60

areas between years. Such behavior complicates the verification problem, and provides further justification for the use of cross-over in preference to fixed target-control sampling designs.

6.3. Sampling Errors in Measurement of Mean Precipitation

Evaluation of cloud-seeding activities has been attempted frequently in the past through comparisons of the mean precipitation recorded on seeded and nonseeded areas. Consequently, sampling requirements for the accurate measurement of this precipitation parameter is essential to reliable evaluations.

Evaluation of sampling requirements under various meteorological conditions has been hampered greatly in the past by the lack of dense rain-gage networks to provide suitable data for this purpose. Several investigators have made limited studies in which sampling error was related to storm mean rainfall and size of sampling area to obtain estimates of gage density requirements. Among these were Light [22], Linsley and Kohler [23], Huff and Neill [24], and McGuiness [25].

In a more recent study briefly summarized here [26], the effects of season, precipitation type, synoptic weather type, and storm duration on the sampling error of storm mean rainfall estimates were investigated, in addition to the usual parameters, mean precipitation and size of sampling area. Also, the study was extended to include rainfall rate, monthly, and seasonal relations.

Data from the East Central Illinois and Little Egypt Networks (Fig. 1) collected during the 1955–1966 period were used in the storm, monthly, and seasonal analyses. Data from the 1952–1953 Goose Creek Network, equipped to measure 1-min rainfall amounts, were used in the rainfall rate analyses. Data from the Shawnee and Little Egypt combined networks (Fig. 1) were available for a three-year period, 1965–1967, and these were used to obtain first approximations of sampling requirements for areas of the order of 1000 miles^2.

Storm sampling error was related to mean precipitation, storm duration, gage density, and size of sampling area with the network data grouped according to season, synoptic storm type, and precipitation type. Approximate normalization of the storm data was achieved through use of logarithmic transformations; and equations, such as the following for May–September storms, were then derived from the network data:

(6.1) $$\log E = -1.5069 + 0.65P + 0.82G - 0.22T - 0.45A$$

where E is the average sampling error in inches, P is the areal mean precipitation in inches, G is gage density in miles^2 per gage, T is storm duration in hours, and A is area. E represents the difference between the best estimate of the true mean P, obtained from the maximum density of raingages on each network, and the sample mean precipitation P_g calculated from the gage amounts for a given value of G.

The average sampling error in storms was found to increase with increasing areal mean precipitation and to decrease with increasing sampling density (gages per unit area) and storm duration, as illustrated in Fig. 22. Relatively large differences were found frequently in the sampling errors between storms of apparently similar characteristics. This emphasizes the difficulty

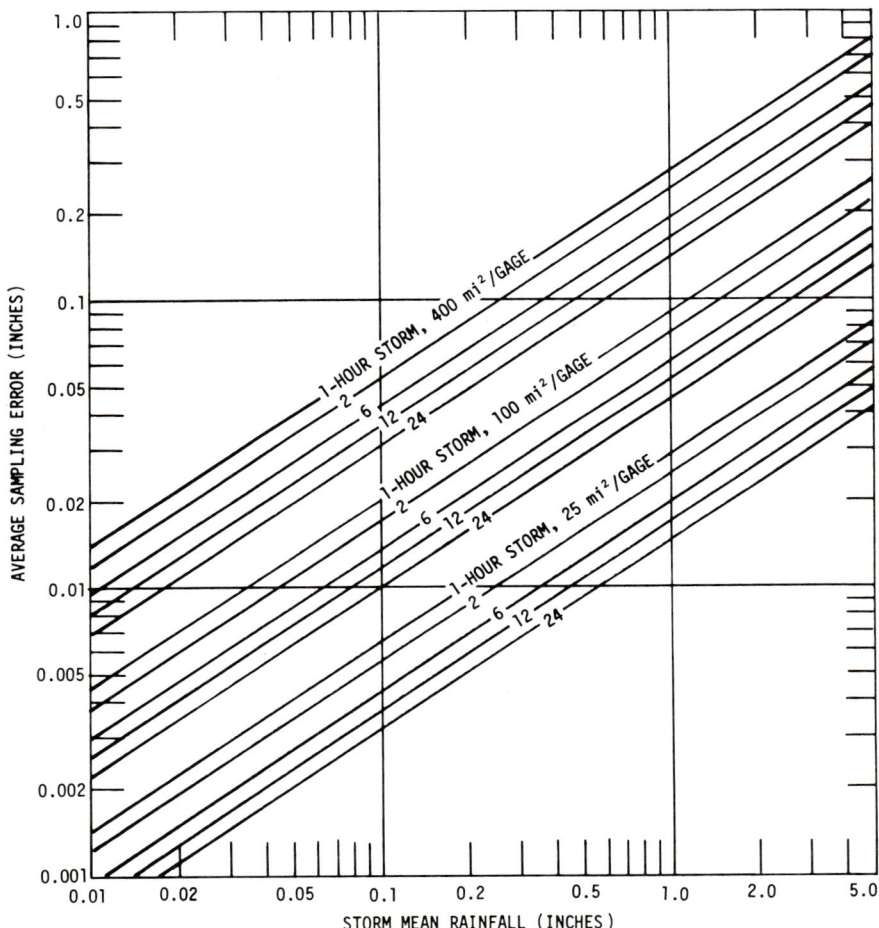

FIG. 22. Average sampling error relations on 400 miles^2.

in predicting sampling errors for specific storms with a given sampling density. The interstorm variability results from the dependency of the sampling error upon numerous factors, some of which can not be readily expressed in mathematical terms. The great amount of variability about the average sampling error is illustrated in Fig. 23. Here, sampling error (percent) has been plotted against network mean rainfall (inches) for all May–September storms in the ten-year period 1955–1964 in which the areal mean was 0.01 in. or more and storm duration was 3 hr or less. Since sampling error changes slowly with storm duration, this duration grouping should contribute only a

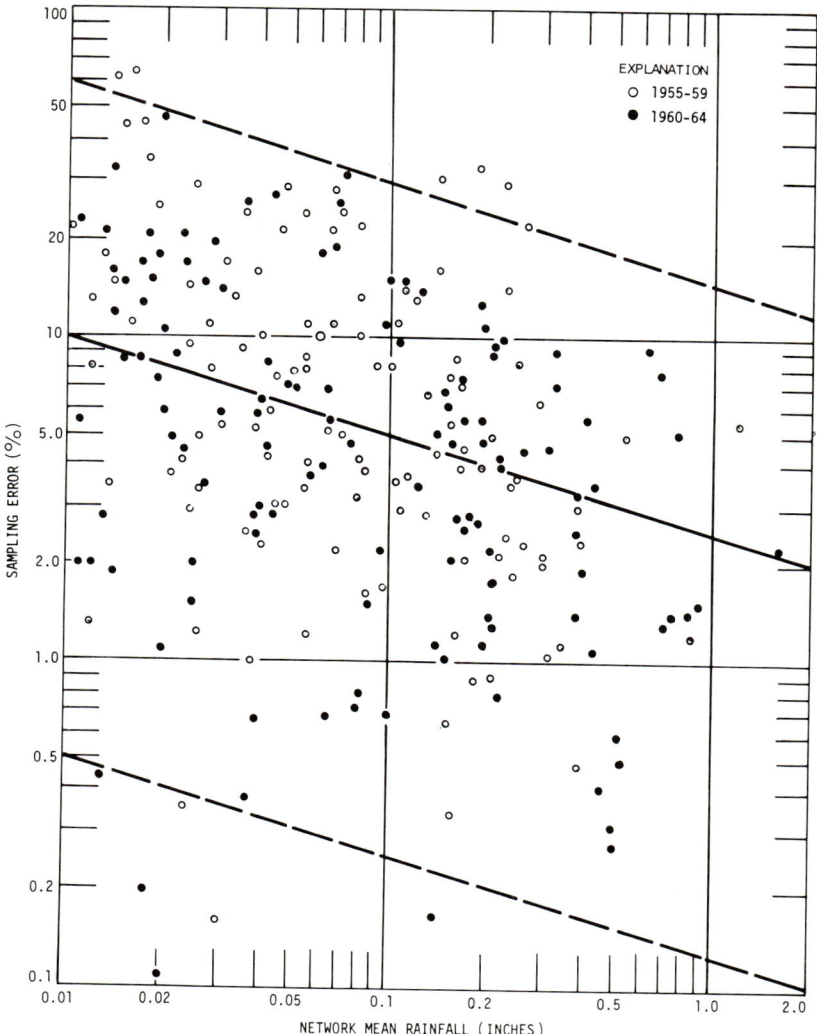

FIG. 23. Sampling errors with gage density of 16 miles^2/gage in May–September storms on 400 miles^2.

small portion of the interstorm variation shown in the graph. The sampling errors are for a gage density of 16 miles^2/gage (25 gages) on the 400-miles^2 network. The solid line represents the median and the dashed lines encompass 95% of the observations. Although the absolute sampling error increases with increasing mean rainfall (Fig. 22), the percentage error decreases as shown in Fig. 23.

Figure 23 shows the sampling error ranging from less than 1 to 60% in the 95% envelope for a storm mean of 0.01 in. Similarly, for a 1-in. mean the sampling error ranges from near zero to 14% with a median of 2.5%. The difficulty in defining the sampling error accurately in specific storms is emphasized by this data plot. As the gage density becomes less, the variation about the mean becomes greater. For example, a similar plot for a gage density of 50 miles^2/gage showed the 95% envelope of sampling errors for a 1-in. mean ranging from less than 1 to nearly 40%.

Other factors being equal, air mass storms were found to require the greatest sampling density among synoptic storm types to maintain a given error level. In the warm season, rainshowers and thunderstorms require nearly twice as many gages as steady rain for a given measurement reliability. Also, the May–September sampling density requirements were two to three times those needed in the October–April period.

Considerable difference in the magnitude of storm sampling errors during the May–September period was found between consecutive five-year periods on the same network. Analyses indicated that the difference resulted partially, at least, from a much higher number of air mass storms in one period. Consequently, one must interpret sampling error relations with caution when they are based upon observational periods of five years or less.

Sampling requirements for the measurement of monthly and seasonal rainfall are less stringent than for storm precipitation. With seasonal rainfall and a relatively sparse density of 200 miles^2/gage, the average sampling error was found to be less than 5% for the entire range of seasonal rainfalls. For a 6-hr rainfall with a mean of 0.5 in., the average error is 14%. Thus, substantial decreases in sampling requirements occur as the time integration of rainfall increases on a fixed sampling area. Conversely, exceptionally dense sampling networks are required for the accurate measurement of areal mean rainfall rate. Thus, for a moderate 1-min rate of 0.5 in./hr, the average sampling error in warm-season storms on a sampling area of 100 miles^2 was found to increase from 33% with 25 miles^2/gage to 66% with 100 miles^2/gage. The effect of time integration of precipitation is illustrated in Table XV taken from [26]. A comparison of average sampling errors in May–September storms is shown for selected time intervals ranging from 1 min to 5 months on a sampling area of 100 miles^2.

6.4. Detection of Storm Precipitation

Previous hydrometeorological studies have indicated that convective storms of small areal extent, such as summer air mass storms, are occasionally not recorded by sampling densities several times greater than those of the normal climatic network [27]. Therefore, as part of the Illinois investigation

TABLE XV. Comparison of average sampling errors in May–September rainfall on 100 miles² for selected time intervals

Gage density (mi²/gage)	Sampling error (%) for given 1-min mean rate (in./hr)			
	0.1	0.5	1.0	5.0
25	42	33	29	23
50	59	46	42	33
100	84	66	60	46
	Sampling error (%) for given 3-hr storm rainfall (in.)			
	0.1	0.5	1.0	2.0
25	10	6	4	3
50	17	10	8	6
100	30	17	13	11
	Sampling error (%) for given monthly rainfall (in.)			
	1.0	2.0	4.0	8.0
25	4	3	3	2
50	6	5	5	4
100	10	9	7	6
	Sampling error (%) for given seasonal rainfall (in.)			
	10	15	20	30
25	2	1	1	<1
50	3	2	2	1
100	5	4	3	2

of precipitation sampling requirements, a study was made of the detection capability of raingage networks of various densities. Results of this study should be particularly useful in cloud seeding experiments in which warm season, air mass showers are seeded and rainfall measurements used in the evaluation. Analyses were accomplished through use of data from areas of 10 to 550 miles² in the dense raingage networks of Fig. 1.

Curves were developed relating probability of storm detection to size of sampling area, rain-gage density, and areal mean rainfall (storm intensity) on an annual and seasonal basis. Some of the more pertinent findings are presented in Fig. 24 abstracted from [27]. Here, detection capabilities of 80 to 99% are shown for appropriate mean precipitation groups in summer and winter. Analyses indicated that the detection problem maximizes in summer and minimizes in winter. The summer storm sample ranged from 350 to 500 among the networks and the winter sample from 210 to 265.

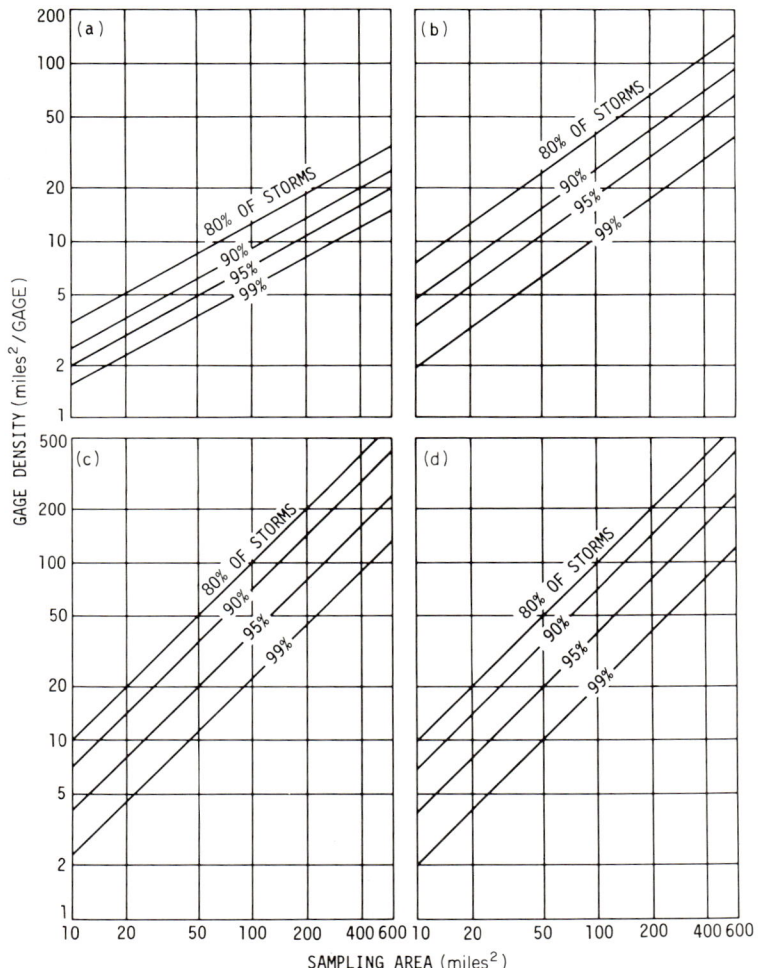

FIG. 24. Sampling requirements for precipitation detection. (a) Summer, mean = trace; (b) summer, mean = 0.01–0.05 in.; (c) summer, mean 0.06–0.10 in.; (d) winter, mean = 0.01–0.05 in.

Curves such as shown in Fig. 24 can be used as guides in determining sampling requirements for precipitation detection under midwestern climatic conditions. Gage density requirements for a detection capability below 80% have been omitted, since it is extremely doubtful that potential users would find a network acceptable that left 20% of the storms undetected. Only one curve is shown for winter since the number of "trace" storms was too few to derive a reliable probability curve, and only one storm with mean exceeding 0.05 in. was undetected on any of the six sampling areas.

Application of Fig. 24 can be illustrated by an example. Assume that a cloud-seeding experiment is being conducted during the summer in Illinois and that the investigators specify 99% detection of all rainstorms in a target area of 400 miles^2 as one of their verification needs. Then, referring to the family of summer "trace" curves, one finds that a density of one gage per 12 miles^2 would be required. If the user would be satisfied with detecting 90% of all storms, the gage density would lower to one gage per 20 miles^2. Now, if the user could limit his detection requirements to 99% detection of storms with areal means exeeding 0.05 in., the summer curves for 0.06–0.10 in. show a required density of only one gage in 90 miles.^{2} However, this density is still considerably greater than the normal climatic network density which averages approximately one gage per 225 miles^2 in Illinois. The foregoing examples illustrate how sharply the sampling requirements lower as the minimum storm intensity is increased.

The winter curves in Fig. 24 show how much sampling requirements are lowered in the cold season when storms tend to be more widespread and the precipitation much more frequently of the stable type compared with summer. Thus, if one is satisfied with 99% detection of all storms with means exceeding 0.01 in. in winter, the 400 miles^2 density requirement is one gage per 80 miles^2 compared with one gage per 30 miles^2 in summer under the same detection requirement. The winter curves for 0.01–0.05 in. are nearly identical with the summer curves for 0.06–0.10 in.

6.5. *Areal Extent of Storm Precipitation on Fixed Sampling Areas*

This study was an extension of the detection study and utilized the same set of network data. It was undertaken to provide additional information on measurement requirements for weather modification experiments, particularly those involving light storms of small areal extent.

The areal extent of storm precipitation was expressed in probability terms also. Minimum percent of area enveloped was related on an annual and seasonal basis to size of sampling area and areal mean precipitation in development of the probability distributions. An example is shown in Fig. 25, in which minimum percent of area enveloped is plotted against size of sampling area for various percentages of all storms sampled annually with mean precipitation of 0.01–0.05 in. Thus, Fig. 25 shows that for a sampling area of 100 miles^2, 20% of these storms, on the average, will envelop 100% of the area, 50% will envelop 75% or more of the area, and 95% will enclose at least 21% of the 100 miles^2. Data used, analytical procedures, and results are presented in much greater detail in [4].

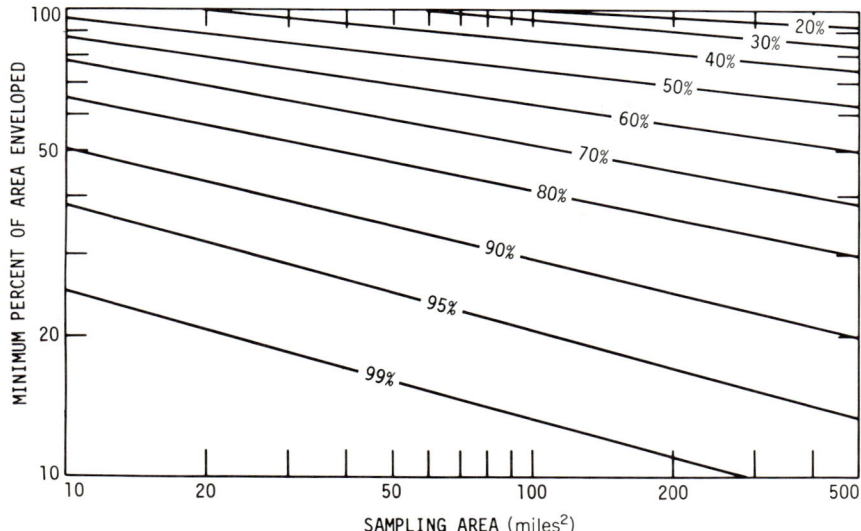

FIG. 25. Relation between storm areal extent and size of sampling area for mean precipitation of 0.01–0.05 in.

7. Statistical Evaluation Techniques

In this section, results of an investigation of statistical evaluation techniques [4] will be reviewed briefly to summarize the more salient features of the findings. In this study, attention was given to data fitting methods, statistical sampling designs, types of verifying parameters, and data stratification effects.

7.1. Use of Areal Precipitation Measurements in Evaluating Seeding Experiments

The verification methods discussed herein are based strictly upon the use of standard statistical designs applied to surface precipitation measurements in the manner used by past investigators. A technique which could expedite significantly the verification of cloud-seeding results has been proposed and tested to some extent [28,29]. Basically, the technique involves use of a cloud model to predict seeding effect upon target clouds. Control clouds (unseeded) are also measured in their tests and used as a check on the applicability of the cloud model. Some of the problems involved in use of this technique have been discussed by Cunningham [30] and Brown and Glass [31]. Obviously, evaluation of the cloud-model prediction techniques was beyond the scope of the Illinois studies involving the natural distribution characteristics of surface precipitation.

The Illinois study of sampling requirements was based upon data from the 12-yr sample of storms on the East Central Illinois Network of 49 raingages in 400 miles^2 (Fig. 1). Areal mean and areal maximum precipitation were selected as the verifying parameters. Storm rainfall was stratified according to season, precipitation type, and synoptic type.

Theoretical frequency distributions were fitted to the storm data, and the best-fit distribution was used to obtain sample size for five statistical designs. Designs included: (1) *random experimental* which involves randomization of days over a single target area into seeded and nonseeded days with nonseeded days being the control; (2) *random historical* in which a random choice is made of days to be seeded over a single target with the historical record as control; (3) *continuous historical* in which all rain days within a given stratification are seeded, with the historical record as control; (4) *cross-over* which requires seeding a target chosen at random with another area being the control (random interchange of target and control); and (5) *fixed target control* in which all potential rain days are seeded in a fixed target area with a nearby area serving as a fixed control. In the statistical evaluations, both sequential and nonsequential analyses were employed.

Results of the various analyses indicated that areal mean and areal maximum precipitation have nearly equivalent power in detecting seeding-induced precipitation changes. In most of the stratifications, the log-normal distribution provided a satisfactory fit of the precipitation data and was superior to the gamma distribution. For a given seeding-induced percentage change, the cross-over design was superior and the random-experimental design required the longest verification time. Stratification of storms by mean rainfall, rain type, and synoptic type produced more homogeneous subsamples; this facilitates fitting the data to theoretical distributions and makes statistical testing easier and more reliable.

Verification time varies considerably between storm and precipitation types because of differences in the degree of natural variability and in frequency of occurrence. Relatively long sampling periods are required with summer air mass storms compared with winter low systems. Despite greater inherent spatial variability, changes in warm-season thunderstorms in Illinois could be detected easier than those induced in cold-season steady rains because of greater frequency of occurrence.

Figure 26 shows the average number of years required to obtain significance in areal mean rainfall at the 95% confidence level on the 400 miles^2 network for various seeding-induced percentage increases, four statistical designs, two rainfall types, and three synoptic storm types. The designs used are cross-over (C), random historical (RH), fixed target control (TC), and random experimental (RE). Rainfall types are warm season thunderstorms and cold-season steady rains. Warm-season air mass and cold frontal storms are shown,

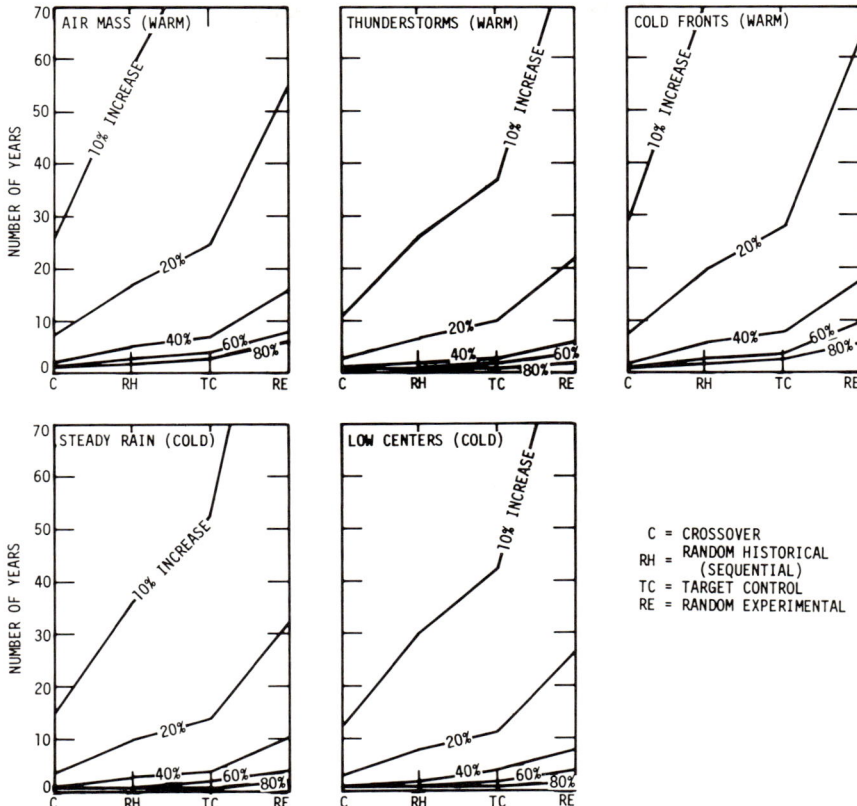

Fig. 26. Number of years to obtain significance for various increases in areal mean precipitation according to synoptic type, precipitation type, and experimental design.

along with cold-season low centers. All storms with areal means exceeding a trace were used in deriving the curves.

Large differences between the cross-over and random-experimental designs are indicated for small percentage increases with each data stratification. However, the difference between designs becomes less important as the seeding-induced percentage increases. Also, note the large decrease in sampling time when the seeding-induced change increases from 10 to 20%. The difficulty in verifying a seeding increase of the order of 10% is clearly illustrated. For example, with the cross-over design applied to air mass storms, 25 years would be required for verification of a 10% increase at the 95% confidence level compared with 7 years with a 20% increase and 2 years with a 40% increase. Also, it is interesting to compare sampling requirements between the several rainfall and synoptic storm types. Thus, for a 20% increase with the cross-over design, the sampling period decreases from 8 years for air mass storms to 3 years with low centers. Similarly, thunderstorms with their greater frequency of occurrence require less verification time than steady rains, 3 years compared with 4 years for a 20% average increase from seeding.

At this time it should be pointed out that the length of sampling periods discussed above is based upon seeding of all storms within a given stratification. Obviously, this would not normally occur in most seeding experiments. The estimates in the graphs can be readily modified to provide sampling lengths under any estimated seeding efficiency. For example, if an investigator estimates only 50% of the naturally occurring storms will be included in his seeding experiment, then the sampling length is doubled. If only 25% are to be included, the sampling time is four times the number obtained from the graphs.

From determination of verification times for 50, 100, 200, and 400 miles^2 it was found that the required sampling time decreased rapidly as the sampling area increased from 50 to 100 miles^2, but changed very slowly between 200 and 400 miles^2. That is, the advantage of increasing the sampling area becomes small at 400 miles^2 unless, of course, very large changes in area are made. Then, sampling homogeneity becomes a problem that could more than offset the advantage of a larger sampling area.

7.2. Use of Area-Depth Curves in Evaluating Precipitation Modification Experiments

Area-depth curves, frequently used in hydrologic analyses, provide a simple mathematical expression of the spatial distribution of precipitation in a rain-gaged area. With a recording gage network, a measurement of the time variation in the spatial distribution within a storm can be obtained. The curve provides a measure of the mean and maximum precipitation, rainfall gradient, and skewness of the distribution. As shown by Huff [32], the above properties make it potentially useful in weather modification research, in which quantitative assessment of cloud-seeding effects upon the spatial and temporal distributions are essential for proper evaluation of benefits. Also, area-depth analyses provide information indirectly on physical changes produced in precipitating cloud systems.

The study involved testing of five statistical designs (listed in the previous section), evaluation of methods for obtaining the best-fit area-depth curve, and determination of the appropriate distributions to describe the three important area-depth parameters (maximum rainfall, mean rainfall, rainfall gradient). The storm data were stratified according to season, storm type, and precipitation type as has been done in nearly all of the Illinois studies. The sample consisted of all recorded storms in which areal mean rainfall was 0.50 in. or more and/or one or more network gages recorded amounts equaling or exceeding 1 in. Analyses indicated that an equation relating rainfall to the square root of the area provides an excellent fit of the area-depth curves for the areas investigated (50 to 400 miles^2), and a log-normal distribution satisfactorily describes the distribution of the two regression constants and rainfall depth.

Study results verified application of the technique for moderate to heavy rainstorms and indicated that the target-control cross-over is the optimum design, followed by the random historical utilizing a sequential analysis approach [4]. The three rainfall parameters were found to agree consistently in prediction of required sampling times with the various data stratifications. Thus, they provide three supporting measures, and thereby can increase the reliability of findings beyond that obtained with the usual single-parameter evaluation (point or areal mean rainfall) employed in cloud-seeding evaluations.

Although evaluation of the area-depth technique was based upon analyses of moderate to heavy rainstorms, there is no reason to believe the method will not be applicable throughout the intensity spectrum of storm mean precipitation. This assumption is lent support by the findings in the previous section in which two of the three area-depth parameters, mean and maximum rainfall, were tested for all intensities of storm precipitation and found to have approximately equal power in detecting seeding-induced increases.

7.3. Use of Rain Cell Measurements in Precipitation Modification Experiments

A limited investigation was made of the characteristics of individual rain cells in warm season storms and how measurements of cell parameters might be used as a verification tool in rain modification experiments. A rain cell was defined as a closed isohyetal system in the surface rainfall pattern with a definite break in time and space continuity from other isohyetal systems in the sampling network. The investigation was accomplished with recording rain-gage data from the combined Little Egypt and Shawnee Networks (Fig. 1) during the summer of 1965, and was based upon 335 rain cells from 21 rainy days. A typical group of rain cells is illustrated in Fig. 27 in which isohyetal patterns, cell movement, and average speed are indicated. A paired-storm design was used in the evaluation. In this design, a pair of rain cells with similar characteristics is selected, and one member of the pair is chosen at random to be seeded.

It was assumed that any two cells with start times within 30 min of each other were likely to have been derived from clouds that had similar meteorological characteristics prior to their production of rain. This selection resulted

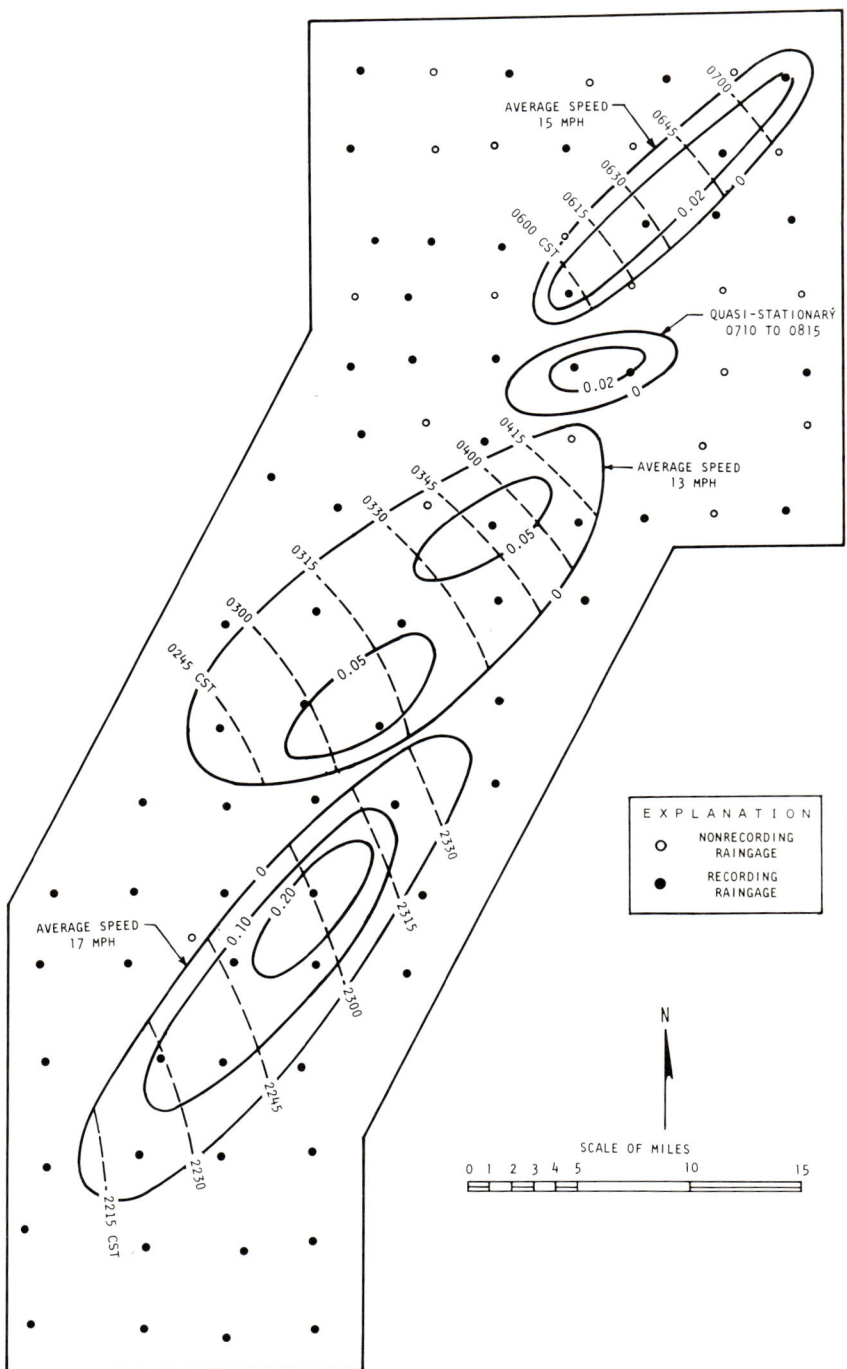

Fig. 27. Examples of rain cells on 6–7 June 1965.

in a sample of 143 pairs. Next, differences between members of each pair were obtained for mean rainfall, maximum rainfall, cell duration, areal extent, and rainfall volume produced. Cumulative ogives for the empirical distributions of differences were then formed and these were designated as the natural distributions. Next, the values of the hypothetically seeded cell of each pair was increased by 20, 40, 60, and 80% and the respective cumulative ogives formed. The differences between the natural and hypothetically seeded curves were assumed to be the effect that seeding would have on the natural differences. The Wilcoxon matched-pairs signed-ranks test was employed in the significance calculations. To derive the number of pairs to obtain significance for the various assumed increase in each rain cell parameter, sample values were generated through use of a Monte Carlo procedure.

From the results it was concluded that the cell comparison method could be quite useful from the standpoint of statistical verification in those experiments concentrating on the seeding of convective rain cells during the warm season. The method would be especially advantageous if multiple pairs of cells could be seeded on operational days. Then, it would approach the cross-over technique with storm rainfall (discussed previously) in achieving significance in relatively short sampling periods. However distinct problems are inherent in the cell sampling technique. These include the need for detailed measurements to identify cells accurately, the difficulty in isolating individual cells in the surface rainfall analyses, and the problems inherent in objectively selecting the proper pairs of clouds to seed. Furthermore, a recording gage density of approximately one gage per 5 miles^2 extending over hundreds of square miles would be required to encompass cell rainout, properly identify treated and untreated cells, and permit accurate isohyetal analysis of cell rainfall.

7.4. *Estimating Natural Distribution of Storm Precipitation in Target Areas from Control Area Data*

The feasibility of using storm precipitation in control areas to predict target rainfall patterns in the verification of cloud seeding experiments was investigated by Nason and Lopez [33] through use of Oklahoma data. Essentially, the technique involves fitting a plane or some higher-order mathematical surface to the control area data and extending this surface through the target area. The target predictions represent the rainfall expected with no seeding, and these values are then compared with measured amounts to determine seeding effects.

In the Illinois studies, further evaluation of this method and two simpler extrapolation techniques was made on a limited scale through use of data from the East Central Illinois Network (Fig. 1). Two 20-storm samples of steady and unstable types of precipitation were selected. First, the surface fitting technique was employed using plane and quadratic surfaces. Second, target predictions were obtained by extension of the adjacent control area isohyetal pattern. Third, target mean precipitation was estimated from a simple averaging of the control area observations.

Initially, the target consisted of 100 miles^2 at the downwind edge of the 400-miles^2 network, with the rest of the network being the control. The target area in each storm was assigned on the basis of wind flow under the assumption that such knowledge would be available in seeding experiments. This target-control arrangement provides a measure of target prediction capability under optimum sampling conditions, since the control was immediately adjacent to the target. Because of the contamination problem, laterally coinciding target and control areas would not be practical in field experiments. Obviously, target prediction errors would then tend to increase with increasing distance between target and control.

TABLE XVI. Comparison between three methods of predicting mean precipitation on 100 miles^2 target

Cumulative percent of storms	Error (%) equaled or exceeded for given method			
	Surface fitting		Isohyetal extension	Gage averaging
	Plane	Quadratic		
Unstable precipitation				
10	44	80	53	95
20	29	44	29	53
30	21	29	19	35
40	17	19	14	25
50	13	14	10	18
60	10	10	7	13
70	8	7	5	9
80	6	4	3	6
90	4	2	2	3
Steady precipitation				
10	18	23	19	22
20	11	14	13	12
30	8	10	10	8
40	5	8	8	6
50	4	6	6	4
60	3	4	5	3
70	2	3	4	2
80	1	2	3	1
90	<1	1	2	<1

Results of the 40-storm study are summarized briefly in Table XVI which shows a comparison between the prediction errors of target mean precipitation for each of the three methods on the 100 miles^2 target. With unstable precipitation, the plane-fitting and isohyetal methods are nearly equal in prediction capability, and both do somewhat better than the quadratic fitting or simple averaging. Overall, prediction errors were considerably smaller with stable precipitation, and differences between methods were relatively small. As expected, prediction errors were found to be larger with air mass than frontal storms in the sample. As the target area was reduced progressively from 100 miles^2 to a point, the target prediction error decreased considerably with unstable precipitation, but little change occurred with the stable type.

Overall, it was concluded that with adequate separation of target and control to eliminate contamination, the method would not be acceptable with unstable types of storm precipitation which normally show great spatial variability. With steady precipitation, the technique may be useful.

8. Downwind Seeding Effects

The mounting interest in all aspects of weather modification makes it essential that the effect of seeding on rainfall in areas directly downwind from seeding experiments be established. The 1960–1964 Missouri seeding experiments, Project Whitetop [9], in which careful attention was given to randomization of seeding operations, has provided an excellent set of data to study this effect. Furthermore, two Illinois dense raingage networks were located where they could assist materially both in evaluating the reality of downwind effects, a controversial subject in the weather modification field at this time, and in demonstrating some of the problems imposed by natural rainfall variability in downwind studies. Results from a study of Whitetop downwind effects, which include an excellent example of the natural variability problem, are summarized in the following paragraphs.

A circular sampling area of 300 miles radius (approximately 283,000 miles^2) about the center of the Whitetop target area in southern Missouri was studied. First, daily rainfalls for seeded and nonseeded days from precipitation stations in an area downwind and out to a distance of 300 miles were used to evaluate the differences between rainfall on seeded and nonseeded days. Secondly, plume analyses using wind trajectories in the lower 500 mb were performed for the year 1961 to obtain a better estimate of the area which could have been affected on each operational day. This permitted a more detailed study of downwind areas for that year in which seeding apparently reduced the rainfall substantially in the Whitetop research area encompassing a circle of 60-miles radius about West Plains, Missouri, where the experiment was centered. Next, data from available recording raingages were used to obtain average areal rainfall for each 3-hr period on operation days in the 1961 evaluation in order to facilitate interpretation of results. Finally, hourly rainfall on two dense raingage networks (Little Egypt and East Central Illinois in Fig. 1), located approximately 175 and 290 miles downwind from the center of the Whitetop area, were examined (1) for evidence of extensive downwind effects in the 1960–1964 period of Whitetop operations, and (2) to aid in evaluating the degree of natural variability polluting the analytical results.

8.1. General Conclusions on Downwind Effects

Among individual years, evidence was found for both positive and negative effects in isolated downwind areas, and a few of these were statistically significant. However, lack of continuity of the significant differences imply these were the result of random sampling. Possible negative effects were observed most frequently in two years, 1961 and 1962, whereas evidence of positive effects was found most often in downwind areas in the other three years. The detailed plume analyses for 1961 indicated a statistically significant probability of a negative seeding effect in an area 100–200 miles downwind; overall, however, the evidence for downwind effects derived from the plume analyses was very weak.

No statistically significant seeding effects were present in the downwind areas when data for all five years were combined. Upon consideration of all analyses performed, it was concluded that the evidence for downwind effects from the Whitetop experiment is weak.

8.2. Analyses of Dense Raingage Network Data

Previous analyses of dense raingage network data have indicated that large differences in the time distribution of rainfall may occur between seeded and nonseeded days solely as a result of natural rainfall variability and thus lead to spurious conclusions if not thoroughly evaluated. The Whitetop study provided an excellent example of how the problem can arise in an actual seeding operation.

Network diurnal distributions of 3-hourly mean rainfall for June–August 1960–1964, the Whitetop seeding period, in Fig. 28 illustrate the problem encountered. Distributions are shown for seeded and nonseeded operational days on both networks and for the remaining nonoperational days on the most distant network. Immediately, attention is drawn to the early morning climatological maxima on the East Central Illinois curves. The maxima are strikingly similar for the nonseeded and nonoperational curves (no-effect hours), but both show much heavier rainfall than does the curve for seeded days. This is especially interesting since the early morning peak occurs at

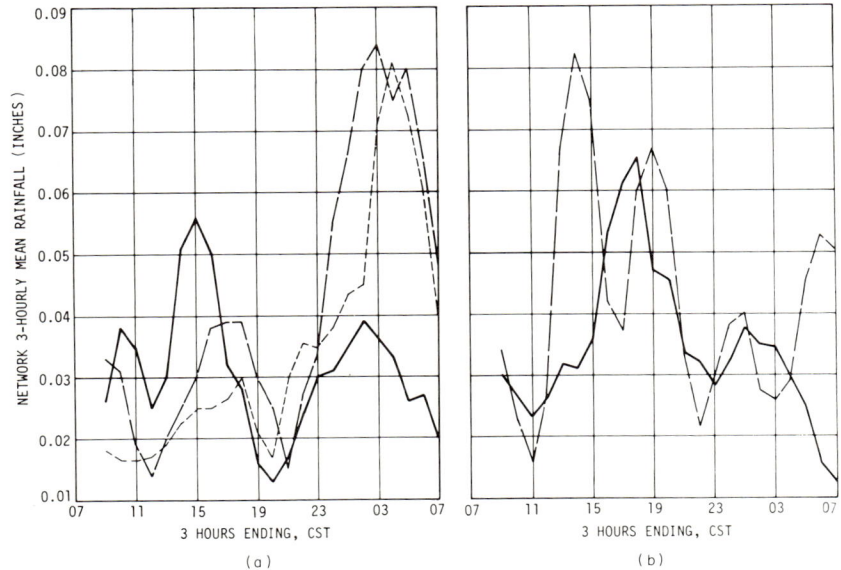

FIG. 28. Diurnal distribution of network mean rainfall. (a) East-central Illinois network; (b) Little Egypt network. ———, Seeded days (58); — — nonseeded days (58); – – – nonoperational days (58).

a time when Whitetop seeding material, or other seeding-induced effects, would most likely have reached the East Central Illinois Network (ECI), based on wind speed climatology. Thus, the ECI curves indicate a strong possibility that a relatively large percentage reduction in rainfall occurred nearly 300 miles downwind as a result of Whitetop seeding.

However, the curves for Little Egypt (LEN) show rainfall differences of a similar magnitude in the early afternoon during the 3–4 hours immediately following the starting time of Whitetop seeding. Such a pronounced downwind seeding effect at 175 miles is unlikely to occur so rapidly or consistently enough to produce this anomaly. Thus, it is concluded that it resulted from natural variability between the two samples drawn from the 1960–1964 rainfall population. The LEN anomaly then raises a question as to whether the ECI early-morning differences merely reflect a sampling vagary.

Results of the 1961 plume analyses were used to investigate further the probability of the ECI anomaly being natural or seeding induced. The 1961 analyses for flight level to 500 mb showed that there was a likelihood of a Whitetop plume reaching ECI on only five of 37 operational days (14%) and on two of these days no rain occurred on ECI. However, with its closer and more strategic position, LEN could have been reached by plumes on 76% of the 1961 operational days. Consequently, it would be expected that if a seeding effect was present it would be more evident on LEN and should be most prominent between 1500 and 0300 CST, when winds would be most likely to bring Whitetop seeding material, clouds, or seeding-induced mesosystems in to the network. However, the two LEN curves in Fig. 28 are very similar during this period, thus indicating little or no seeding effect.

Assuming that the 1961 plume distributions, in which the calculated areal extent of the plumes was doubled in the network areas to allow for computational error, are reasonably representative of plume movements in the 1960–1964 period, it is very difficult to conclude that downwind seeding effects were the cause of the early morning anomaly shown by the ECI curves. It is more likely a consequence of natural rainfall variability such as produced the early-afternoon anomaly on LEN.

The Wilcoxon ranked pairs test was employed in conjunction with the hourly data on LEN and ECI to evaluate the statistical significance of the differences between the hourly averages for seeded and nonseeded days' and between the hourly averages for seeded and nonoperational days. Results are summarized in Table XVII. First, tests were made on a 25-hr period

TABLE XVII. Probabilities obtained from Wilcoxon paired ranks test for hourly rainfall differences on seeded, nonseeded, and nonoperational days

	East Central Illinois			Little Egypt
Time (CST)	Seeded versus Non-seeded	Seeded versus Non-operational	Time (CST)	Seeded versus Non-seeded
25-hr (06–07)	0.06	0.47	25-hr (06–07)	0.25
06–17	0.06	0.01	06–15	0.37
17–07	0.006	0.001	15–07	0.41

starting at 0600 CST which included hours of potential effect and hours of no effect from the Whitetop seeding. Next, the network time distributions were divided into two parts (based on wind climatology) to include the overall range of hours in which a Whitetop seeding-induced effect may have reached the networks and those hours during which it was unlikely for any seeding effect to have been present. The no-effect hours were 06–17 on ECI and 06–15 on LEN, whereas the potential effect hours were 17–07 on ECI and 15–07 on LEN.

Assuming that any probability of 0.05 or less is significant, Table XVII shows no indication of downwind seeding effect on LEN, approximately 175 miles downwind of Whitetop. However, highly significant probability values (0.01 or less) were obtained not only for the hours of potential downwind effect on ECI, 290 miles from Whitetop, but for the no-effect period (06–17) in the seeded versus nonoperational comparison. Also, a value close to significance (0.06) was obtained in both the no-effect period and in the 25 hr period bracketing effect and no-effect hours in the seeded versus nonseeded comparisons.

Table XVII provides an excellent example of the high degree of statistical significance that may be obtained with differences between two samples drawn from the same rainfall population. One may insist, despite evidence to the contrary presented here, that the highly significant test statistics for 17–07 (effect hours) on ECI were caused by downwind seeding effects. However, there still remains the highly significant 0.01 probability for the no-effect hours in the seeded versus nonoperational comparisons on this network.

9. Design of a Precipitation Modification Experiment

Up to this point, an effort has been made to describe the natural distribution characteristics of precipitation in terms useful in the design and evaluation of weather modification experiments and to clarify the problems imposed by natural variability. In the following pages, results from various Illinois studies will be used to illustrate the adaptability of such accumulated knowledge in establishing design criteria and verification procedures for potential precipitation modification experiments.

Although primary emphasis in this review has been upon applications of precipitation climatology in weather modification, other factors must enter into the selection of experimental sites or establishment of operational procedures and verification methods. For example, regional needs and uses for additional precipitation should be an important consideration in selecting a site for an experiment. Regional physiography will be a factor also in site selection, its importance depending upon the primary objective of the seeding project (enhancement of orographic rainfall, modification of convective storms, etc.). Availability of support facilities, such as radiosonde stations, radar facilities, and airports will influence location of an experiment. Factors such as equipment costs, aircraft capabilities, and major purpose of a planned experiment all enter into determining design, operational, and verification procedures employed. The factors listed above will be given consideration in the recommendations which follow.

The proposed Illinois design is based upon the assumption that surface precipitation measurements will be the primary method of evaluation. Furthermore, the design is based upon presently available knowledge (published in technical journals and reports) on the efficacy of cloud seeding and upon accepted procedures, techniques, and capabilities in weather modification.

From a practical standpoint, economic benefits (and damages) to agricultural, hydrological, recreational, and other affected interests must be evaluated from what happens at the ground when clouds are seeded to increase or decrease rainfall. Scientifically, of course, there is mutual interest in atmospheric and surface effects. Certain features of any experimental design will, therefore, vary with the primary purpose of the experiment.

9.1. Selection of Sites

Both central and southern Illinois provide acceptable sites for cloud seeding experiments from both scientific and practical considerations [3]. From the standpoint of existing support facilities, central Illinois is the better choice. However, economic benefits from seeding are likely to be greater in the southern part of the State [34].

Potential experimental areas are shown in Fig. 29. In central Illinois, areas A, B, and C or B, E, and D were selected for an experiment employing a cross-over design, whereas the larger dashed area (Central Illinois Network) would be used for an experiment employing a single area, such as the random experimental design. Similarly, in southern Illinois, areas 1, 2, and 3 are for a cross-over design and the larger dashed area is a recommended location for a single-area design.

The central Illinois sites are near existing weather radar facilities at the Illinois Water Survey (Urbana), flight facilities at the University of Illinois and Chanute Air Force Base, radiosonde observations at Peoria, and they incorporate the Central Illinois Network of 196 recording raingages in 1600 miles^2. A 15-yr record of precipitation climate (1955–1970) obtained from dense raingage networks is available for assistance in the design and evaluation of a seeding experiment.

In southern Illinois, rainfall has been shown to have a greater effect upon major crop yields than in the central and northern portions [34], and the area is more dependent upon surface water supplies. Climatologically, opportunities for seeding would normally be more frequent in cold season operations, since average precipitation is greater and precipitation types more diverse than in the central and north. In summer, the primary diurnal maximum occurs in mid to late afternoon in southern Illinois, a convenient period for seeding operation, whereas the primary maximum occurs in the early morning hours (darkness) over most of the rest of the state. Thus, an added advantage is the possibility of using seeding-induced effects on this primary diurnal maximum as one of the verification tools. Also, southern Illinois experiences a greater number of air mass storms in the warm season than the

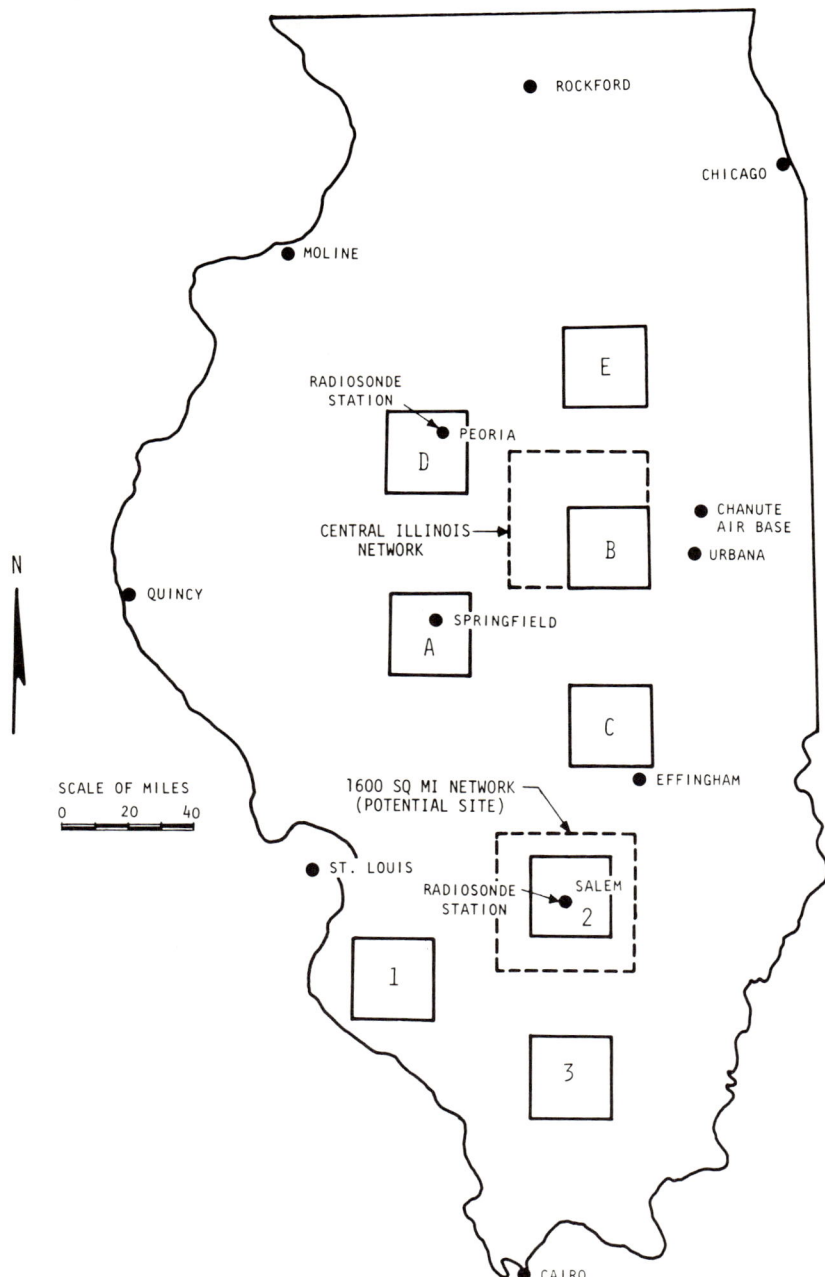

FIG. 29. Recommended locations for Illinois experiment.

central and northern regions [2]. These are frequently the primary target in scientific seeding experiments, since they are relatively simple storms to treat and observe because of their scattered distribution in the atmosphere.

Both areas 1–2–3 and A–B–C are in exceptionally flat areas so that topographic influences on precipitation should be minor and not interfere with the evaluation of seeding experiments [3]. Area D of the B–E–D combination may be affected to some extent by the Illinois River bluffs in the Peoria region; therefore, this combination becomes second choice in central Illinois.

9.2. Sampling Design and Verification Methods

Two types of sampling designs recommended for central and southern Illinois are shown in Fig. 29. The three-area sampling design, such as areas 1, 2, and 3 in southern Illinois, would be used with the cross-over target-control evaluation technique. The triple sampling area allows operations with storm movement from any direction; that is, two usable areas can be chosen for randomization regardless of storm movement.

The three areas in the above designs are 30 miles apart at the shortest distance. Spatial correlation studies [14] indicate that this separation will maintain the average correlation coefficient at 0.75 or greater between target and control. At the same time, the separation should minimize possibilities of contamination from the target seeding when individual convective cells or storms are being treated.

Each of the three unit areas encompasses 600 miles^2, near the average county size in Illinois. Selection of this size is based upon several considerations. These include initial costs of equipment and installation, manpower requirements for efficient operations, and data reduction requirements. Furthermore, areas of county size are frequently exposed to dry spells or minor droughts which would be amenable to alleviation from seeding, so that from practical considerations these are reasonable areal sizes. The target-control areas are square-shaped, since it is anticipated that any seeding experiment would not be directionally oriented.

The three-area design is also suitable for experiments employing a fixed target, such as recommended by Spar [35] for the Northeastern United States experiment. For example, using Area 2 in southern Illinois as the fixed target, Areas 1 and 3 would serve as controls, depending upon storm movement. In this way, individual convective elements moving from nearly any direction could be sampled with small danger of contamination.

Although the cross-over sampling will provide verification of seeding results quicker than other statistical methods, its application in Illinois is questionable when the characteristics of the precipitation distribution and prevalent types of storm systems are considered. This is especially true if investigators are involved in a practical evaluation of seeding benefits in this state. The majority of the Illinois precipitation does not result from scattered convective cells or showers but from extensive storm systems of relatively long durations associated with fronts and squall lines [12]. Mesoscale circulations associated with such systems and interaction between neighboring convective elements would make the contamination problem difficult to overcome. However, from a practical standpoint, these storm systems should be subjected to seeding if substantial benefits are to be obtained from a seeding program.

Unless one is restricting his experiment to the isolated convective storm (air mass storms), it may be necessary to use a design employing randomization over a single area, such as employed in the Arizona experiments [8]. However, the random experimental design results in the longest verification time of the several investigated. To partially overcome this problem, the random experimental method of verification could be supplemented by a random historical analyses. This approach is especially applicable in Illinois with its network historical records. In this approach, the seeding would be done with the random experimental design and the nonseeded samples compared with available historical records to check for trends. With no trend indicated during the experimental period, the seeded sample could also be compared with the historical record and the verifying time would be reduced by 50% from that with the random experimental design used alone [3]. In any case, the combination of verifying methods should decrease the verifying time somewhat. In central Illinois, the existing Central Illinois Network of 196 recording gages in 1600 miles^2 would serve the purpose of a single-area seeding experiment admirably. In southern Illinois, a similar network could be installed about the radiosonde station at Salem, as shown by the dashed outline in Fig. 29.

Examination of the literature indicates a general consensus among those with expertise that aircraft seeding is the most desirable method of cloud inoculation, especially in warm-season operations on convective clouds. Adequate flight facilities are available in the proposed experimental areas, especially in the central Illinois region. However, much of central and western Illinois is crossed by major airways so that heavy aircraft traffic would be a problem in the proposed central Illinois experiment.

9.3. Precipitation Measurement Requirements

9.3.1. Mean Precipitation. Taking into consideration measurement accuracy [26], equipment costs, operational requirements, and data processing, a gage density of 10 miles^2/gage is considered suitable for the measurement of areal mean rainfall in storms on the sampling areas involved in the

Illinois design. Also, this gage density is considered very satisfactory for deriving storm area-depth relations and adequate for estimating storm maximum precipitation.

With a sampling density of 10 miles^2/gage, the average sampling error in a large sample including all types of storms should not exceed 2–3%, even in short-duration storms of light to moderate intensity which normally exhibit the highest spatial relative variability. In fact, the above percentage error is based upon calculations for an average 1-hour storm with mean rainfall of 0.10 in. occurring in the warm season [26]. One must understand, however, that the sampling error may exceed 5% in approximately 5% of the storms and, occasionally, increase to 10% or more.

The above error estimates are for all storms combined. The sampling error will maximize, on the average, in air mass storms. If only this storm type is involved in a seeding experiment, the average sampling error rises to 5–6% for a gage density of 10 miles^2/gage. Similarly, the error can be expected to exceed 10% in 5% of the storms and reach 15–20% occasionally.

9.3.2. Pattern Analysis. From [21], it has been determined that the spacing of one gage every 3.3 miles (10 miles^2/gage) recommended for areal mean precipitation measurements is generally satisfactory for isohyetal pattern analysis. With this spacing, the variance explained, on the average, is approximately 90% for all warm season storms combined. It decreases to 75% for storms with durations less than three hours; otherwise, it is greater than 80% for all storm durations, synoptic weather types, and precipitation types.

9.3.3. Detection and Areal Extent. Studies have shown that the gage spacing of 3.3 miles in the recommended experimental network is satisfactory for detecting all storms with significant precipitation and to measure accurately their areal extent [27].

9.3.4. Rain Cell Analysis. Based on the rain cell study described in [4], a gage density of 10 miles^2/gage, generally satisfactory for all other surface precipitation measurements discussed earlier, is considered usable although not entirely satisfactory for measurement of rain cell mean rainfall. This is because approximately two-thirds of those cells analyzed in the Illinois rain cell study had means less than 0.10 in. From gage density studies [26], it is estimated that the average sampling error with 10 miles^2/gage will increase from 3–5% at 0.10 in. to 8–10% at 0.01 in. It would be most desirable to increase the density to 5 miles^2/gage, especially if pattern analyses are to be performed in the seeding evaluation.

9.3.5. Type of Raingages. All gages in the target and control areas should be of the recording type to provide space-time distributions for partial storm, total storm, or daily precipitation amounts. To provide data on possible downwind effects, the recording gage networks could be augmented by standard nonrecording gages or by the much cheaper wedge gage (fence-post type) evaluated by Huff [36].

9.3.6. Radar Instrumentation. Radar observations are a basic requirement for a scientifically oriented seeding experiment. Radar is needed to identify and track seeded and nonseeded storms. With the cross-over target-control design where contamination is a problem, radar would be useful in determining whether contamination had affected a seeding operation. In all cases, radar is very useful for determining the growth characteristics of target and control clouds, changes in direction and speed of cloud systems, changes in intensity of rain-producing clouds, duration of precipitation, location of maximum seeding effect in clouds with respect to location of treatment, and other factors pertinent to the assessment of seeding operations. A narrow-beam, high-powered 10-cm set with rapid scan capabilities is needed for this purpose [4].

9.4. Time of Year

The time of year to conduct a precipitation modification experiment depends to some extent upon the purpose to be served. If primary interest is in agricultural application, the experiment should be conducted during the warm season. For the major Illinois crops, corn and soybeans, July and August rainfall are the most important [34]. Therefore, it would be logical to concentrate efforts to determine the efficacy of cloud seeding for agricultural applications in the above two months.

If primary interest is in the use of cloud seeding to help replenish water supplies, the cold season during the period from October through March becomes very important. As shown by Hudson and Roberts [37], the general trend is for a withdrawal of water from storage in the April–September period in Illinois, whereas it enters into storage in the October–March period when transpiration and land evaporation minimize. Although cloud seeding could be undertaken to augment both surface and groundwater supplies during the warm season, losses to evapotranspiration would be large, especially during dry periods when the greatest requirement for rainfall augmentation exists. If a seeding experiment is to be confined to a single season in Illinois, the warm season is the most logical choice considering both agricultural and water supply needs.

9.5. Time of Day

The optimum time of day to carry out seeding experiments is dictated to a considerable extent by operational capabilities. Aircraft seeding during the night has usually been avoided. In southern Illinois, the optimum time of day during the warm season would be the afternoon and early-evening hours, since a climatological maximum in the diurnal distribution occurs in mid-afternoon. Afternoon seeding operations in the southern area have two advantages—more frequent seeding opportunities and the possible use of seeding-induced changes on the magnitude and/or time of the afternoon maximum as a verification tool. In most other regions of the state, the primary maximum occurs in the early-morning hours, but there tends to be a secondary maximum in late afternoon or early evening.

9.6. Weather Types

Synoptically, air mass storms are the easiest to seed by aircraft and to observe because of their tendency to be scattered in the atmosphere. However, these storms produce only a small percentage of the Illinois precipitation. For example, on the East Central Illinois Network, they account for 17% of the warm-season rainfall compared with 77% from frontal associated precipitation [12]. Therefore, any comprehensive scientific seeding experiment aimed toward practical application of the results must evaluate the efficacy of seeding all types of synoptic weather.

Thunderstorms and rainshowers account for 85–90% of the warm season rainfall [4]. Consequently, any seeding experiment during this period should concentrate on these rain types. In the cold season, steady rain accounts for approximately 33% of the total precipitation, so that seeding effects of this type of precipitation should be studied in addition to the convective types.

Less than 20% of the annual precipitation in Illinois occurs in storms of relatively short duration (three hours or less). Consequently, a comprehensive seeding experiment should evaluate seeding effects on the more extensive storm systems. The same conclusion is reached from a consideration of storm or daily rainfall totals. In Illinois, approximately two-thirds of the annual precipitation occurs on days with amounts exceeding 0.50 in.

The above discussion leads to the general conclusion that a comprehensive evaluation of the efficacy and economic benefits of cloud seeding in Illinois should involve study of both frontal and nonfrontal weather, stable and unstable types of precipitation, and both short-lived and extensive storm systems. This is not to say that useful information can not come from the usual type of experiment that restricts seeding to the isolated, air mass system, but these storms are not a major water producer. At this time, no adequate proof has appeared in the literature to show that seeding efficacy and efficiency in one type of storm is applicable to all types.

9.7. Length of Experiment

The necessary length of an Illinois experiment can not be specified exactly, since there are many complicating factors that would vary between experiments, depending upon the needs, interests, and operational capabilities of the investigating group. If the investigator's primary interest was to determine the overall benefits to agriculture or water supply from Illinois seeding operations, the information compiled in the Illinois studies indicates he should endeavor to seed all types of storms [12]. This would result in a much larger statistical sample in a given period of operation than that obtained when the primary interest was in seeding a particular type of storm. However, in certain scientific experiments the investigator might have greater interest in observing the microphysics and dynamics of clouds exposed to seeding than in evaluating the surface rainfall output. Then, he might very well concentrate on the isolated convective clouds typical of air mass storms, as done in the Missouri experiments [9]. Any attempt to evaluate the contribution to surface rainfall from seeding, with a high degree of reliability through standard statistical analyses, would require a long time with air mass storms unless the seeding effect was very large [15].

The length of an experiment will also depend upon the percentage of available storms that can actually be seeded. Operational problems and forecasting limitations will prohibit seeding of the total number of storms available during an experimental period. Furthermore, if seeding operations are limited to a particular period of the day, as frequently done in aircraft seeding experiments, the number of seeded storms is further reduced. The diurnal distribution of seasonal precipitation provided in [4] can be used to estimate opportunities lost when seeding is restricted to a given period of hours. For example, assume seeding is to be carried on only during the 6-hr period, 1200–1800 CST in a central Illinois experiment. On the average, only 22% of both the total summer precipitation and number of hours with measurable rainfall occur in this 6-hr period. Add three hours (1800–2100) for postseeding effects, and the normal 9-hr total is 35%.

If one incorporates a "rest" period in the seeding operations to minimize possible residual contamination effects, as suggested by Spar [35], the available number of seeding days is reduced substantially. Estimates of the reduction can be obtained from the wet day sequences in [4]. Assuming a 24-hour rest period after each day of seeding, the number of possible seeding days would be reduced by approximately 30%. Estimated length of experiments are provided in [4] for various stratifications of precipitation data by season, synoptic type, and precipitation type, and for various seeding-induced precipitation increases on county-size areas.

The effects of selective or fractional seeding of available storms discussed above are illustrated in Table XVIII abstracted from [4]. Here, length of

TABLE XVIII. Length of experiment required to obtain significance at 95% confidence level on East Central Illinois Network

Sampling design	Years required for verifying given increase (%)					
	Warm Season			Cold season		
	20	40	60	20	40	60
	All storms combined					
Target-control cross-over	2	1	1	2	1	1
Random-experimental	16	5	2	13	4	2
Random-historical non-sequential	8	2	1	6	2	1
	Warm season air mass storms			Cold season low centers		
Target-control cross-over	7	2	1	3	1	1
Random-experimental	54	16	8	26	8	4
Random-historical non-sequential	27	8	4	13	4	2

experiment required to obtain significance in warm- and cold-season experiments on the East Central Illinois Network is shown for the three sampling designs discussed in the Illinois design, three assumed seeding-induced increases, and three storm categories. Estimates are based upon seeding all storms occurring in each stratification. However, estimates of length of experiment under any assumed seeding efficiency can be readily calculated. For example, with any specific design and storm grouping, if the investigator estimates 75% of the naturally occurring storms will be included in the experiment, then the experimental period is increased by 25% from the indicated value. If only 50% of the naturally occurring storms could be included in the study, the experimental period would be doubled. Sampling requirements with the cross-over design in Table XVIII are based upon a correlation coefficient of 0.75 between target and control mean rainfall. If the correlation decreases, the sampling time naturally lengthens. Thus, if the correlation coefficient decreases from 0.75 to 0.50, the sampling time is approximately doubled.

Acknowledgments

Much of the research upon which this paper is based was supported by the National Science Foundation under Grant GA-1360. Major assistance in the various statistical analyses was provided by my associate, Dr. Paul Schickedanz.

References

1. McDonald, J. E. (1956). Variability of precipitation in an arid region: a survey of characteristics for Arizona. Inst. Atmos. Phys., Univ. of Arizona, 88 pp.
2. Huff, F. A. (1966). The effect of natural rainfall variability in verification of rain modification experiments. *Water Resources Res.* **2**, 791–801.
3. Huff, F. A. (1970). Rainfall evaluation studies. I. Final Rep., NSF Grant GA-1360, Ill. State Water Survey, 53 pp.
4. Huff, F. A., and Schickedanz, P. T. (1970). Rainfall evaluation studies. II. Final Rep. NSF Grant GA-1360, Ill. State Water Survey, 224 pp.
5. Stall, J. B. (1964). Low flows of Illinois streams for impounding reservoirs. *Ill. State Water Survey, Urbana, Bull.* 395 pp.
6. Grant, L. O., Chappell, C. F., and Mielke, P. W., Jr. (1968). The recognition of cloud seeding opportunity. *Proc. 1st Natl. Conf. Weather Modification, Boston, Massachusetts*, pp. 372–385. Am. Meteorol. Soc.
7. Braham, R. R., Jr., and Flueck, J. A. (1970). Some results of the Whitetop experiment. *Preprint, 2nd Natl. Conf. Weather Modification, Boston, Massachusetts*, pp. 176–179. Am. Meteorol. Soc.
8. Battan, L. J., and Kassander, A. R. (1960). Design of a program of randomized seeding of orographic cumuli. *J. Meteorol.* **17**, 583–590.
9. Braham, R. R., Jr. (1966). Project Whitetop: design of the experiment. Final Report to NSF, Parts 1 and 2, Univ. of Chicago, 156 pp.
10. Gabriel, K. R. (1967). The Israeli artificial rainfall stimulation experiment. *Proc. 5th Berkeley Symp. Math. Statist. Probability, Univ. of California, Berkeley* pp. 91–113.
11. MacDonald, G. J. F. (1966). Weather and climate modification, problems and prospects. Final Report. Panel on Weather and Climate Modification, National Academy of Sciences–National Research Council, Washington, D.C., 198 pp.
12. Huff, F. A. (1969). Climatological assessment of natural precipitation characteristics for use in weather modification. *J. Appl. Meteorol.* **8**, 401–410.
13. Conrad, V., and Pollak, L. W. (1950). "Methods in Climatology," 459 pp. Harvard Univ. Press, Cambridge, Massachusetts.

14. Huff, F. A., and Shipp, W. L. (1968). Mesoscale spatial variability in midwestern precipitation. *J. Appl. Meteorol.* **2**, 886–891.
15. Decker, W. L., and Schickedanz, P. T. (1965). The evaluation of rainfall records from a five-year cloud seeding experiment in Missouri. *Proc. 5th Berkeley Symp. Math. Statist. Probability, Univ. of California, Berkeley* pp. 55–63.
16. Huff, F. A. (1967). Rainfall gradients in warm season rainfall. *J. Appl. Meteorol.* **6**, 435–437.
17. Huff, F. A., Shipp, W. L., and Schickedanz, P. T. (1969). Evaluation of precipitation modification experiments from precipitation rate measurements. Final Report to U.S. Dept. of Interior, Bureau of Reclamation, Ill. State Water Survey, 122 pp.
18. Huff, F. A. (1967). Time distribution of rainfall in heavy storms. *Water Resources Res.* **3**, 1007–1019.
19. Changnon, S. A., Jr. (1969). Climatology of severe winter storms in Illinois. *Ill. State Water Survey, Urbana, Bull.* **53**, 45 pp.
20. Bergeron, T. (1960). "Operation and Results of Project Pluvius," Monograph No. 5, Physics of Precipitation, pp. 152–157. Am. Geophys. Union, Washington, D.C.
21. Huff, F. A., and Shipp, W. L. (1969). Spatial correlations of storm, monthly, and seasonal precipitation. *J. Appl. Meteorol.* **8**, 542–550.
22. Light P. (1947). Hydrologic aspects of thunderstorm rainfall. Hydrometeor. U.S. Weather Bureau-Corps of Engineers, Rep. No. 5, 260–268.
23. Linsley, R. K., and Kohler, M. A. (1951). Variations in storm rainfall over small areas. *Trans. Amer. Geophys. Union*, **32**, 245–250.
24. Huff, F. A., and Neill, J. C. (1957). Rainfall relations on small areas in Illinois. *Ill. State Water Survey, Urbana, Bull.* **44**, 61 pp.
25. McGuiness, J. L. (1963). Accuracy of estimating watershed mean rainfall. *J. Geophys. Res.* **68**, 4763–4767.
26. Huff, F. A. (1970). Sampling errors in measurement of mean precipitation. *J. Appl. Meteorol.* **9**, 35–44.
27. Huff, F. A. (1969). Precipitation detection by fixed sampling densities. *J. Appl. Meteorol.* **8**, 834–837.
28. Simpson, J., Brier, G. W., and Simpson, R. H. (1967). Stormfury cumulus seeding experiment, 1965: statistical analysis and main results. *J. Atmos. Sci.* **24**, 508–521.
29. Weinstein, A. I., and Davis, L. G. (1968). A parameterized numerical model of cumulus convection. Rep. No. 11 to NSF GA-777, Dept. of Meteorology, Penn State University, 43 pp.
30. Cunningham, Robert M. (1970). Problems in evaluating effects of seeding cumulus clouds. *Preprint, 2nd Conf. Weather Modification, Boston, Massachusetts*, pp. 193–197. Am. Meteorol. Soc.
31. Brown, H. A., and Glass, M. (1970). The use of a cumulus model in a cloud modification experiment. *Preprint, 2nd Conf. Weather Modification, Boston, Massachusetts*, pp. 8–13, Am. Meteorol. Soc.
32. Huff, F. A. (1968). Area-depth curves—a useful tool in weather modification experiments. *J. Appl. Meteorol.* **7**, 940–943.
33. Nason, C. K., and Lopez, M. E. (1967). A test of certain evaluation designs for cloud seeding. Tech. Rep. B-3662, W. E. Howell Associates, Lexington, Massachusetts, 68 pp.
34. Changnon, S. A., and Neill, J. C. (1966). Areal variations in corn-weather relations in Illinois. *Trans. Ill. Acad. Sci.* **60**, 221–230.

35. Spar, J. (1968). Design study for a weather modification experiment in northeastern United States. Final Rep., Grant E22-112-67(G), ESSA, New York Univ., 85 pp.
36. Huff, F. A. (1955). Comparison between standard and small-orifice raingages. *Trans. Amer. Geophys. Union* **36**, 689–694.
37. Hudson, H. E., Jr., and Roberts, W. J. (1955). 1952–1955 Illinois drought with special reference to impounding reservoir design. Ill. State Water Survey, 52 pp.

FOR A SPACE PROBER*

From Time's obscure beginning, the Olympians
   Have, moved by pity, anger, sometimes mirth,
Poured an abundant store of missiles down
   On the resigned, defenceless sons of Earth.

Hailstones and chiding thunderclaps of Jove,
   Remote directives from the constellations:
Aye, the celestials have swooped down themselves,
   Grim bent on miracles or incarnations.

Earth and her offspring patiently endured,
   (Having no choice) and as the years rolled by
In trial and toil prepared their counterstroke—
   And now 'tis man who dares assault the sky.

Fear not, Immortals, we forgive your faults,
   And as we come to claim our promised place
Aim only to repay the good you gave
   And warm with human love the chill of space.

Thomas G. Bergin
Professor of Romance Languages
* and Master of Timothy Dwight College*
Yale University
New Haven, Connecticut

* Prior to the launch of the Navy TRAAC Satellite by the Johns Hopkins University Applied Physics Laboratory in 1961, G. F. Pieper, Head of the Space Research and Analysis Group at that laboratory, requested of Professor Thomas Bergin that he be kind enough to compose and dedicate a poem to the newly developing field of space research. The above poem was received, inscribed upon an instrumentation panel, and launched aboard the TRAAC Satellite 1961 $\alpha \eta$ 2 on November 15, 1961. Members of the group launching the poem and signing their names to the effort were G. F. Pieper, C. O. Bostrom, G. Bush, H. Lie, F. Swain, and D. J. Williams.

CHARGED PARTICLES TRAPPED IN THE EARTH'S MAGNETIC FIELD

Donald J. Williams

Space Environment Laboratory, Environmental Research Laboratories,
National Oceanic and Atmosperic Administration, Boulder, Colorado

	Page
1. Introduction	137
1.1. Geomagnetic Trapping	137
1.2. General Comments	145
2. Magnetic Field Considerations	146
3. Particle Survey	156
3.1. Protons	156
3.2. Electrons	160
3.3. Alpha Particles and $Z \geq 3$	168
4. Sources, Losses, and Transport	170
4.1. Inner Zone Protons	170
4.2. Inner Zone Electrons	177
4.3. Outer Zone Protons	184
4.4. Outer Zone Electrons	191
5. Concluding Remarks and Future Directions	206
List of Symbols	207
References	208

1. INTRODUCTION

1.1. Geomagnetic Trapping

Although early descriptions of charged particle motion in the geomagnetic field [1–11] were applied mainly to cosmic ray propagation, geomagnetic storms, and auroral phenomena (understandably the idea of a permanent intense trapped particle population was not expressed), the concept of charged particle trapping, the overall trapping region geometry in relation to a dipole field, and trapped particle motion within such a field were quantitatively well understood before the discovery of the earth's trapped radiation [12]. Verification of the interpretation that the newly observed charged particles above the atmosphere were geomagnetically trapped [12] came from a number of experiments, all of which demonstrated the direct control exerted by the earth's magnetic field on the observed particle intensities. As an early example of this, Fig. 1 shows the work reported by Yoshida *et al.* [13] wherein the ordering of the data using magnetic field coordinates can be clearly seen.

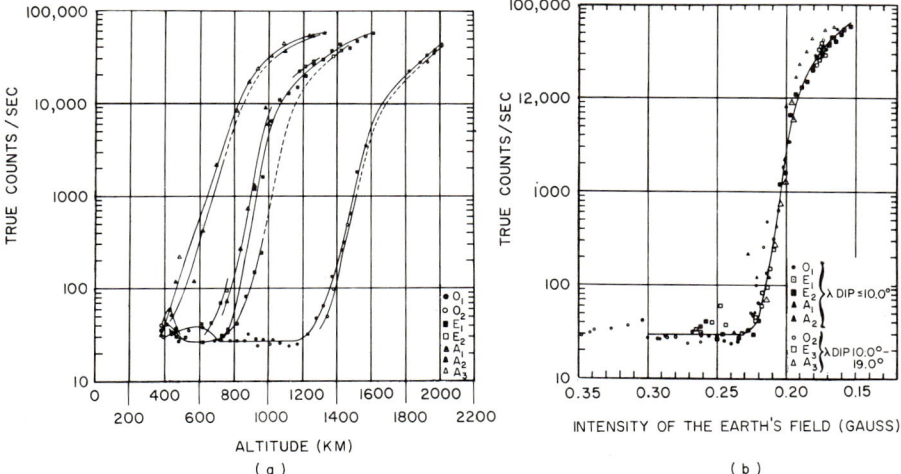

FIG 1. Example of an early observation clearly demonstrating the control of the earth's magnetic field over the observed charged particle population. (a) Count rate versus altitude curves for several different longitudes near the magnetic dip equator; O_1 and O_2 cover the span 93.4°E to 113.9°E; E_1, E_2, and E_3 cover 12.7°E to 3°W; and A_1, A_2, and A_3 cover 60.9°W to 84.6°W. (b) Same data in (a) but now plotted versus the intensity of the earth's field (from Yoshida et al. [13]).

The most useful approach for describing the motion of charged particles in the earth's magnetic field has been the guiding center approximation and subsequent development of adiabatic invariant concepts [8,14]. In the guiding center approximation, the instantaneous position, $\mathbf{r}$, of a particle moving in a magnetic field is broken down into its circular motion of radius $\boldsymbol{\rho}$, and the motion of the guiding center $\mathbf{R}$.

(1.1) $$\mathbf{r} = \mathbf{R} + \boldsymbol{\rho}$$

A general expression for the motion of the guiding center can be obtained [15] by substituting (1.1) into the equation of motion

(1.2) $$m \frac{d^2 \mathbf{r}}{dt^2} = m\mathbf{g} + \frac{e}{c} \frac{d\mathbf{r}}{dt} \times \mathbf{B} + \mathbf{E}e$$

where m = particle mass, e = electronic charge, c = velocity of light, $\mathbf{g}$ = acceleration of gravity, $\mathbf{B}$ = magnetic field, and $\mathbf{E}$ = electric field. This yields the nonrelativistic guiding center equation

(1.3) $$\frac{d^2 \mathbf{R}}{dt^2} = \mathbf{g} + \frac{e}{m} \left\{ \mathbf{E} + \frac{1}{c} \frac{d\mathbf{R}}{dt} \times \mathbf{B} \right\} - \frac{\mu}{m} \nabla \mathbf{B} + O\left(\frac{\rho}{x}\right)$$

where μ = particle's magnetic moment due to gyration, ρ = cyclotron radius, x = scale length over which magnetic field changes appreciably, and $O(\rho/x)$ = terms of order ρ/x. In this approximation the particle's motion in the earth's field is broken down into three components: gyration about a field line, bounce back and forth along the field line between mirror points, and a slow longitudinal drift around the earth. While these motions are not strictly separate from one another, the vast difference in the time scales associated with them makes such a separation possible and leads directly to the consideration of adiabatic invariants. These motions are illustrated in Fig. 2.

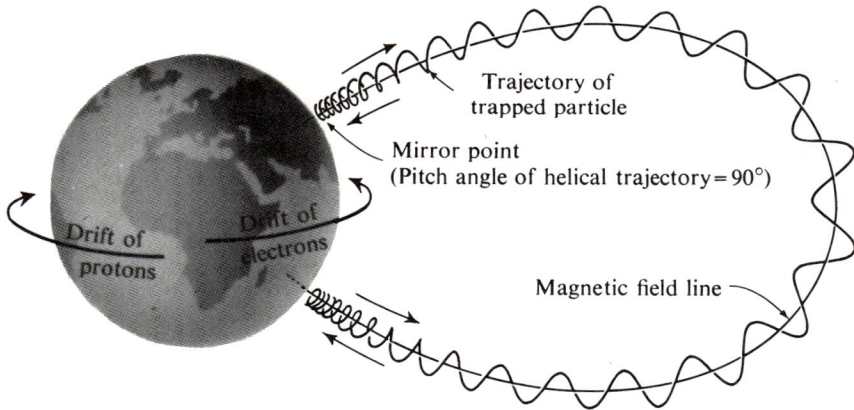

FIG. 2. Illustration of the motion of a charged particle trapped in the earth's magnetic field.

The adiabatic invariants may be considered constants of the particle's motion provided that magnetic field variations are small compared with the time and spatial scales associated with the particle's motion. The first of the adiabatic invariants is the magnetic moment generated by the particle as it gyrates around the field line,

(1.4) $$\mu = m v_\perp^2/2B = mv^2 \sin^2 \alpha/2B = E_\perp/B$$

where α = angle between the field line and the velocity vector. If the changes in the field are small over one radius of gyration ρ and during one gyroperiod τ_c, then $\mu \simeq$ constant, i.e., if

(1.5) $$|\nabla B/B| \ll 1/\rho$$

and

(1.6) $$dB/dt \ll B/\tau_c, \qquad \tau_c = 2\pi mc/eB$$

Then $\mu \simeq$ constant.

This leads directly to the mirror equation and definition of the particle's mirror point.

$$(1.7) \qquad \mu = \frac{E \sin^2 \alpha}{B} = \text{constant}$$

In a static field, $E = $ constant and

$$(1.8) \qquad \sin^2 \alpha_1 / B_1 = \sin^2 \alpha_2 / B_2 = \text{constant}$$

As a particle moves into a converging field, α increases until $\sin \alpha$ reaches the value 1.0. At this point the particle reverses its motion and travels back along the field line in the direction from which it came. Resolution of the $\mathbf{v} \times \mathbf{B}$ force acting on particles traveling toward higher B values in a converging magnetic field yields a component which is always directed away from the higher field, resulting in eventual mirroring as described by (1.8). Using the earth's equatorial magnetic field as a reference, the mirror point B value on a given line of force is determined by the particle's pitch angle at the equator,

$$(1.9) \qquad B_M = B_{eq} / \sin^2 \alpha_{eq}$$

In a dipolelike field configuration such as the earth's, having a minimum B value at the equator, particles will simply bounce back and forth between conjugate mirror points located in the northern and southern hemispheres (cf. Fig. 2).

A range of pitch angles (the loss cone) can be defined from Eq. (1.9) in which trapping is not possible because particles with such pitch angles impact the atmosphere and are absorbed. Using an altitude of 100 km as the effective edge of the dense atmosphere, the atmospheric loss cone is given by

$$(1.10) \qquad \alpha_{eq} = \arcsin (B_{eq}/B_{100})^{1/2}$$

where B_{100} is the field value at 100 km. All particles with equatorial pitch angles in the range 0 to α_{eq} will enter the atmosphere.

The second adiabatic invariant, obtained from the action integral and associated with the particle's bounce back and forth along a field line, is given by

$$(1.11) \qquad J = 2 \int_M^{M^*} P_\| \, ds$$

where $P_\| = mv_\|$ is the particle momentum along the field line and the integral is taken along the field line between the two conjugate mirror points

M and M^*. Using $v_\| = v(1 - \sin^2 \alpha)^{1/2}$ and (1.8), (1.10) can be used to define the integral invariant I

(1.12) $$I = \frac{J}{2mv} = \int_M^{M^*} [1 - (B/B_M)]^{1/2}\, ds$$

The convenience of using the I integral is that it is conserved in a static field ($\partial B/\partial t = 0$) and depends only on the particular magnetic field configuration being considered. If $\partial B/\partial t \neq 0$, the actual adiabatic invariant J, and not I, must be used in describing the particles' motion.

Forces due to the gradient of the earth's magnetic field and field line curvature cause a longitudinal drift across field lines with electrons drifting eastward and protons drifting westward. In an ideal dipole field this effect produces a drift surface which is simply the figure of azimuthal revolution of a line of force. Associated with this drift motion is the third adiabatic invariant, the flux invariant

(1.13) $$\Phi_M = \int B\, dS$$

Φ_M, the magnetic flux linked in the drift orbit of the particle, is the weakest of the three invariants μ, J, Φ_M since it has associated with it the largest spatial and temporal scales. Therefore the conditions for adiabatic invariance are most easily violated for Φ_M.

In Table I we show for reference characteristic times associated with charged particle motion in the magnetosphere.

Direct experimental verification of the trapping ability of the geomagnetic field was obtained from the Argus series of high-altitude nuclear detonations [16,17]. These explosions produced electron intensity enhancements that

TABLE I. Characteristic times associated with particles trapped in the earth's magnetic field[a]

Particle	Energy (keV)	Gyroperiod ($\propto R^3$) (sec)	Bounce period ($\propto R$) (sec)	Drift period ($\propto 1/R$) (hr)
electron	10	$9.4(10)^{-6}$	0.64	36.7
	100	$13\ (10)^{-6}$	0.23	4.1
	1000	$80\ (10)^{-6}$	0.13	0.54
proton	10	$17\ (10)^{-3}$	27	36.7
	100	$17\ (10)^{-3}$	8.6	3.4
	1000	$17\ (10)^{-3}$	2.7	0.35

[a] Values correspond to $\alpha_e = \pi/2$ and $R = 2R_E$.

were spatially confined at all longitudes to a particular magnetic shell. These peaked intensity distributions were observed for time periods much greater than the particle drift times and thus directly displayed the high trapping efficiency of the geomagnetic field.

Related to these Argus results was the development of the B, L coordinate system [18] based on the adiabatic invariants which describe the motion of a charged particle trapped in the earth's magnetic field. The value of the magnetic field B and the integral invariant I are obtained for any observational point in space using the earth's magnetic field as represented by a polynomial expansion of the surface field. The equatorial crossing distance L of a dipole field line passing through the point in space having these values of B and I is then obtained. Along most field lines in the earth's field at low altitudes, L varies by $\leq 1\%$. This transformation of each observational point in the earth's field to an idealized dipole field yields a coordinate system not only based on the adiabatic invariants but also allows a convenient spatial conception of the trapping regions. Each point is characterized by its position along a field line B and by the equatorial crossing distance of its field line L. As the particles populating a field line drift in azimuth, they will remain on a drift surface which is simply the azimuthal figure of revolution of the field line characterized by a given L value, the L shell. The success of this coordinate system in ordering trapped particle data from all spatial locations within the geomagnetic field has further substantiated the concept of stable geomagnetic trapping. An example of the ordering power of the B, L coordinate system and an indication of its great usefulness is shown in Fig. 3 (from McIlwain [18]). Here are shown both the position in L value of the Argus bursts as a function of longitude and the first of the now well-known B-L representations of trapped particle intensities.

However, the B, L coordinate system fails to order charged particle data when magnetosphere distortions, due to magnetopause, tail and ring current effects, become significant. This occurs at $L \gtrsim 5\text{-}6$ R_E during geomagnetically quiet periods. The initial observation of the effect of such magnetospheric distortions on charged particles was that of a large diurnal variation in low-altitude ≥ 40-keV electron intensities [19], whereby a given electron intensity was found at a lower latitude at midnight than at noon. However, the use of invariant theory within realistic geomagnetic field models has been able to order much of the high-latitude data and indicates that the concept of stable trapping is also supported in the distorted geomagnetic field [20]. Figure 4 shows the ordering of energetic electron data obtained at 1100 km in a distorted field. The amount of latitude shift $\Delta \Lambda$, obtained from the latitude profiles shown, is plotted versus the noontime latitude Λ_D of observation. The predictions of adiabatic motion within a model distorted geomagnetic field are also shown in Fig. 4 and indicate that these

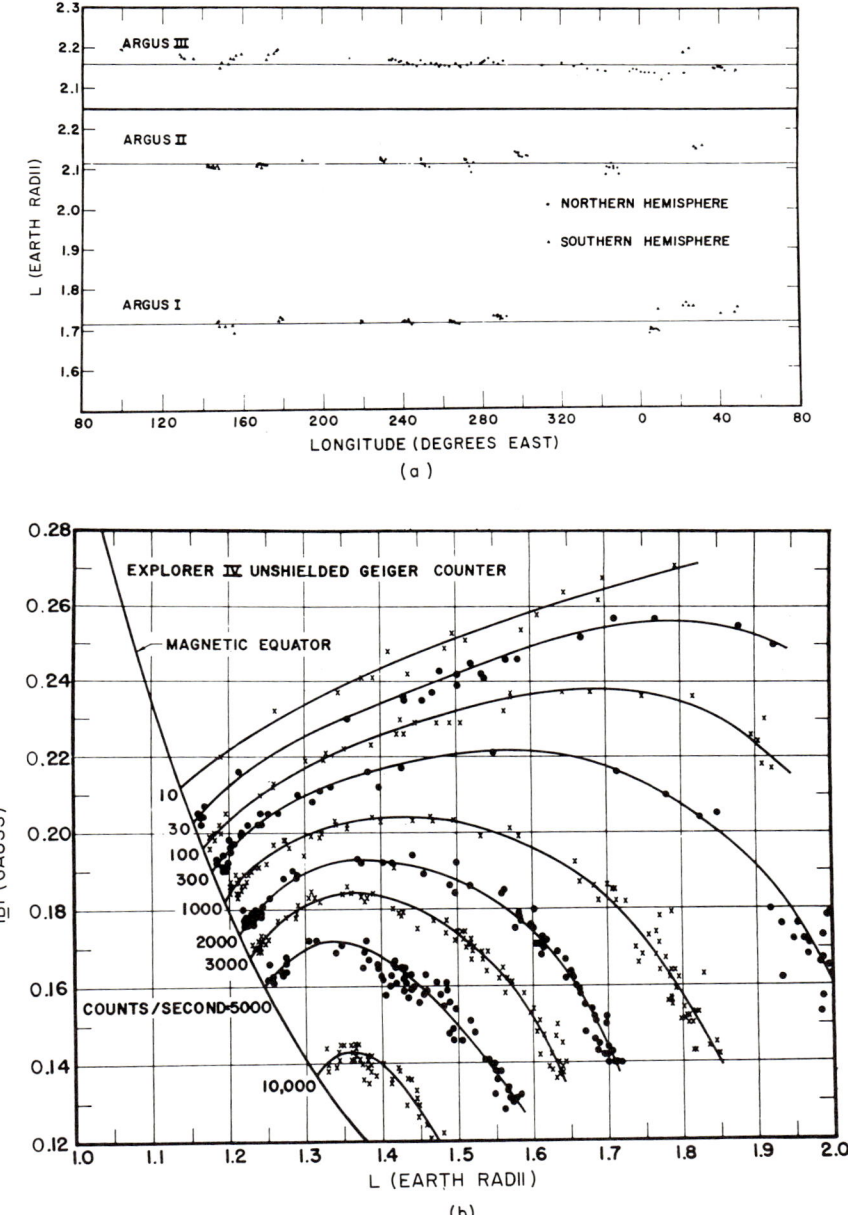

FIG. 3. (a) Observed L values of three Argus shells plotted versus longitude. (b) Contours of constant intensity plotted in B, L coordinate system (from McIlwain [18]). These initial examples of the B, L system clearly display the natural ordering ability of this coordinate system for charged particle observations in the earth's magnetic field. Since the B, L system is based on the concepts of particle trapping, its success graphically demonstrates the high trapping efficiency of the geomagnetic field.

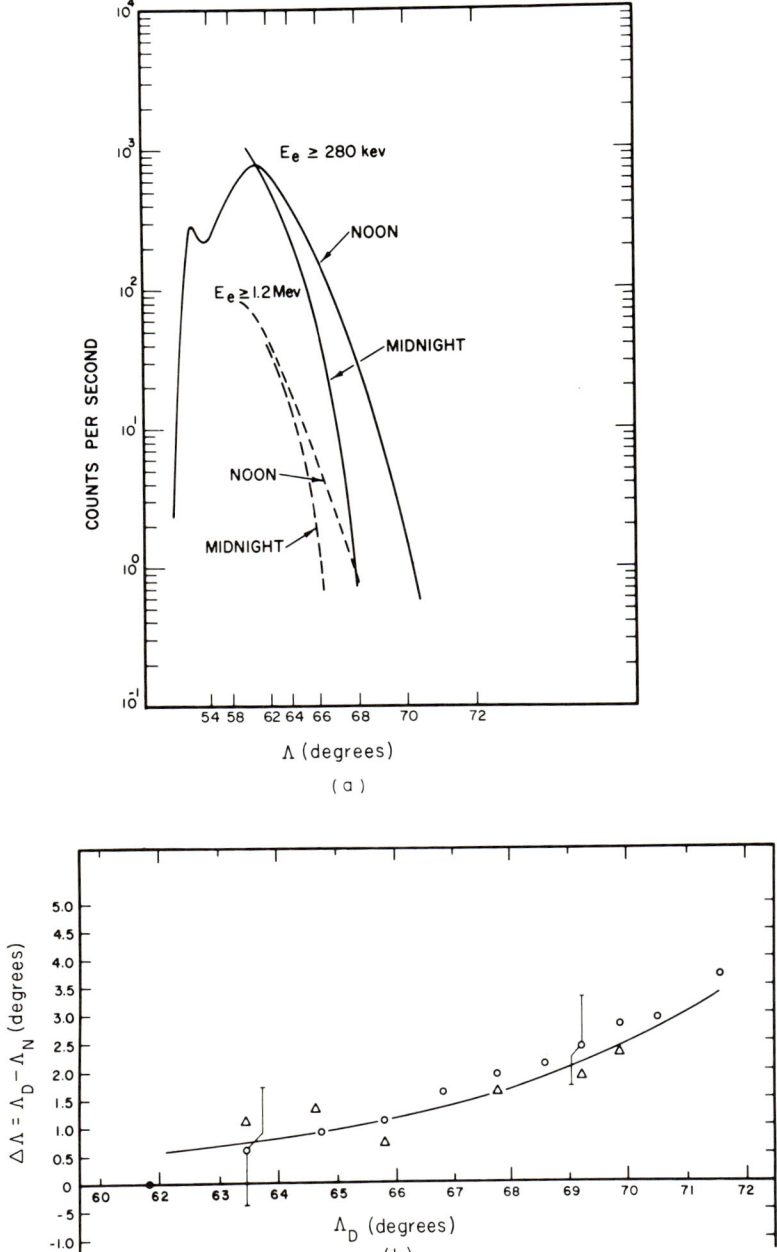

Fig. 4. Example of the maintenance of charged particle trapping even in the highly distorted geomagnetic field. (a) Invariant latitude ($\Lambda = \cos^{-1}L^{-1/2}$) profiles at 1100 km on the noon–midnight meridian for outer zone energetic electrons. (b) Diurnal variation, $\Delta\Lambda = \Lambda_D - \Lambda_N$ (Λ_D, Λ_N = dayside and nightside latitude of observation, respectively)

energetic electron spatial distributions are consistent with stable trapping in such a field. The model field used by Williams and Mead [20] is shown in Fig. 5.

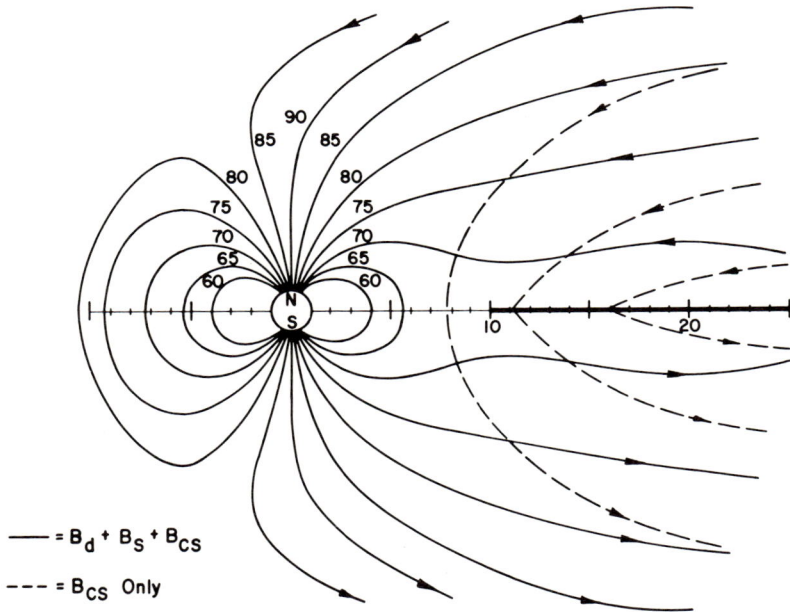

Fig. 5. Geomagnetic field model used to explain trapped electron distributions shown in Fig. 4. Solid lines show the distorted geomagnetic field as made up of components due to the earth's dipole B_d, surface currents at the magnetopause B_s, and a current sheet in the nightside hemisphere B_{cs}. The dashed lines show the field due to the current sheet alone B_{cs} (from Williams and Mead [20]).

We therefore come to the seemingly obvious conclusion that stably trapped particles occupy a wide range of altitudes and latitudes within the earth's magnetic field. Stably trapped particles are those able to complete all three motions characteristic of geomagnetic trapping, gyration, bounce, and drift; i.e., particle lifetimes equal or exceed drift period. In a later section this conclusion will be slightly modified.

1.2. General Comments

Since its discovery [12], reviews of the geomagnetically trapped particle population and its behavior in the geomagnetic field have been numerous

---

from data in (a) plotted versus dayside latitude, Λ_D. Solid curve is expected variation based on adiabatic motion of electrons trapped in the distorted geomagnetic field; 0, trapped electrons $E_e \geq 280$ keV-matched pass data; Δ trapped electrons $E_e \geq 280$ keV — daily averages (from Williams and Mead [20]).

and extensive (e.g., see [21–35]). Such summaries and proceedings have done an excellent job in documenting and advancing the growth in our knowledge and understanding of the trapped particle population and its role in the field of magnetospheric physics. Characteristics of the trapped particle distributions, their temporal behavior, and many ideas, models, and theories have been presented and reviewed in detail. From these reviews it is clear that our morphological knowledge of the trapped radiation is extensive. Quantitative models of most aspects of the trapped radiation, including long-term time variations, now exist for purposes of radiation dose calculations and are being continually updated with the most recently available data (e.g., see [36,37]). However, our understanding remains more limited than our knowledge. While particle types, their distributions, and their behavior are known their sources, losses, and transport are less certain.

This chapter presents a survey of the time-averaged trapped particle distributions in the magnetosphere with primary emphasis on steady state sources, losses, and transport. Although it is assumed herein that the time-averaged distributions represent the steady state case, it is recognized that great time variations occur in these populations, which may make this a poor approximation.

The chapter is divided into three sections: magnetic field considerations; particle survey; and sources, losses, and transport. In order to place trapped particle observations into an appropriate perspective, the first section will deal with a consideration of the overall magnetospheric configuration. The trapping, pseudo-trapping, and nontrapping regions are described within a realistic magnetic field configuration. The spatial relationship of the observed low-energy electron distributions to these regions is also described. This is followed in the second section by a presentation and discussion of the trapped electron, proton, and alpha-particle populations. Finally, in the last and largest section, the current status of specific mechanisms contributing to trapped particle sources, losses, and transport is reviewed.

For descriptions of and references to the large body of magnetospheric particle observations obtained in the short history of this field of study, reference is made to the many comprehensive summaries listed earlier in this introduction and, in particular, to the conference proceedings edited by McCormac [33–35] and Williams and Mead [32].

2. Magnetic Field Considerations

Calculations using model distorted magnetospheric configurations [20, 38–41] have shown the importance of such distortions in the determination of the spatial distributions of the trapped particle population. Not only can many features of these spatial distributions be explained using the simple assumption of adiabatic drift in a realistic model field [20,38,41–43], but

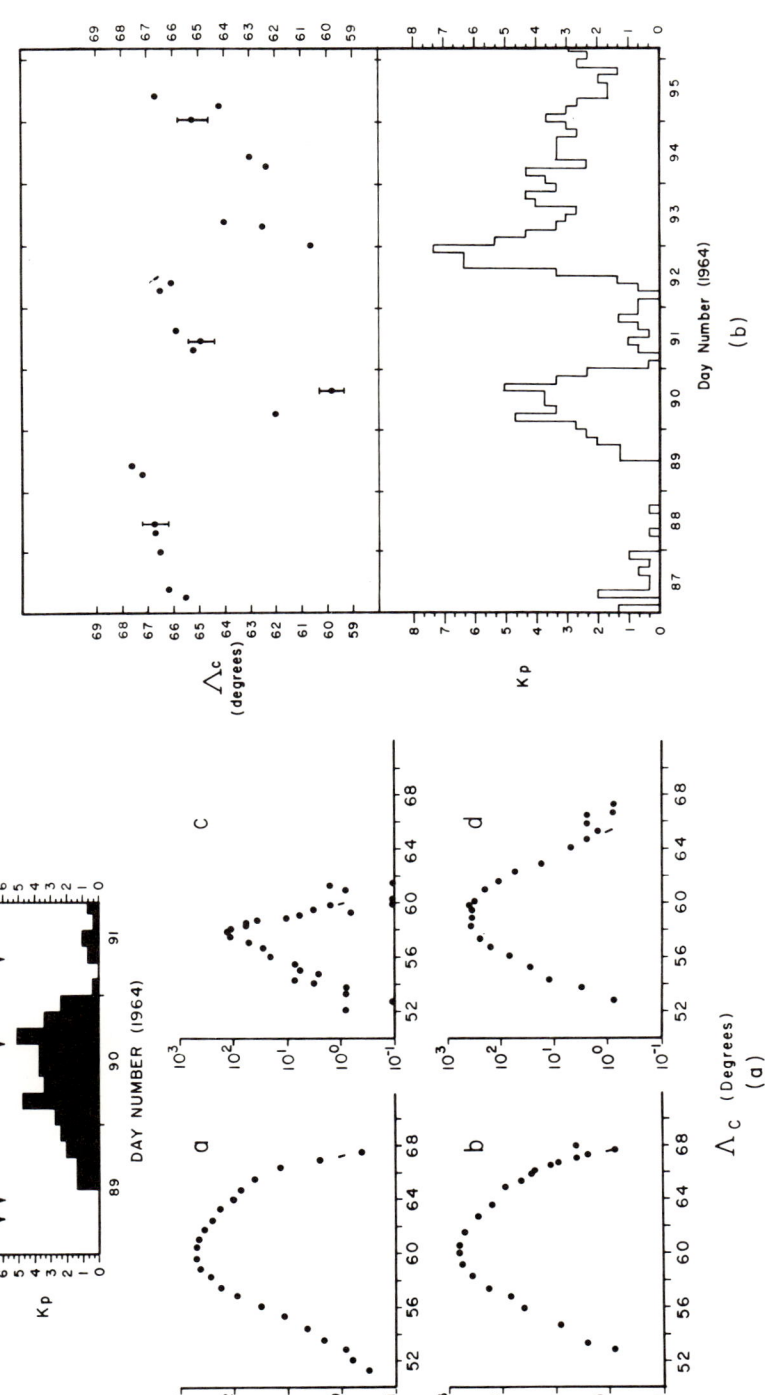

FIG. 6. Motion to lower latitudes during geomagnetic storms of trapped electron low-altitude, high-latitude boundary. (a) Four passes displayed as count rate versus Λ plots obtained near the midnight meridian before, during, and after geomagnetic storm shown in Kp versus time plot. (b) Behavior of outer boundary (defined at one count per second) during period 29 March–2 April, 1964 and its correlation with geomagnetic activity as measured by Kp (from Williams [46]).

certain time variations in these distributions and in the magnetic field itself can be explained [44,45]. For example, during geomagnetic storms the high-latitude edge of the low-altitude energetic ($\geq$280-keV) trapped electron intensity profile displays a characteristic collapse toward low latitudes [46]. This is shown in Fig. 6a where four passes through the outer zone $\geq$280-keV trapped electron region are plotted at various times during a geomagnetic storm. In Fig. 6b is shown the close correlation of this boundary collapse with magnetic activity for a several day period. The boundary collapse during the storm and its subsequent recovery can be seen. A possible cause for such a collapse is the stretching of additional field lines on the nightside hemisphere into the geomagnetic tail to the extent that they no longer can support an energetic trapped particle population. If these field lines were previously populated, this effect would produce the boundary collapse (disappearance of electrons) observed at low altitudes. In addition, an increase in field strength should be observed in the geomagnetic tail.

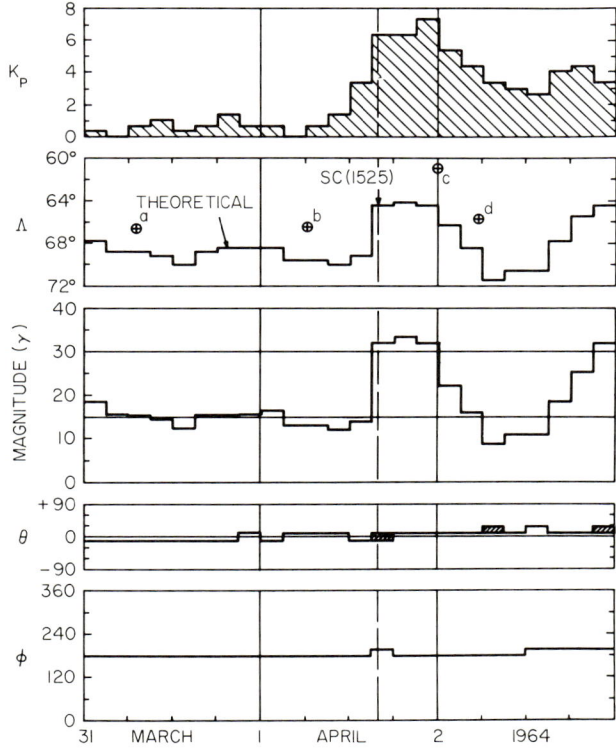

FIG. 7. Plot showing Kp variation, tail field magnitude and direction, and the predicted and observed outer trapping boundary throughout the disturbance of 1 April, 1964 (from Ness and Williams [47]).

Figure 7 shows one result of a study utilizing simultaneous data on trapped electron intensities at low altitudes and the magnetic field strength in the geomagnetic tail during magnetic storms [44,47]. Note the increase in tail field intensity during the magnetic disturbance, consistent with additional field lines being pulled into the tail. Using the last closed nightside field line in the model of Williams and Mead [20] as the high-latitude boundary for trapped electrons, the observed tail field variations in Fig. 7 were used to predict the high-latitude boundary during the course of the storm. These predictions are also shown and compared with the measured boundary in Fig. 7. Note that while the absolute values of the predicted and measured high-latitude boundary differ by a small amount, the time variation of this boundary can be explained using measured variations in the distorted field and assuming stable trapping with conservation of the adiabatic invariants.

From calculations using realistic geomagnetic field models have come the concepts of L-shell splitting and pseudo-trapping [40,48] The term "L-shell splitting" refers to the imposition of a strong pitch angle dependence on invariant drift shells caused by the removal of longitudinal symmetry in the field due to boundary and tail currents (to date L-shell splitting calculations have not included ring current effects). Pseudo-trapping follows immediately from L-shell splitting and refers to particles, conserving μ and J, whose longitudinal drift carries them out of the magnetosphere.

In Fig. 8 the results of shell splitting calculations of Roederer [40] are shown as applied to the model field presented by Williams and Mead [20]. The shell-splitting effect of magnetospheric distortions is shown both for particles beginning on given field lines on the noon meridian and drifting to the midnight meridian and for particles initially on given field lines on the midnight meridian drifting to the noon meridian. Particles mirroring near the equator will follow contours of constant B value in their azimuthal drift. Particles mirroring with small α_e (i.e., at low altitudes) not only maintain constancy of their mirror point B value B_M but also through conservation of I maintain a nearly constant field line length between conjugate mirror points.

From Figure 8 it is seen that particles mirroring near the equator at altitudes $\gtrsim 8\ R_E$ on the midnight meridian will drift to near equatorial altitudes of $\gtrsim 10\ R_E$ on the noon meridian. Since this is outside the magnetopause boundary for this model field, these particles are pseudo-trapped in the sense that their drift will carry them into the magnetopause, and they will be lost from the trapping regions. These particles are unable to complete the azimuthal drift motion characteristic of stably trapped particles. Likewise particles beginning at high altitudes on the noon meridian with small α_e will drift to nightside field lines which extend into the geomagnetic tail and will also be lost from the trapping regions.

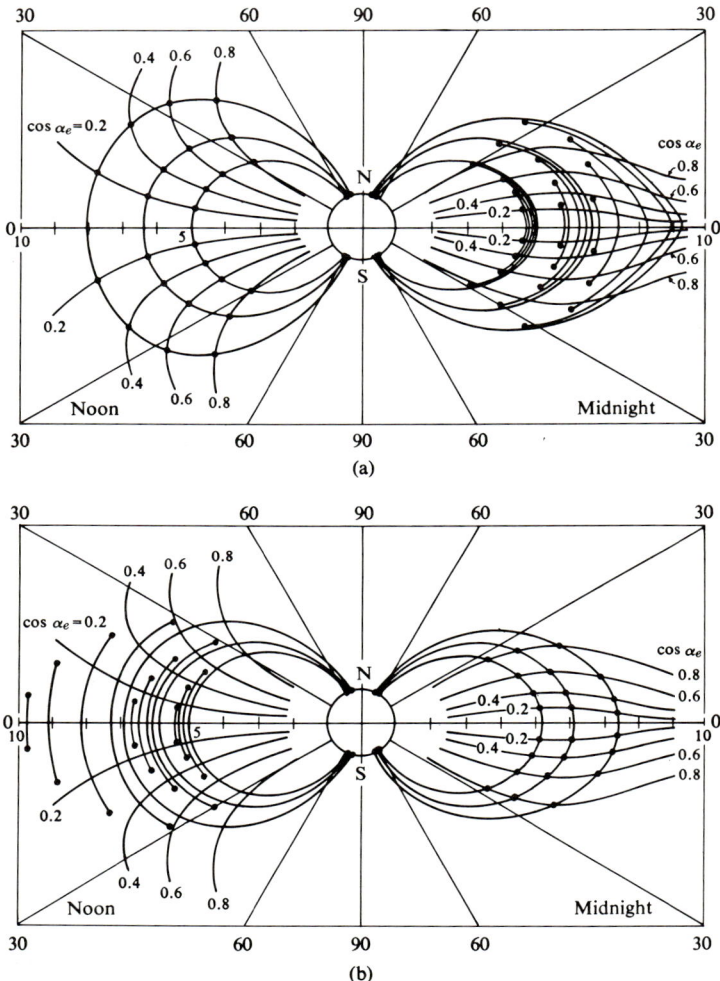

FIG. 8. Computed shell splitting (a) for particles beginning on common field lines on the noon meridian and (b) for particles beginning on common field lines in the midnight meridian. Dots represent particles' mirror points and curves giving position of mirror points for constant equatorial pitch angle α_e are shown (from Roederer [40]).

These effects are demonstrated in Fig. 9 where the stably trapped and pseudo-trapped regions are shown in a plot of equatorial pitch angle α_e versus altitude R in geocentric earth radii. Noon and midnight boundaries are assumed to be at 10 R_E. The atmosphere loss cone is shown and for 10 R_E altitude, the various portions of the equatorial pitch angle distribution are diagramed on the noon and midnight meridian.

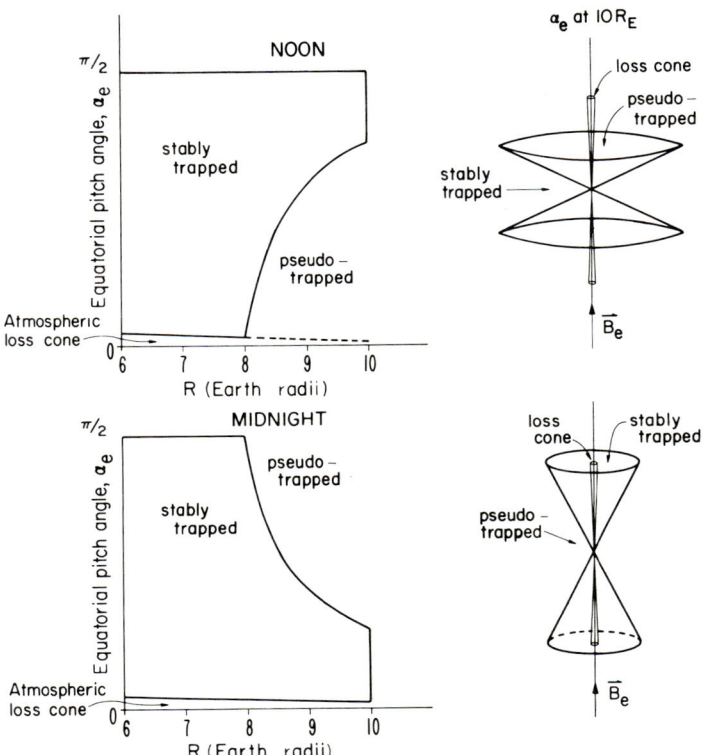

FIG. 9. Diagrammatic view of equatorial pitch angle versus altitude showing stably trapped and pseudo-trapped realms. Atmospheric loss cone also shown. Noon boundary assumed at 10 R_E. These various pitch angle regions are also shown schematically at the equator B_E at 10 R_E.

The above description is admittedly oversimplified by neglecting many effects which influence the motion and lifetime of a trapped particle (e.g., electric fields, pitch angle scattering, etc.). However, these effects are energy dependent and generally are more important at low energies (e.g., less than a few tens of kilovolt electrons). Moreover, since all trapped particles are strongly influenced by the geomagnetic field, the trapping regions will be considered in more detail and an attempt made to show what is known of the spatial relationship between these regions, the magnetosphere, and the magnetospheric plasma.

Figure 10 presents experimental observations of the earth's magnetic field referred to the equatorial plane [49]. Contours of constant magnetic field intensity and contours designating the latitude and local time of the

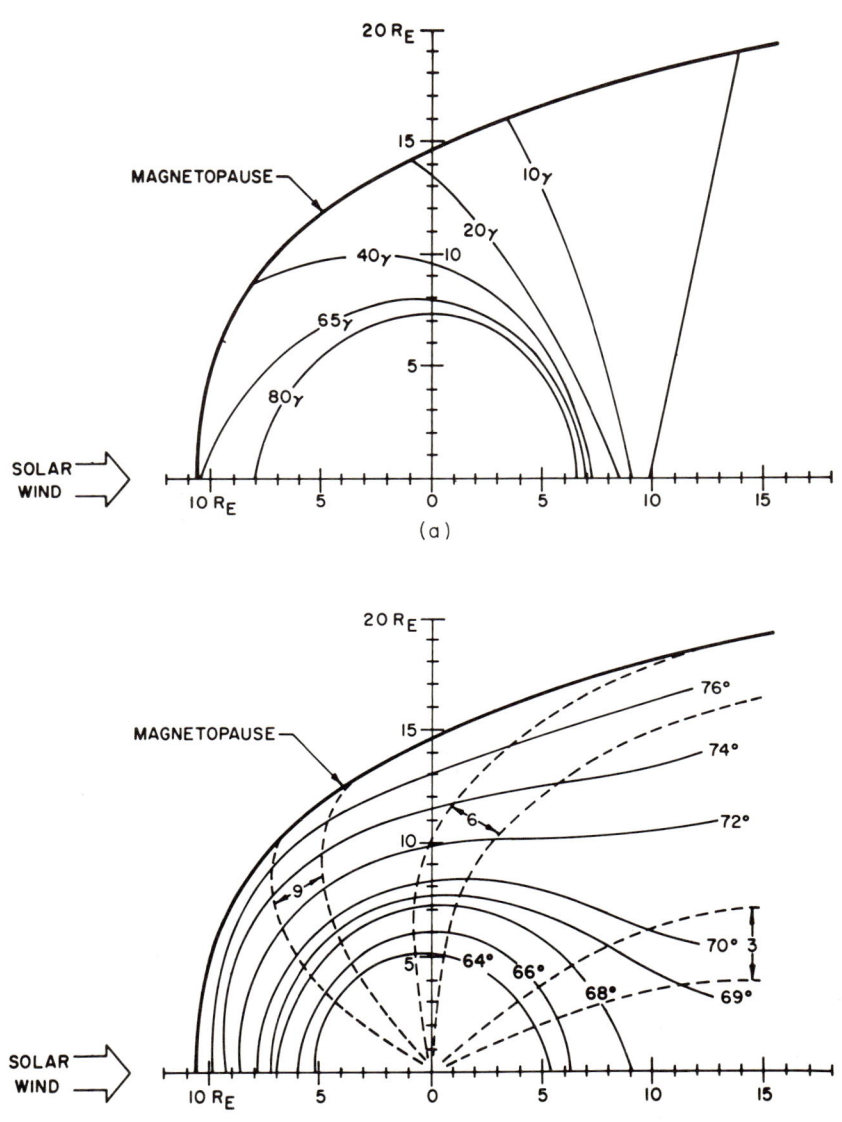

Fig. 10. Average magnetic field characteristics projected to the equatorial plane (from Fairfield [49]). (a) Contours of constant magnetic field intensity. (b) Contours of latitude and local time of the point of intersection of a field line with the earth's surface.

point of intersection of a field line with the earth's surface are shown. Many details of trapped particle motion may be obtained from Fig. 10. Equatorially mirroring particles follow contours of constant B and thus form a stable trapping region within the 65γ ($\gamma = 10^{-5}$ G) contour. Those outside this contour ($\gtrsim 7$ R_E at midnight) will drift into the magnetopause and be lost, thus defining a pseudo-trapping region for equatorially mirroring particles. The pseudo-trapping region is bounded by the 65γ contour and a line defining the limit of sustained bounce motion on a given field line (presumably due to significant extension into the tail region). This outer boundary of pseudo-trapping can vary over a large spatial extent and is arbitrarily drawn, for reference, as the solid line beginning at ~ 10 R_E in the tail in Fig. 10. This line also forms the outer boundary of the pseudo-trapping region for low-altitude (1000 km) mirroring particles.

From Fig. 10 it can be seen that particles mirroring at 1000 km and drifting to field lines on the nightside hemisphere which intersect the earth at latitudes $\gtrsim 68°$–$69°$ will be lost into the tail regions. Such particles would start at noon at latitudes $\gtrsim 71°$–$73°$ [20,40] or from Fig. 10, equatorial crossing altitudes $\gtrsim 8$–9 R_E. The inner boundary of the pseudo-trapping region for 1000 km mirroring particles can thus be visualized as roughly a circular contour running from ~ 8 R_E at noon to ~ 10 R_E at midnight in Fig. 10.

The high-latitude nightside trapping boundary is probably energy dependent and controlled by field line curvature. Therefore, such boundaries are not strict measures of "last closed" field lines but rather a measure of significant field line extension into the tail for a particular particle momentum. This effect yields a large altitude variation at the equator and a rather small ($\sim 1°$) latitude variation at low altitudes for electron energies from ~ 40-keV to ~ 1 MeV.

Note that the pseudo-trapping regions (and thus the stable trapping regions) have a marked pitch angle dependence, and the two extremes discussed here ($\alpha_e = \pi/2$ and $\alpha_e \ll \pi/2$, where $\alpha_e =$ equatorial pitch angle) overlap considerably. As a result, an omnidirectional detector will respond to both trapped and pseudo-trapped particles at high equatorial altitudes.

Using the above results and the calculations reported by Roederer [40], Figs. 11 and 12 show the spatial extent of the trapping and pseudo-trapping regions in an experimentally observed magnetospheric configuration. The noon–midnight magnetospheric projection shown in Fig. 11 is that reported by Fairfield [49] based on IMP 1, 2, and 3 satellite observations. Figure 12 shows an equatorial section through the magnetosphere based on the magnetic geometry of Fig. 10.

The shaded pseudo-trapping regions in Fig. 11 apply only to particles mirroring in those regions. Stably trapped particles mirroring at low altitudes

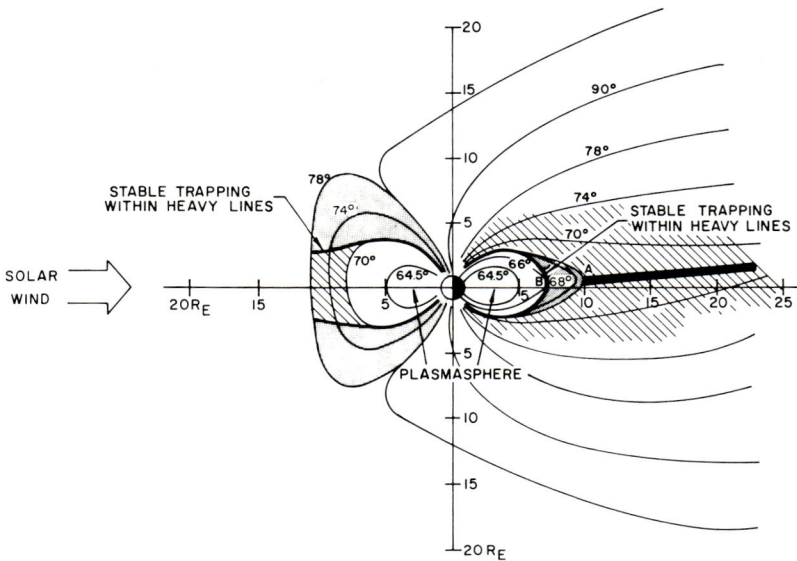

FIG. 11. Noon-midnight projection of the stable trapping, pseudo-trapping, and observed plasma sheet regions within the average magnetospheric configuration reported by Fairfield [49]. Pseudo-trapping region (dotted area); low energy electrons (cross hatched area); neutral sheet (solid area).

can exist in the pseudo-trapping region shown on the midnight meridian in Fig. 11 and apparently have been observed [50,51].

Conversely, the stably trapped regions apply in general to particle mirror points and not their total motion. Note that the stable trapping regions do not follow field lines.

The low-energy electron distribution shown (taken as representative of the plasma sheet) is based mainly on the observations reported by Gringauz et al. [52,53], Freeman [54], Bame et al. [55] and Vasyliunas [56,57] (see also review by Gringauz [58]). No plasma sheet associated intensities are shown at high latitudes on the dayside in Fig. 11, since no comprehensive observations in this region have yet been made. Gringauz [58] has reviewed existing high-latitude dayside plasma observations and has associated the observed intensities with neutral line (point) injection. Within the trapping regions, a quiet time plasmasphere [59] is shown at ~ 5 R_E.

The trapping, pseudo-trapping, plasmasphere, and plasma sheet regions are shown in equatorial cross section in Fig. 12. The dayside high-latitude pseudo-trapping region has been projected to the equator along field lines. A region lacking general survey measurements of plasma sheet intensities can be seen in the postmidnight–early-morning sector.

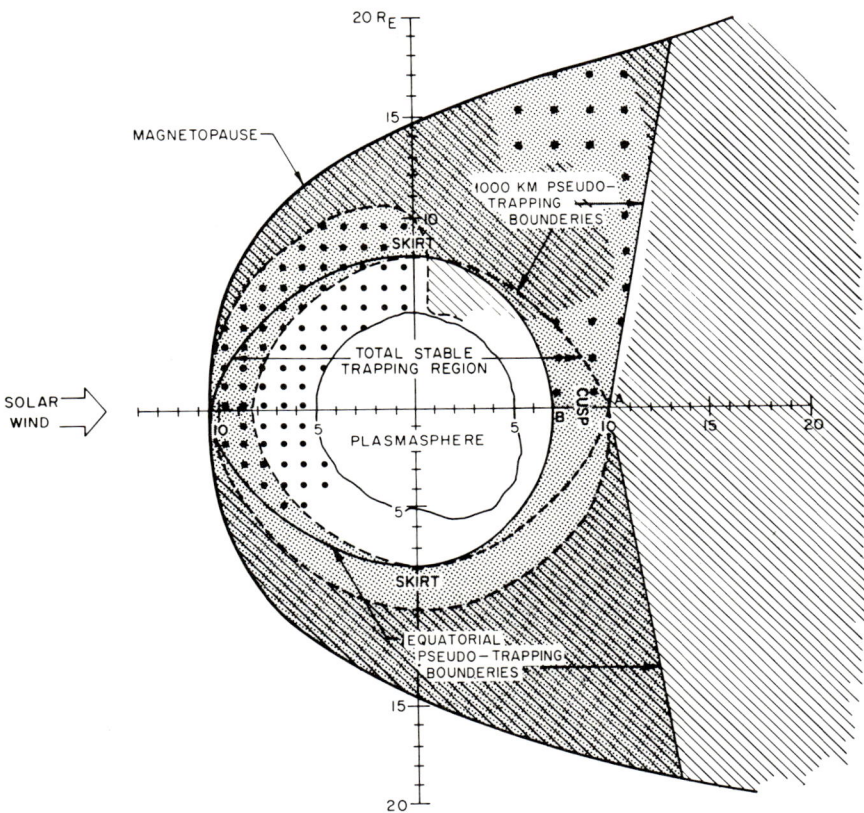

FIG. 12. Projection to the equatorial plane of the stable trapping, pseudo-trapping, and observed plasma sheet regions within the average magnetospheric configuration reported by Fairfield [49]. Total pseudo-trapping region (fine dots); intense low energy electron region (cross hatched area); low intensity low energy electron region (coarse dots); no low energy electron observation available (starred area).

The pseudo-trapping region shown in Figs. 11 and 12, originally called "the distant trapping region" by Anderson [60], contains the familiar skirt and cusp regions [61–63]. The cusp can be identified with the near midnight region between points A and B which denote, respectively, the "last closed" field line and the equatorial limit of trapping. The skirt region extends along the trapping boundary in Fig. 12 from the predawn and postdusk region to the subsolar point.

All the features shown in Figs. 11 and 12 have significant time variations, and, in particular, points A and B and the distance between them may vary over several earth radii. Of further interest are the sharp boundary at midnight between the plasma sheet and the low-altitude trapping boundary and the protrusion of the plasma sheet to near plasmasphere altitudes in the early morning sector.

Finally, we note that without additional forces and interactions (e.g., electric fields, wave–particle interactions) the trapping and pseudo-trapping regions depend only on the magnetic field configuration and are well represented in Figs. 11 and 12 for the field configuration shown. Demonstrably the addition of interactions, particularly wave–particle interactions, significantly changes the extent of these regions for electrons $\gtrsim 40$ keV.

3. Particle Survey

This section is concerned with time-averaged intensity distributions for protons and electrons. Data for these time-averages, obtained from a number of satellites, have been compiled by the National Space Science Data Center at Goddard Space Flight Center and results made available as a series of NASA special publications (see also Vette [64]). Additional data are also presented in order to emphasize structural features and characteristic behavioral patterns that tend to be eliminated by the long-term averaging process. For example, during slow magnetic variations, particle populations remain essentially in a steady state condition, although the overall distribution changes significantly by adiabatically adjusting to the field variation. Corrections for this effect are required in order to compare properly particle distributions during different time periods as well as being necessary for identifying, measuring, and locating nonadiabatic effects (acceleration, injection, and loss). To complete the particle survey, a brief summary of trapped alpha-particle observations in the magnetosphere is also included.

3.1. Protons

Figure 13 presents radial profiles of equatorial integral omnidirectional fluxes for several proton energy thresholds [64]. This trapped proton population covers a wide range of intensities and energies and displays a rather featureless structure as a function of altitude. However, three characteristic features can be noted. These are:

(1) peak intensities occur at lower L values for progressively higher energies, leading to a much harder spectrum at low altitudes;

(2) not only does the high-altitude spectrum become very soft, but the intensity gradient at low energies also becomes very small at high altitudes;

(3) the proton distributions are much more stable in time than the corresponding electron distributions.

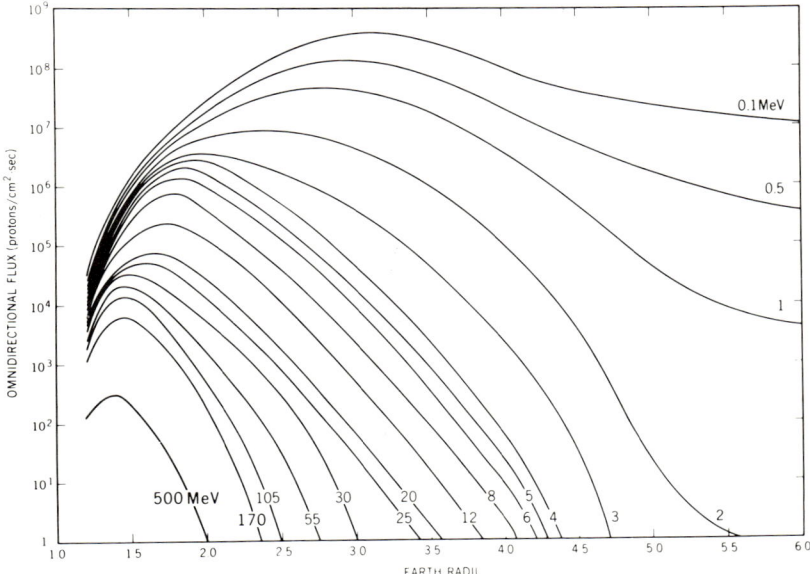

FIG. 13. Time-averaged radial intensity profiles of the integral equatorial omnidirectional proton flux for several energy thresholds (from Vette [64]).

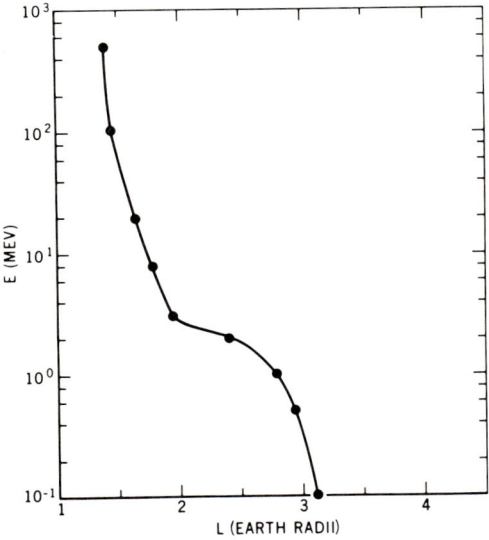

FIG. 14. A plot of the L value of the peak of the intensity distributions, $J(>E)$, shown in Fig. 13, versus E.

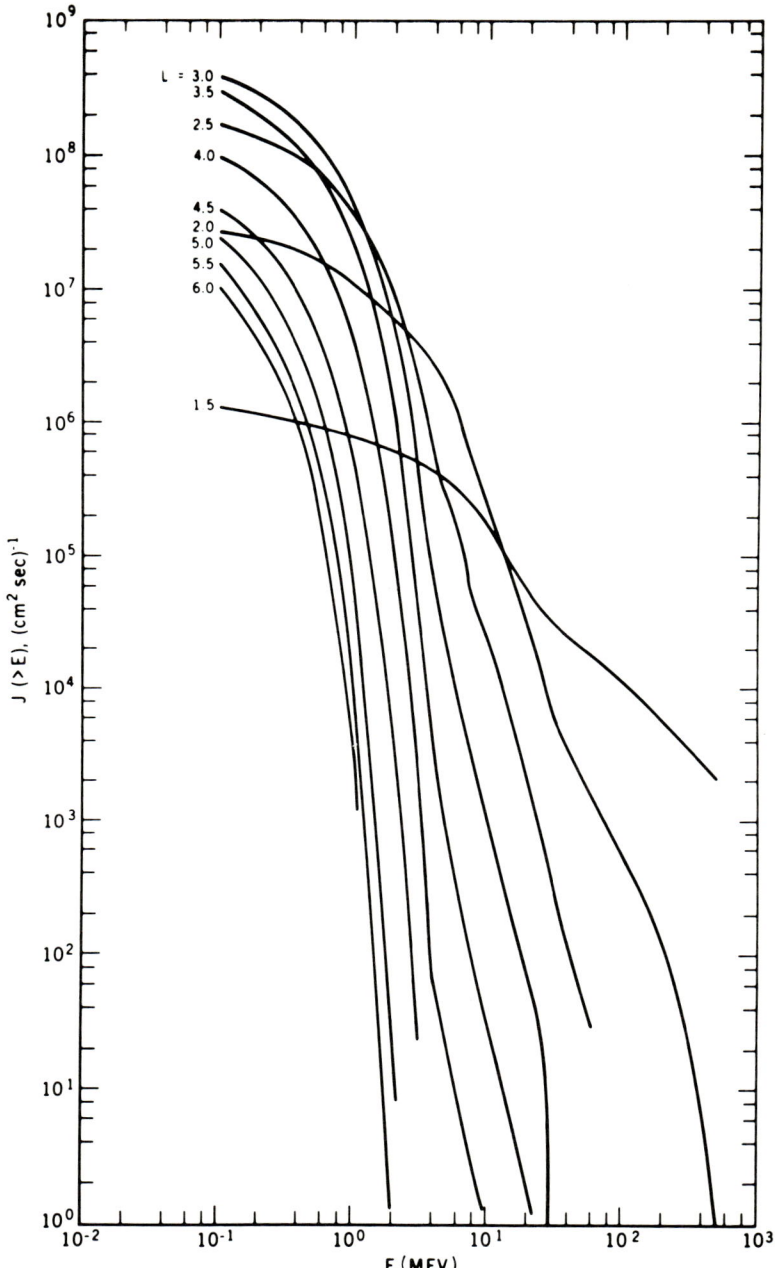

FIG. 15. Time-averaged absolute spectra throughout the trapping regions for the integral equatorial omnidirectional proton flux distributions shown in Fig. 13. A clear hardening of the spectrum is seen at low L values and is indicative of an additional source of protons existing at low altitudes.

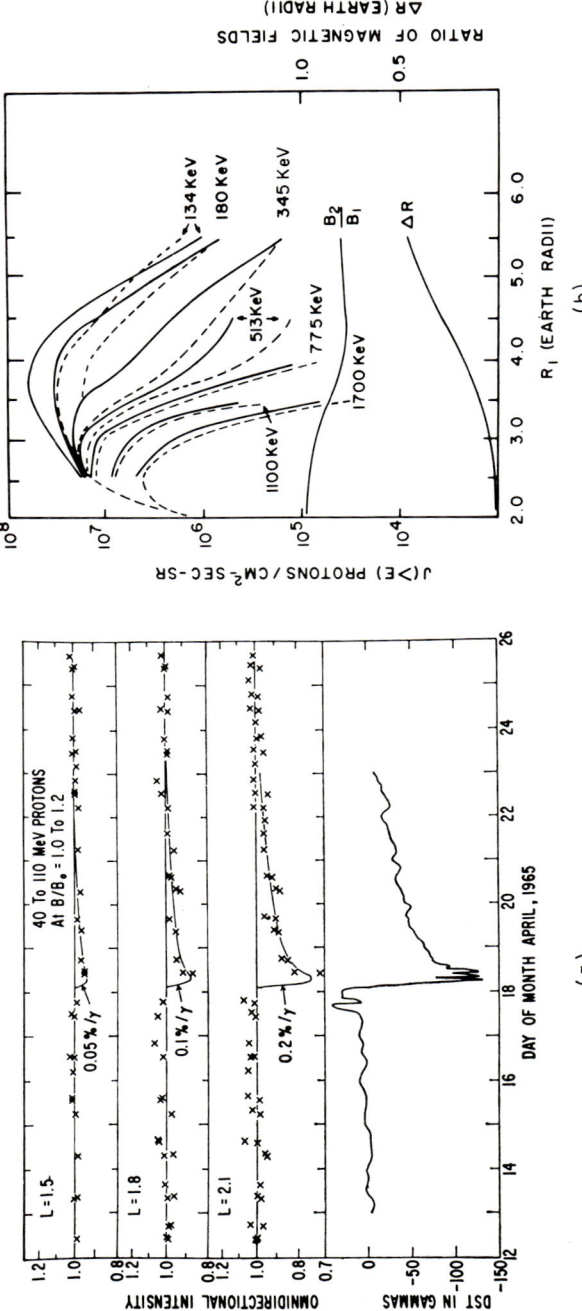

Fig. 16. Two views of adiabatic readjustments of the trapped proton distribution during the 18 April, 1965, magnetic storm. (a) Time history of 40–110 MeV proton intensities throughout the event. The solid lines through the data are the adiabatic intensity variations expected from the observed D_{ST} values (from McIlwain [65]). (b) Readjustments of the low-energy proton radial intensity profiles due to adiabatic effects on 19 April, 1970, when $D_{ST} = -47\gamma$. Ratio of the field strength at the time of the observation B_2 to the prestorm field B_1 and the radial movements ΔR of the particles are also shown. - - - Uncorrected data; —— corrected data (from Söraas and Davis [66]).

Figure 14 shows the peak of the integral proton distribution above a given energy E as a function of L and presents in a more quantitative way the progression to lower L values of the higher-energy peaks. The trend shown in Fig. 14 has no simple physical description, inasmuch as it is a result of all the source loss and transport mechanisms which produce the proton distributions shown in Fig. 13.

Figure 15 shows the absolute energy spectra of the proton distributions given in Fig. 13 for several different values of L. It is readily seen that the spectra harden dramatically at low L values and are indicative of the appearance of a source for low-altitude protons which is separate from that responsible for high-altitude protons.

Time variations observed in trapped proton intensities have been both adiabatic (invariant conserving) and nonadiabatic (invariant violating) in nature. The adiabatic variations are simply redistributions of the steady state population owing to a slowly varying magnetic field configuration and can result in significant intensity changes at a given spatial location. Examples of these effects, covering a wide energy range, are shown in Fig. 16. Figure 16a (from McIlwain [65]) shows variations in 40–110-MeV proton intensities on several L shells throughout the magnetic storm of 17 April 1965. The solid lines through the proton intensities are the variations expected from adiabatic effects, as calculated from the observed D_{ST} values.

Figure 16b (from Söraas and Davis [66]) presents another view of adiabatic effects, again during the 17 April 1965, storm. Adiabatic effects are removed from the data by transforming the observed fluxes from the time-dependent field to a reference field where the ring current is effectively zero. The "corrected" proton fluxes resulting from this transformation are shown in Fig. 16b along with the uncorrected fluxes for the intensity versus L distributions from 134 keV through 1700 keV on 19 April when $D_{ST} = -47\gamma$. Note that whether adiabatic redistributions produce intensity increases or decreases at a given point in space depends strongly on the existing proton energy spectra and spatial gradients, as well as the specific field perturbations in effect at the time. Only with such effects accounted for can nonadiabatic effects be quantitatively measured and described.

3.2. Electrons

Time-averaged equatorial omnidirectional integral electron fluxes at several energy thresholds are shown in Figs. 17 and 18 (Vette, personal communication). These averages, shown near solar minimum and maximum, have been obtained from several experiments and include data from all available longitudes. The averaging out of diurnal variations (especially important for $L > 6$), combined with the occurrence of great fluctuations

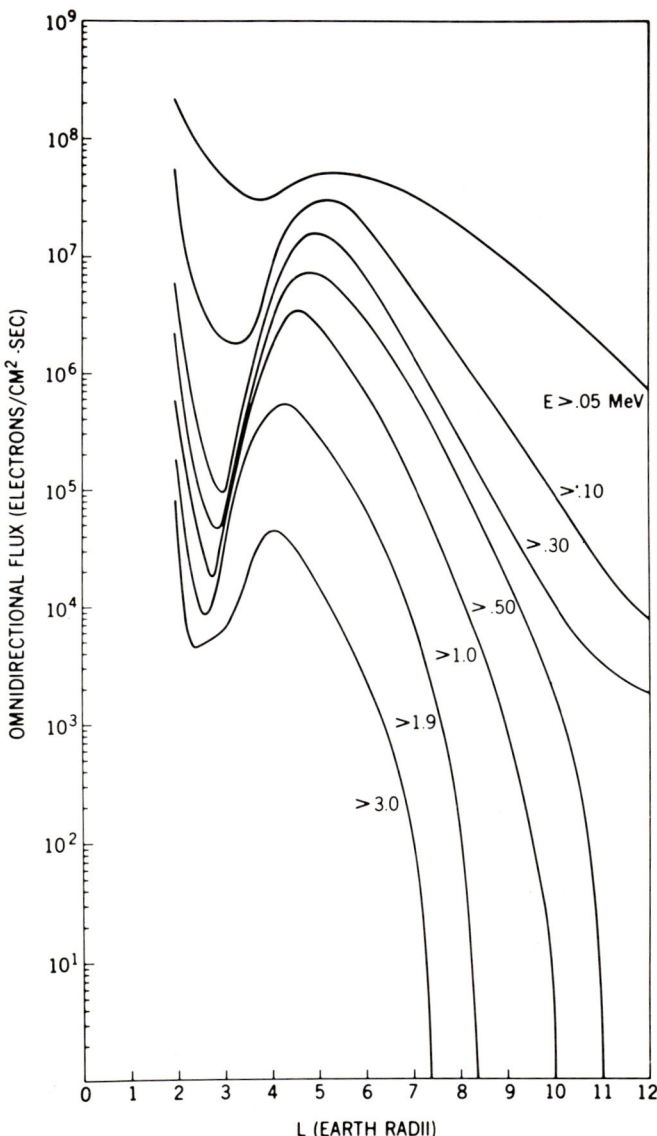

FIG. 17. Time-averaged radial intensity profiles of the integral equatorial omnidirectional electron flux for several energy thresholds. Data obtained near solar minimum and from all available longitudes.

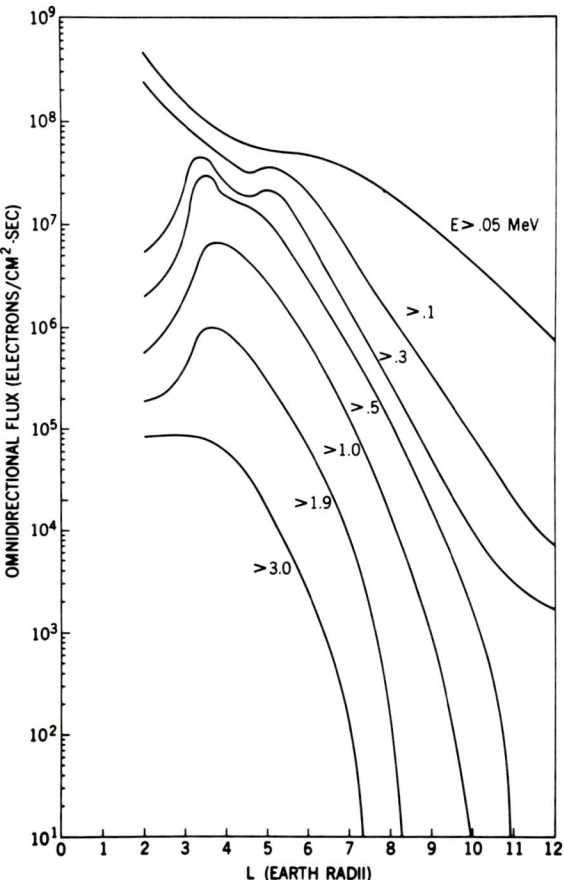

FIG. 18. Time-averaged radial intensity profiles of the integral equatorial omnidirectional electron flux for several energy thresholds. Data obtained near solar maximum and from all available longitudes.

in electron intensities throughout the trapping regions, make the electron time averages much less representative of a steady state situation than the corresponding proton averages. For example, the spatial structure characteristic of the skirt and cusp regions [60,61], the total pseudo-trapping region in Fig. 12, is not seen in the averaged data.

The main difference between the solar minimum and maximum curves in Figs. 17 and 18 is that the slot region, $2 \lesssim L \lesssim 3$, becomes much less pronounced at solar maximum due to a larger flux enhancement there than at lower and higher altitudes. These curves also agree with earlier observations that outer

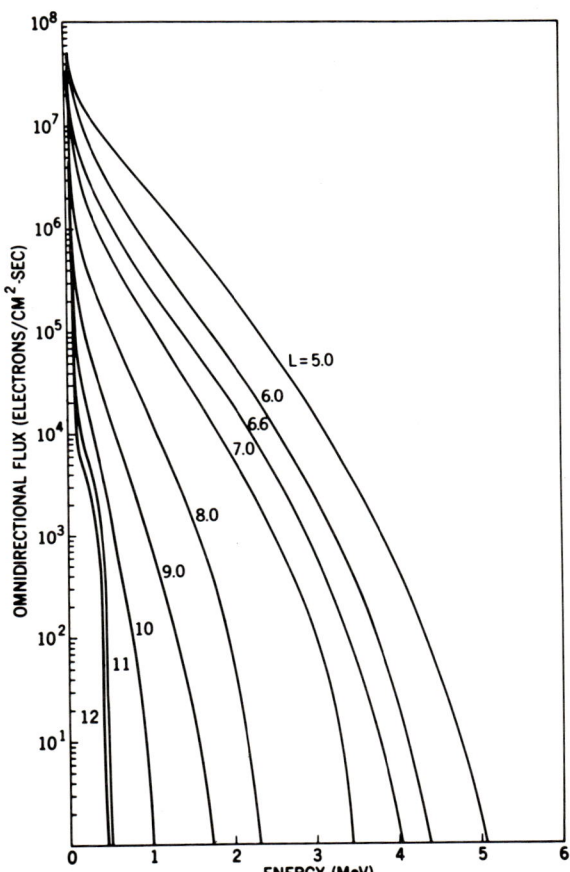

FIG. 19. Time-averaged absolute integral electron spectra throughout the outer trapping regions.

zone maxima and the position of the slot region move toward higher L values as solar minimum is approached [67,68]. This behavior is strongly energy dependent and the slot, maximum, and outer boundary positions can vary over large distances (up to several earth radii) on short time scales (hours to days).

Figure 19 shows, for completeness, averaged integral electron spectra at outer zone altitudes (Vette, personal communication) and shows the rapid softening of the spectrum at altitudes corresponding to the overall pseudotrapping region outlined in Fig. 12.

Probably the most accurate representation of average outer zone electron fluxes is the log-normal distribution used to represent the electron flux distribution at synchronous altitude [69]. Using the mean and standard deviation of the logarithm of the flux F, the Gaussian density $P(F)$, and the probability of observing an integral flux greater than F_x for a given energy threshold E_j, $P_{Ej}(F > F_x)$ can be calculated. Paulikas et al. [70] have applied this technique to an extensive body of ATS-1 data and present results at various local times and different levels of magnetic activity. Figure 20 shows sample plots of $P(F > F_x)$ for four electron energies at

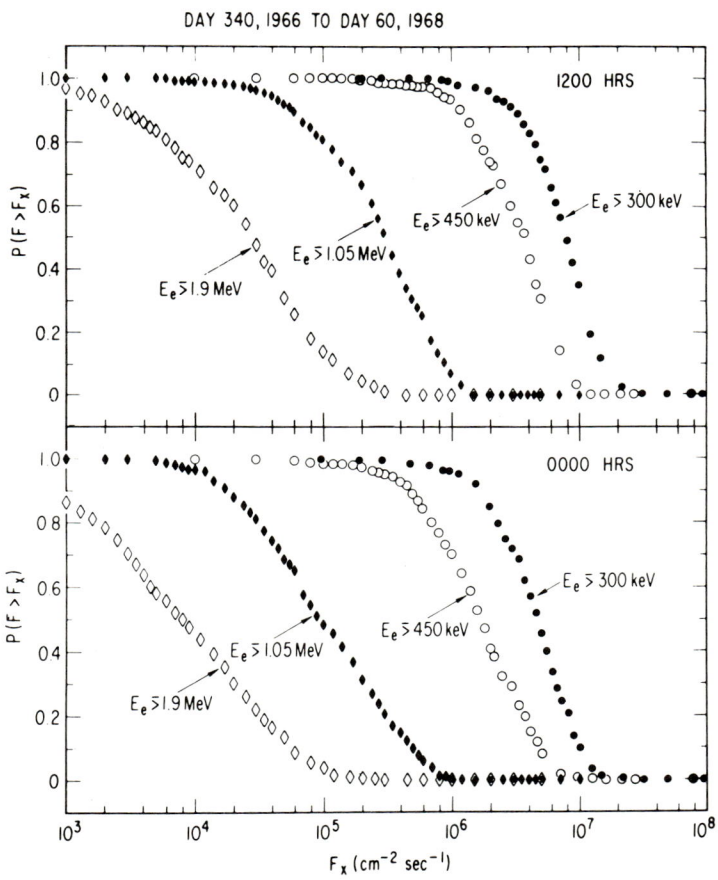

FIG. 20. Log-normal representation of the electron flux at synchronous altitude. The probability of observing an omnidirectional flux greater than F_X, $P(F > F_X)$, is plotted versus F_X for four electron energies and two local times. All magnetic conditions included in the data sample (from Paulikas et al. [70]).

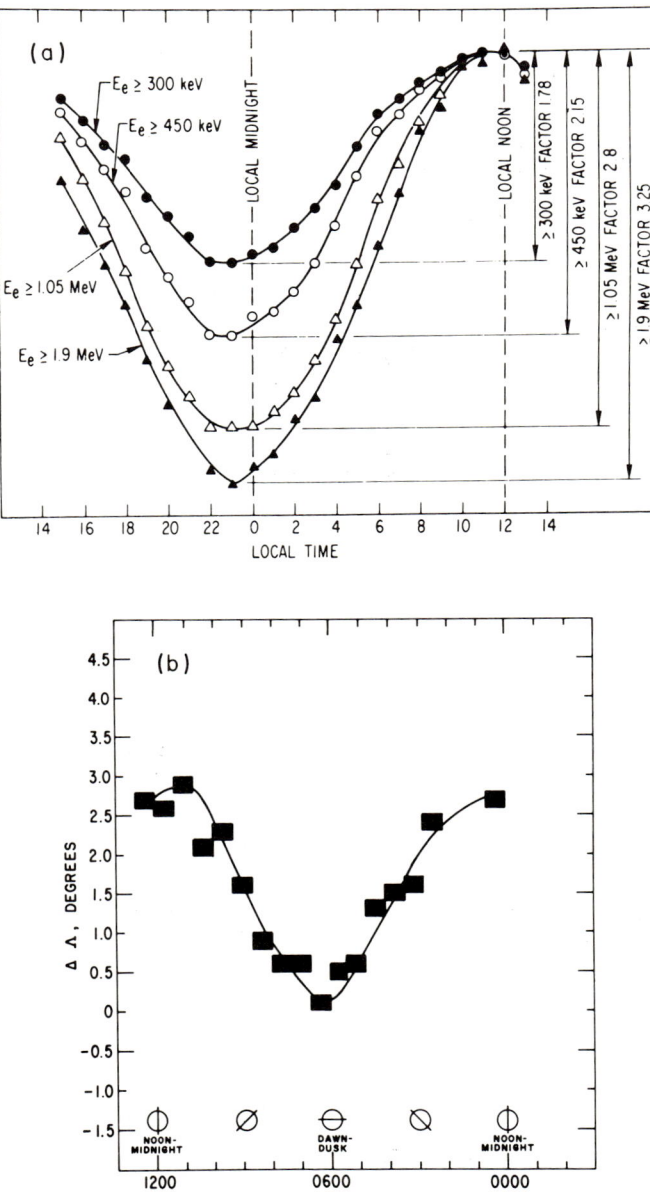

FIG. 21. Examples of diurnal variations in the outer zone energetic electron population. (a) Electron intensity variation as observed at synchronous altitude at all local times (from Paulikas et al. [70]). (b) Variation over all local times of the high-latitude ≥ 280 keV electron boundary at 1100 km during magnetically quiet periods. $\Delta \Lambda$ is the latitude difference between the dayside and nightside high latitude boundaries.

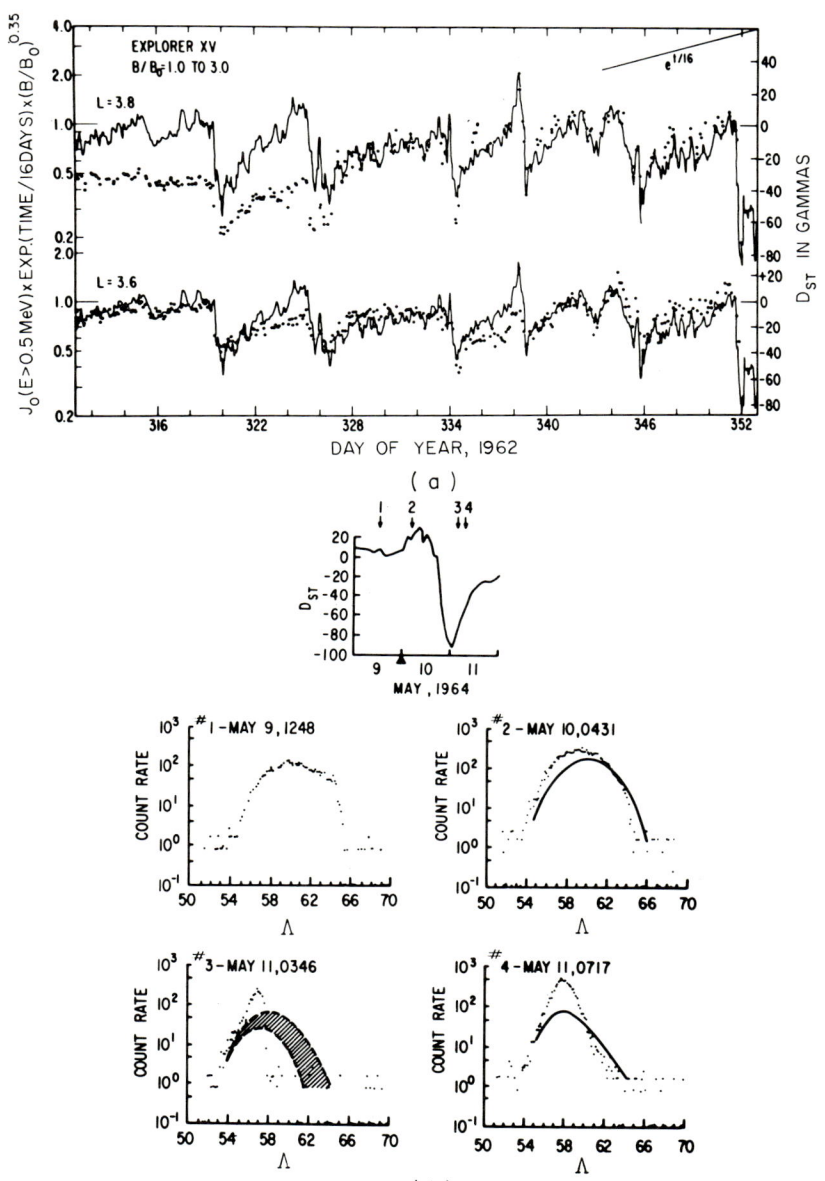

FIG. 22. Examples of adiabatic effects in the trapped electron population. (a) Near equatorial intensities, corrected for steady e-fold decay of ~ 16 days, plotted along with D_{ST}. Close correspondence of electron intensities and D_{ST} is indicative of the adiabatic response of the electron distribution to large scale magnetic variations (from McIlwain [72]). (b) Variations of the low-altitude (1100 km) trapped ≥ 280 keV electron profile

local times of 1200 hr and 0000 hr [70]. All magnetic conditions are included in Fig. 20.

Strong diurnal variations which characterize outer zone electron distributions for $L \gtrsim 5$ and which are averaged out in Fig. 17 can be clearly seen in Fig. 20. These variations, important throughout the high-L trapping region [19,54,61,71], are shown in Fig. 21 for all local times at synchronous altitude (from Paulikas et al. [70]) and at 1100 km during magnetic quiet periods (Williams, unpublished data). The magnitude of the variation shown in Fig. 21 depends not only on magnetospheric distortions [44] but also at synchronous altitudes on existing gradients in the electron spatial distributions.

We have seen that the "steady state" proton distribution may change simply by adiabatically adjusting to a slowly varying magnetic field. Such effects also occur in the electron distributions, although a steady state is much harder to describe because of the many strong nonadiabatic effects occurring.

In Fig. 22 we show examples of adiabatic readjustments in the electron population. Figure 22a shows a very high degree of correlation between variations in D_{ST} and variations in $\geq$500-keV electron intensities at the geomagnetic equator on the $L = 3.6$ and 3.8 shells [72]. The electron data have been corrected for an observed persistent decay, having an e-folding lifetime of ~ 16 days. The close correspondence between D_{ST} and electron intensities shown in Fig. 22a is a strong indication that major adiabatic readjustments take place in the electron population, just as has been described for the proton population. McIlwain [72] has used these data, where adiabatic effects are clearly seen, in one of the initial attempts to identify and catalog nonadiabatic processes operating on energetic electrons.

Figure 22b (from Arens and Williams [73]) shows observed low-altitude energetic ($\geq$280-keV) electron behavior during a geomagnetic storm, as well as predicted behavior due to adiabatic effects alone. The characteristic feature observed in low-altitude electron profiles during storms is their collapse to lower latitudes [46,74]. While Fig. 22b shows that only a portion of the boundary collapse is a result of adiabatic effects, it does point out the importance of these effects, even at low altitudes.

---

during the 10 May, 1964 magnetic storm. Solid lines are expected variations due to adiabatic effects. Shaded area represents spread in the adiabatic response due to available data used in calculation. Note that the low altitude adiabatic effects are significant even though they cannot explain the larger observed variations (from Arens and Williams [73]).

3.3. Alpha Particles and $Z \geq 3$

Positive observations of magnetospherically trapped alpha particles were first reported by Krimigis and Van Allen [75] at an energy of $E_\alpha \gtrsim 0.52$ MeV/nucleon. Previous observations at higher energies had yielded null results and corresponding upper limits on alpha-particles fluxes [76–78].

The initial observations have now been confirmed and extended [79,80], although no measurements of trapped alpha particles have yet been reported in the trapping region at high altitudes. The available results show an alpha to proton flux ratio, j_α/j_p, which varies during magnetic storms at energies, $E_N \lesssim 0.5$ MeV/nucleon, and which is relatively steady at a value of $\sim 10^{-4}$ for energies $E_N \gtrsim 0.5$ MeV/nucleon. Below $E_N \sim 0.5$ MeV/nucleon, the j_α/j_p ratio appears to increase as E_N decreases.

An attempt to summarize what is presently known of the geomagnetically trapped alpha population is shown in Figs. 23 and 24. Figure 23 presents the energy dependence of the j_α/j_p ratio. The low-energy ($E_N \lesssim 0.5$ MeV/nucleon) data (from Krimigis [81]) are obtained after a magnetic storm enhancement of j_α/j_p has returned to prestorm values, which is used here as

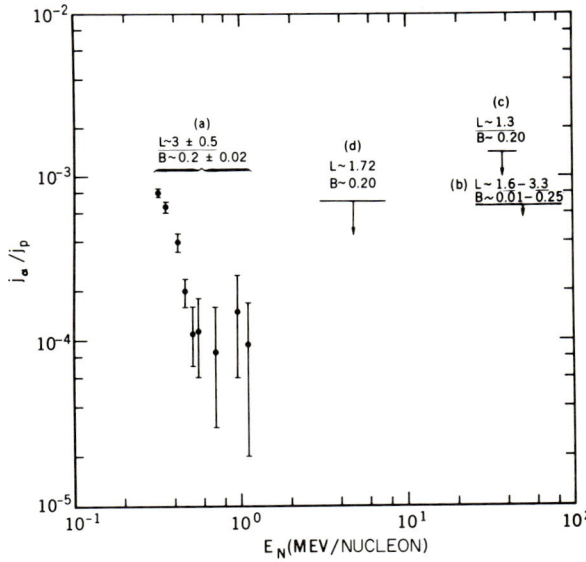

FIG. 23. Summary of alpha particles spectral observations. The alpha to proton flux ratio j_α/j_p is plotted versus energy per nucleon E_N. The data are from (a) Krimigis [81], (b) Fenton [78], (c) Heckman and Armstrong [76], (d) Naugle and Kniffen [77]. The upper limits shown for (c) and (d) have been obtained from the referenced upper limit alpha fluxes in combination with current proton flux information.

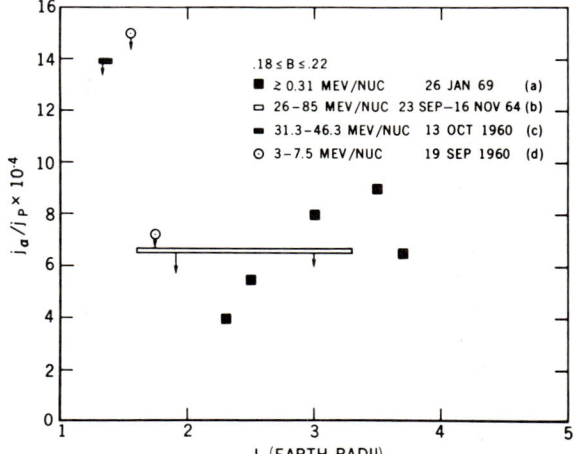

FIG. 24. Summary of available information on alpha-particle spatial distribution. Data used are same as in Fig. 23.

the best available representation of a steady state condition. The upper limits shown for the Heckman and Armstrong [76] and Naugle and Kniffen [77] data were obtained by combining their energy, position, and upper limit information on alpha flux with the proton distributions reported by Vette [82,83]. Figure 24 presents available data concerning the spatial distribution of the j_α/j_p ratio.

Initial observations on $Z \geq 3$ trapped particles have been reported by Krimigis et al. [84] and Van Allen et al. [85] using a single element thin solid state detector. They report a maximum intensity of $Z \geq 3$ nuclei for $0.15 \leq B \leq 0.20$ G at $L \sim 3.2$ with a flux value of 1.1 ± 0.2 $(cm^2 \, sec \, ster)^{-1}$ for $E_N \geq 0.3$ MeV/nucleon. An intensity ratio of $Z \geq 3$ nuclei to alpha particles at $E_N \gtrsim 0.3$ MeV/nucleon of $(2.8 \pm 0.5) \times 10^{-3}$ is found for $3.0 \leq L \leq 3.5$ and $0.15 \leq B \leq 0.20$ G.

Additional measurements of alpha particles and $Z \geq 3$ nuclei, extending both the energy coverage and spatial distribution, are required before a meaningful morphology of these particles can be established. Although existing alpha-particle and $Z \geq 3$ nuclei data do provide new insights on the source of radiation belt particles, it supplies no definitive answers partly because of the limited coverage available. It should be noted, however, that the observation of various particle species in the trapping regions may well prove to be one of the most powerful methods of determining both the long-term and instantaneous sources for these regions [86].

4. Sources, Losses, and Transport

This section is devoted to a survey of major source, loss, and transport mechanisms thought to affect directly the "steady state" trapped particle population. We divide the trapping region into the historical categories, inner and outer zones. This division is made on the basis that large day-to-day flux variations are observed in the outer zone, whereas similar large flux variations are much less frequent in the inner zone. The boundary between the two regions is considered to be at $L \sim 2$ and to be very diffuse.

The most familiar transport mechanism acting on the trapped particle population is their motion through the geomagnetic field under conservation of the adiabatic invariants. In this category we include the intensity redistributions produced by these particle populations adiabatically adjusting to slow magnetic field changes, as described previously. These variations plus the intense nonadiabatic variations which occur in the trapping regions will not be considered directly in this section since they are covered elsewhere in the literature. However, time variations are considered to the extent that they may provide direct information concerning source, losses, and transport.

4.1. Inner Zone Protons

Energetic inner zone ($L \simeq 1.2$) protons were the particles detected by Van Allen et al. [12] in their discovery of the trapped radiation. Soon after these initial observations, it was suggested that the decay of albedo neutrons produced by cosmic-ray interactions in the atmosphere formed the source of these low-altitude energetic trapped protons [87–89]. Thus, at a very early date in the history of trapped particle observations, the first (and for many years the only) source–loss problem amenable to detailed quantitative theoretical study was defined. However, even now after much theoretical and experimental study (see reviews in [31,90–92], a definitive result is still unavailable. The reason is that only one of the three unknowns required to solve the problem has been determined. The following situation exists:

(1) the proton distribution (the result) is known;

(2) source function is unknown—*no* measurements have been made above the atmosphere of neutron intensities and spectra $\gtrsim 10$ MeV;

(3) loss mechanisms are poorly known—pitch angle scattering and invariant breakdown for high-energy protons are not well understood.

This situation points out the impossibility of quantitatively answering large-scale source–loss problems in the magnetosphere unless relevant input (boundary) conditions are known.

4.1.1. Present Situation. Cosmic ray albedo neutron decay (CRAND) is certainly a source of trapped protons. However, its absolute magnitude, spectrum, and effectiveness in different regions of the magnetosphere are either poorly known or unknown, thus prohibiting a quantitative comparison with observed proton intensities, spectra, and spatial distributions. For example, theoretical determinations of the albedo neutron flux and spectra above the earth's atmosphere [93–95] differ by factors of three in magnitude and two in latitude variation and are subject to much larger errors in the critical energy region $\gtrsim 10$ MeV [92]. Until recently, the albedo neutron flux above the atmosphere at $\lesssim 10$ MeV was known experimentally to no better than a factor of 3–5 in magnitude and to a factor of 2–3 in its latitude variation [96–98].

Very recent and probably the best available data for neutrons ≤ 10 MeV above the atmosphere have been reported by Jenkins *et al.* [99,100]. Results from their OGO-6 experiment give a total neutron flux of 0.25 ± 0.03 n/cm² sec in the 1–10-MeV range along with an upper limit to the steepness of the energy spectrum of $E^{-1.15}$, in reasonable agreement with the calculations of Newkirk [94]. Below 1 MeV, the observed flux and latitude variation is similar to that predicted by Lingenfelter [95]. Even though the situation at ≤ 10 MeV is much improved, we emphasize that nothing is known of albedo neutrons ≥ 10 MeV, since no measurements have yet been made above the earth's atmosphere. Extrapolations of the <10-MeV results to the >10-MeV region are subject to large errors, as different processes operate to produce neutrons in these two energy regions above the atmosphere, namely, leakage and knock-on processes.

The spatial variation of the effectiveness of the CRAND source depends critically on loss mechanisms, such as atmospheric collisions, invariant breakdown, and pitch angle diffusion. Although the effects of atmospheric losses on inner zone protons now seem to be understood [101–103], only a beginning has been made in understanding the more complex mechanisms and effects of invariant breakdown and pitch angle scattering [24,104–107].

Figure 25 (from Freden [91]) displays the inner zone proton spectrum at high energies, along with spectrum predictions using the CRAND source. It can be seen that the *shape* of the proton spectra can be satisfactorily described using CRAND plus atmospheric loss and invariant breakdown. It can also be seen that the theory breaks down in the low-energy portion of the spectrum. The precise energy at which this occurs is unknown and probably depends on spatial location. Freden [91] estimates that the proton energy above which CRAND is a dominant source at the equator is ~ 40 MeV at $1.5 < L < 2.5$ and $\lesssim 15$ MeV at $L \lesssim 1.3$.

Figures 26 and 27 (from Freden [91]) present *absolute flux* comparisons of experiment and theory as a function of B for two L shells and show that

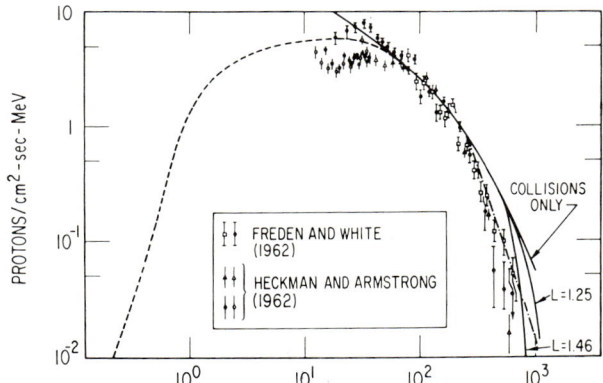

FIG. 25. Experimentally observed omnidirectional proton spectra compared with predictions of cosmic-ray albedo neutron theory as obtained by Haerendel [104]. Solid curves include losses due to atmospheric interactions and magnetic moment breakdown in a dipole field. Dot–dash curve includes third adiabatic invariant breakdown. Dashed curve includes charge exchange losses (from Freden [91]).

(1) fair agreement exists at these energies near the equator, but the shapes of the intensity versus B curves are significantly different;

(2) more protons are seen at high B (low altitude) than predicted by up to a factor of 50.

Historically, these data have been used in arguments both for and against the CRAND theory.

An indirect argument in favor of the CRAND source is the long-term time behavior of low-altitude trapped proton intensities. If the solar cycle variations in trapped proton intensities is governed by atmospheric losses, then the source of these protons is effectively constant in comparison to atmospheric solar cycle variations. This is consistent with the CRAND source since it only varies by $\sim 12\%$ over a solar cycle.

Initial calculations of the relative trapped proton and atmospheric variations [108] used extrapolated atmospheric densities based on the previous solar cycle and predicted proton variations now known to be far too large. Recent work [101–103,109], using actual atmospheric density variations in the past cycle, shows that trapped proton intensities are indeed controlled by the atmosphere at low altitudes. Figure 28 (from Macy et al. [103]) shows measured 55-MeV proton intensities at 440 km on the $L = 1.4$ shell from 1961 into 1969. The low-altitude increase caused by the Starfish detonation in 1962 [110] is clearly visible. The solid curve gives the predicted proton intensity variation assuming a constant source and using the solar cycle variation in atmospheric densities. The dashed curve shows the predicted

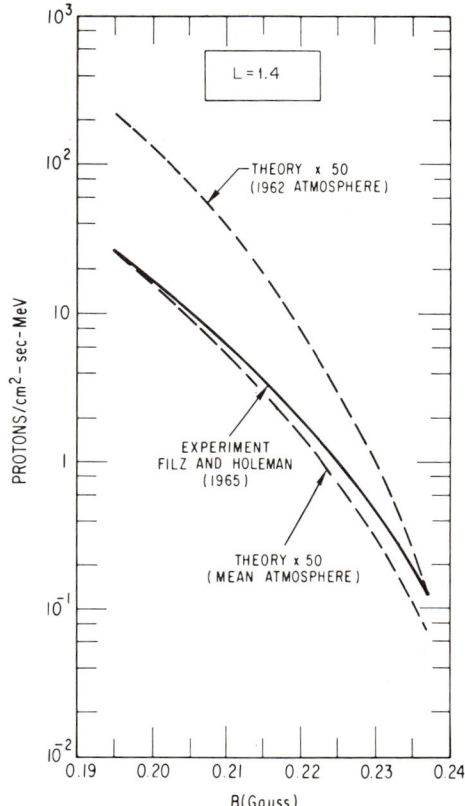

FIG. 26. Comparison of absolute proton (35–63 MeV) flux versus B observations with the albedo neutron theory calculations of Dragt et al. [119] (from Freden [91]).

atmospheric losses for the starfish protons. Similar results were also obtained by Macy et al. [103] at lower altitudes.

Note that this result is not proof of CRAND as a dominant proton source at these altitudes and energies but is only a consistency argument since CRAND is, in effect, a constant source compared with the variation shown in Fig. 28.

The CRAND source is incapable of supplying the large intensities of low energy ($\lesssim 10$ MeV) protons observed in the inner zone (e.g., see measurements of Fillius and McIlwain [111], Bostrom et al. [112], Freden et al. [113], Fillius [107], Gabbe and Brown [114]). The initial observations of these lower-energy protons [115] led to the suggestion that they may be supplied by decay of neutrons produced by solar proton interactions in the atmosphere (solar proton albedo neutron decay, SPAND) [115,116]. Detailed

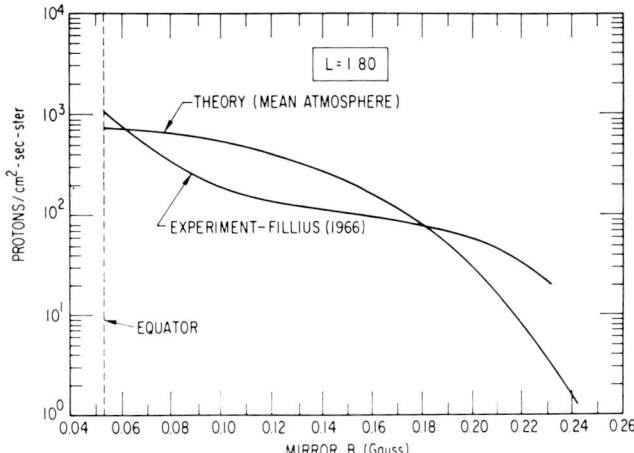

FIG. 27. Further comparison of experiment and theory (Dragt *et al.* [119]). The theoretical values have been multiplied by 50 for comparison (from Freden [91]).

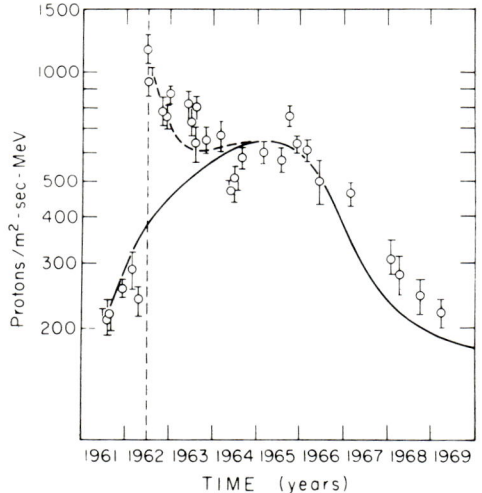

FIG. 28. Comparisons of experimental observations with theoretical predictions using measured atmospheric variations for the 9-yr period shown and assuming a steady source function. $E = 55$ MeV, $L = 1.4$ R_E; $h = 440$ km. The solid line gives the theoretically expected variation. The low-altitude proton enhancement associated with the 1962 Starfish nuclear detonation is clearly seen. The dashed line gives the expected decay from that enhancement, assuming only atmospheric loss processes. It is clear that at this altitude, the atmosphere governs the energetic trapped proton time variation. The source of these protons must therefore have a time variation much less than the variation shown in the figure (from Macy *et al.* [103]).

calculations concerning this source [92,117–119] have led to the conclusion that the SPAND source is totally inadequate to explain the observed fluxes.

While SPAND can be considered a negligible source of trapped protons, it is possible that the decay of solar neutrons (solar neutron decay, SND) may be as important a source for energetic trapped protons as CRAND [120,121].

Since even less is known of the solar neutron spectrum than of the earth's albedo neutron spectrum, the existing comparisons of the relative proton injection rates from the SND and CRAND sources [122] should be considered only as guidelines which emphasize the possible importance of SND.

Another potential source of the inner zone proton population is the diffusion of a lower-energy population from higher L shells. This mechanism, discussed more thoroughly in Section 4.3, Outer Zone Protons, involves the transport of charged particles across L shells under violation of the third adiabatic invariant while conserving the first and second invariants. Much theoretical effort has been expended on this mechanism, particularly concerning the outer zone proton population where it has been shown to be a major source and transport mechanism. While no detailed calculations have been performed comparing cross-L diffusion with the inner zone proton population, some evidence exists indicating the need to further explore this situation.

What appears to be a slow inward motion of a large secondary peak in the energetic trapped proton radial profile [123] has been observed (64,91,114). Figure 29 shows this equatorial radial profile as it appeared in January 1963

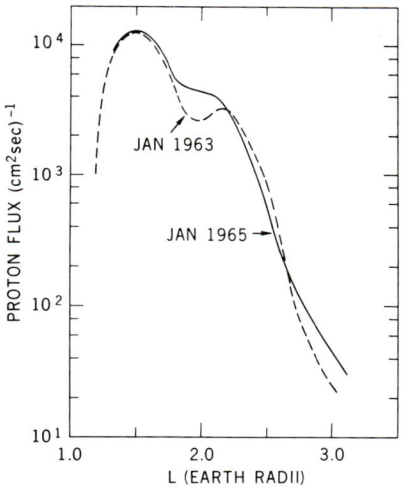

FIG. 29. Equatorial radial profiles of energetic proton fluxes showing apparent inward diffusion of secondary high-altitude peak between January 1963 and January 1965. Data from McIlwain [123] copyright 1963 by the American Association for the Advancement of Science, and Vette [64].

and again in January 1965. These data have been obtained from plots given by Vette [64] using data of McIlwain ([123], personal communication). The behaviour shown in Fig. 29 is qualitatively that of a slow diffusion of the high-L peak inward during the two year interval shown. Similar shorter time scale results have been reported by Gabbe and Brown [114] for 49–145 MeV protons. Recent work of Farley *et al.* [124] indicates that combining the process of radial diffusion with the CRAND source produces a reasonable fit in the intensity and energy spectrum for the observed inner zone proton distribution in the energy range 20–170 MeV.

Figure 30 (from Freden [91]) shows a comparison of experimentally observed proton fluxes in the inner zone with the diffusion results of Nakada and Mead [125]. This comparison suggests that diffusion may be an important

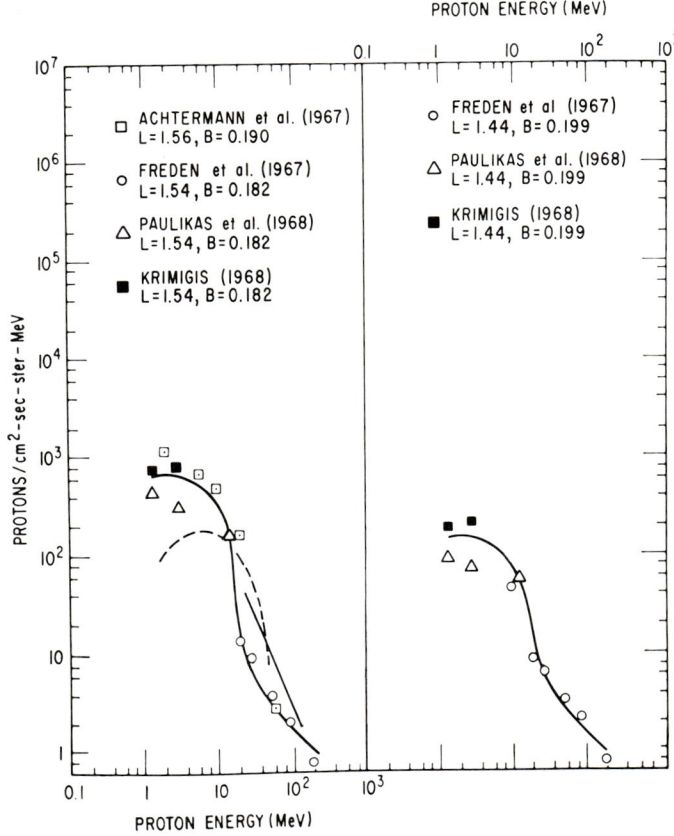

FIG. 30. Comparison of observed low-energy inner zone proton spectra above 1 MeV with cross-L diffusion calculations of Nakada and Mead [125]. Solid curves are average fits through the data. Dashed curve is theoretical prediction (from Freden [91]).

source of the lower-energy inner zone population. While more accurate computations are required before any quantitative estimates can be made, it is clear that diffusion theory fits the 2–50 MeV proton distribution in the inner zone ($L \sim 1.5$) at least as well as CRAND.

We show in Fig. 31 (from Bostrom et al. [126]) the time history of proton fluxes observed at mirror point altitudes (minimum mirror altitude >600 km) where atmospheric effects should be negligible. These results confirm that nonadiabatic effects (redistributions) occur in the normally stable high-energy proton distribution during large magnetic storms and that these effects are most noticeable at higher-L shells [127]. In addition, Fig. 31 shows very clearly that trapped proton behavior at low-L values becomes much more complex at low energies. In fact, Bostrom et al. [126] point out that the spatially restricted enhancement of low energy protons near $L \sim 2.2$ during the May 1967 disturbance is suggestive of a resonance phenomenon.

4.1.2. Summary of Inner Zone Protons. From the preceding discussions conclusions concerning inner zone protons follow:

Input source functions are poorly known—e.g., no measurements have yet been made of neutron intensities and spectra above the atmosphere in the critical range >10 MeV.

Loss mechanisms are not well known—whereas atmospheric effects are now well understood, much work is required to assess accurately the importance of pitch angle scattering and invariant breakdown with resultant loss of trapping.

With these points in mind, it is further concluded:

Neutron decay (CRAND and SND) is the most probable source of protons ≥ 50 MeV at $L \leq 1.5$. In addition this characteristic energy probably decreases as L decreases.

Cross-L diffusion is the most probable source of protons ≤ 50 MeV. Again this characteristic energy probably decreases as L decreases.

For protons ≥ 50 MeV at $L \geq 1.5$, neutron decay and diffusion are both effective sources with diffusion becoming the dominant source by $L \sim 1.8$.

4.2. Inner Zone Electrons

Prior to the Starfish detonation in July 1962, measurements of inner zone electrons were few and showed little agreement in intensity and spectra (91,128]. Although definitive comparisons could not be made, it was thought that a CRAND source for electrons [129] could account for at most only a very small fraction of observed inner zone electrons [128,130,131].

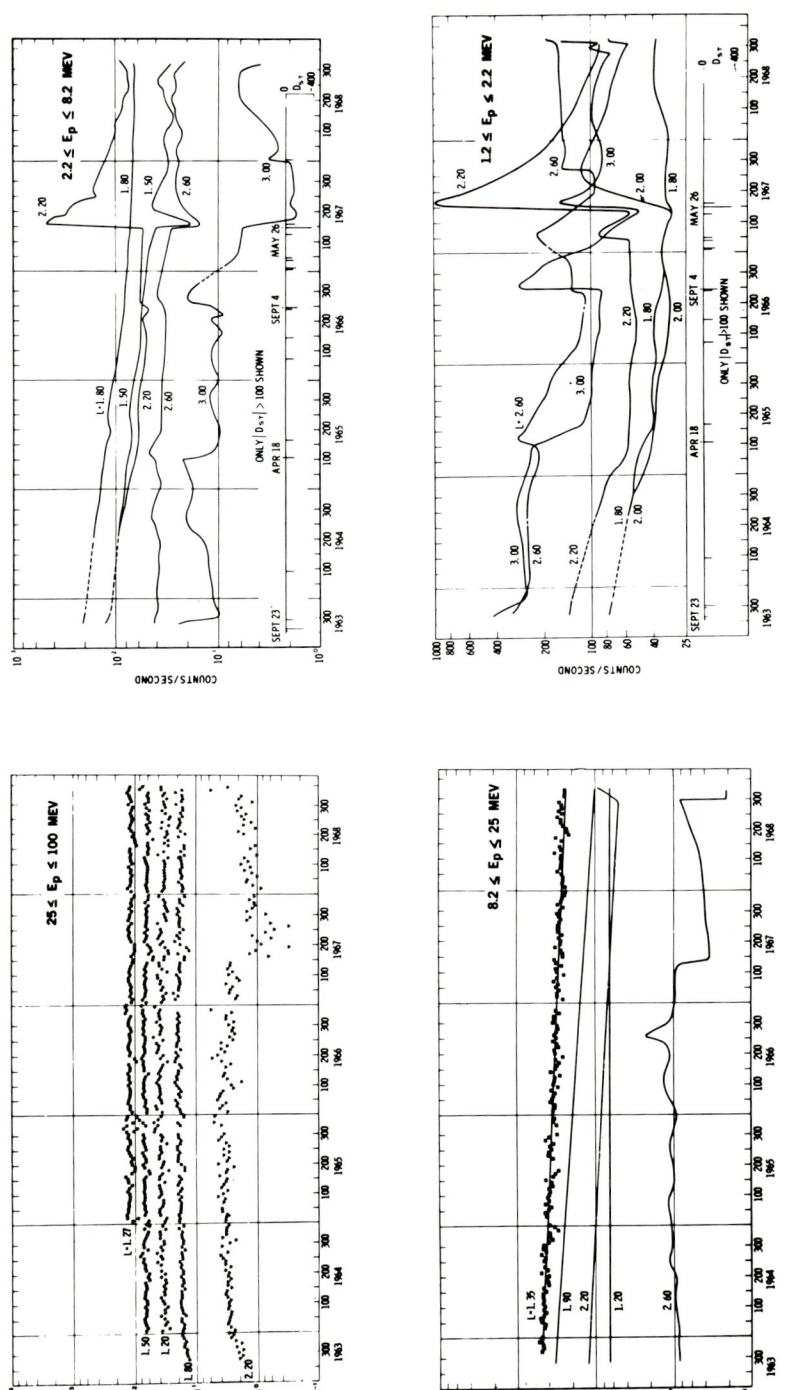

FIG. 31. Five-year time history of trapped proton intensities at 1100 km and at several energies throughout the inner zone. Intensities become very variable at lower energies and higher L values. Nonadiabatic effects can be seen (from Bostom et al. [126]).

It was at this point that the Starfish high-altitude nuclear explosion occurred and overwhelmed the natural inner zone electron intensities with its own fission electrons. This voided for several years the possibility of looking for naturally occurring electrons of energy ≥ 780 keV in the inner zone and thus directly testing albedo neutron decay as an electron source.

A detailed presentation of data obtained shortly after the Starfish detonation has been compiled by Hess [132]. One (and possibly the only) advantage of the Starfish event is that it approximated to high order a delta function injection of electrons in time and space. Due to the strength of this injection, it has been possible for many years to observe electron loss characteristics in the inner zone under the assumption of a zero strength source. Unfortunately, not enough satellite and rocket instrumentation were in space at the time of the explosion to study in detail the spatial development of the artificial belt. However, much postevent data have been made available to describe its time behavior (e.g., see [132–136]).

A thorough study of inner zone electron lifetimes has been reported by Beall et al. [137]. Data from an electron spectrometer aboard the satellite 1963 38C was used to follow the decay of the Starfish electron intensities throughout the inner zone for a 27-month period. Figure 32 (from Beall et al. [137]) shows measured electron lifetimes as a function of B for the integral thresholds indicated. It is seen that these lifetimes decrease as B and E increase.

It was further pointed out by Beall et al. [137] that shorter term increases

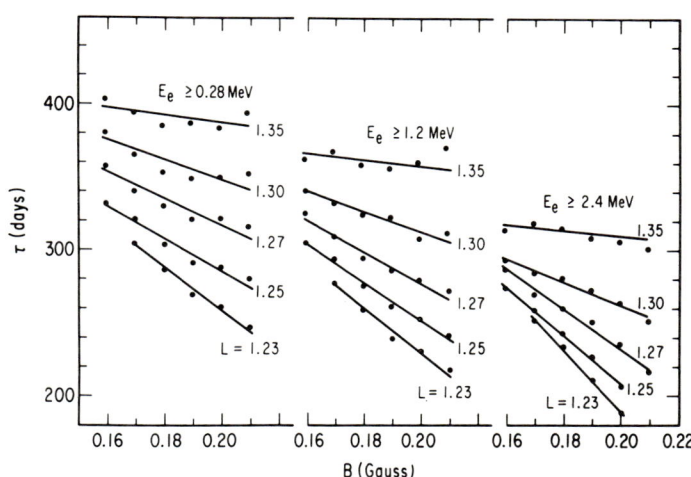

FIG. 32. Electron distribution e-folding lifetimes (τ) versus B for several L shells and energies (from Beall et al. [137]).

of ≥ 280-keV electron fluxes were seen as soon as 10–15 months after the Starfish explosion. Thus, effects of natural inner zone electron events may have been observed at these early post-Starfish times.

Figure 33 (from Walt [136]) shows a comparison of measured decay times with those predicted on the basis of multiple small-angle coulomb scattering

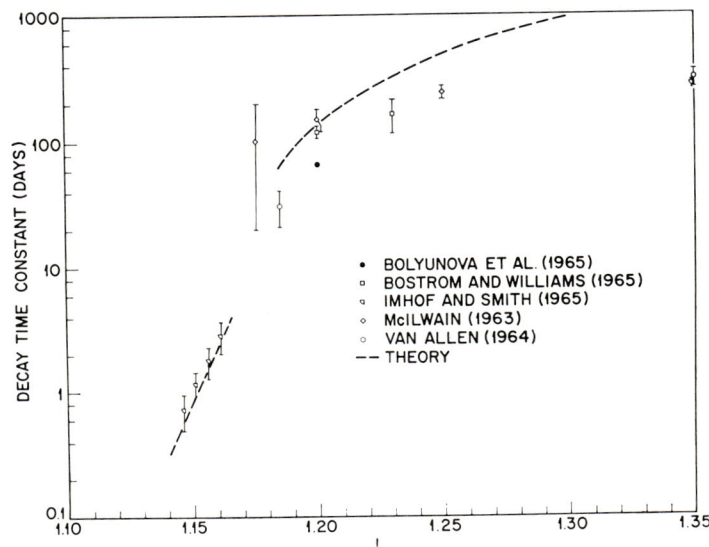

FIG 33. Comparison of measured decay times with predictions using atmospheric losses based on multiple small-angle Coulomb scatterings. Data deviate significantly from theory for $L \gtrsim 1.23$ (from Walt [136]).

within the atmosphere [138]. Clearly this atmospheric loss dominates for $L \lesssim 1.25$ while other loss mechanisms are required at the higher L values. By studying the flux levels and energy spectra just to the west and east of the South Atlantic anomaly, Imhof and Smith [139] concluded that single large-angle coulomb scattering in the atmosphere was responsible for the redistribution and loss of these electrons in the region $1.2 \leq L \leq 1.4$. Thus the range of atmospheric control for inner zone electron loss extends to $L \sim 1.4$ rather than to $L \sim 1.25$, as indicated in Fig. 33. Other mechanisms are required for $L \gtrsim 1.4$.

Direct evidence of the existence of natural processes capable of producing energetic electrons deep in the inner zone comes from the observation of monoenergetic peaks of ~ 1-MeV electrons at $L \sim 1.15$ [140]. In this case, the source (acceleration) mechanism is a fluctuating local magnetic field, produced by modulated ionospheric currents, which resonates with the electron drift frequency and thus imparts energy to the electrons during each drift. The electrons, having gained energy, move to lower L shells (conservation of μ) where they are readily observed [141]. The short-term decay of these monoenergetic peaks is consistent with atmospheric losses and is included in the data of Fig. 33.

In addition to these spatially limited intensity enhancements, naturally occurring electron intensity increases have now been clearly observed throughout the inner zone [126,142]. Many of these enhancements are

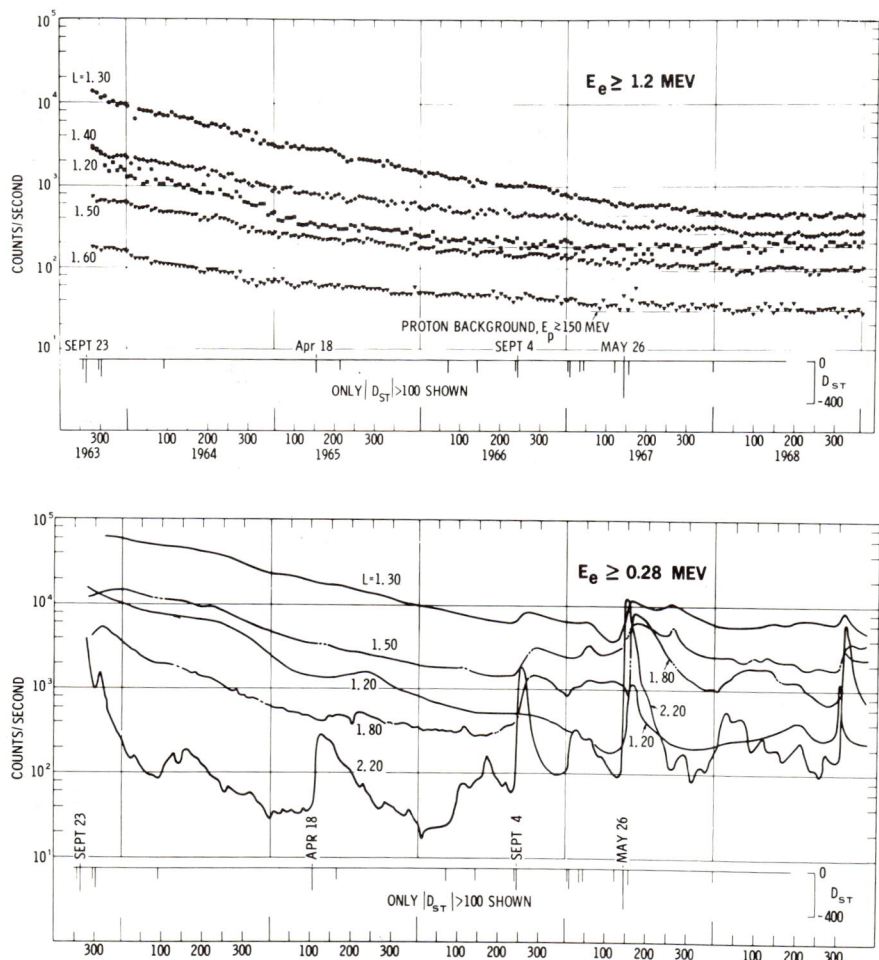

FIG. 34. Five-year time history of trapped electron intensities at 1100 km throughout the inner zone. Note that while ≥ 280 keV electron intensities display large variations throughout this region, the ≥ 1.2 MeV intensities steadily decay to detector background values (from Bostrom et al. [126]).

suggestive of a diffusion process because during a storm, rapid enhancements occur at higher L values with progressively slower, longer-lasting enhancements being observed at lower (down to $L \sim 1.4$) L values. Figure 34 (from Bostrom et al. [126]) shows the long-term behavior of inner zone electrons for several L values and two threshold energies. Note that while increases are seen throughout the inner zone for $\geq$280-keV electrons, no discernable perturbations occur in the $\geq$1.2-MeV electron fluxes. At these higher energies electrons decay to a constant background produced by energetic protons. It appears that it is very difficult for several hundred kilovolt electrons to gain access to or to be injected throughout the inner zone (see also [142]).

Inspection of Fig. 34 shows two regions of electron enhancement separated by a transition region. A higher L response, as described above and similar to that reported by Williams et al. (143) is seen along with a separate response

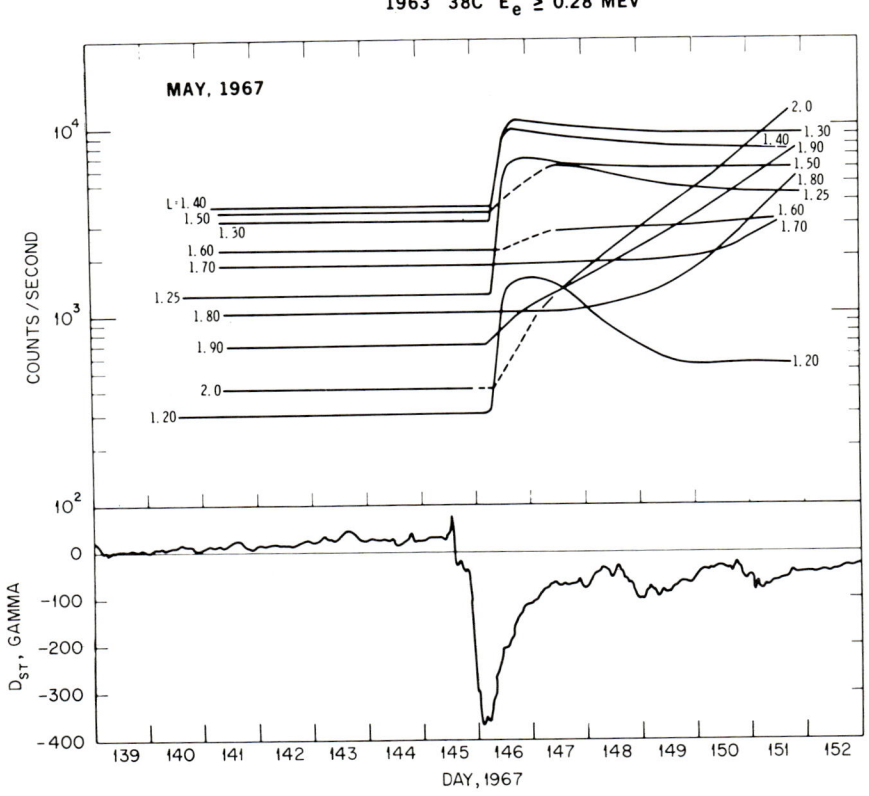

Fig. 35. Inner zone electron intensity enhancement on 26 May, 1967 showing possible resonant acceleration phenomenon (from Bostrom et al. [126]).

at low L values ($L \sim 1.2$). This is shown more clearly in Fig. 35 where the response occurring from $L = 1.2$ through $L = 1.6$ is clearly separate spatially from that occurring at $L \gtrsim 1.7$. Note from Fig. 34 that a rapid increase also occurs at $L = 2.2$ for this storm. This low-L response may be a higher-altitude observation of the resonance acceleration observed by Imhof and Smith [140].

Further evidence of cross-L diffusion deep in the inner zone has been reported by Imhof et al. [144]. As described previously, the observed short-term decay of the monoenergetic electron peaks at $L \sim 1.15$ is consistent with atmospheric loss. However, the long-term time histories of energetic electrons at $1.14 \lesssim L \lesssim 1.26$ show decay times independent of L and much longer than those due to atmospheric loss. It was concluded by Imhof et al. [144] that other sources must exist to maintain these low-altitude electron intensities, and they suggested cross-L diffusion as a likely candidate.

Empirical evaluations of the radial diffusion coefficient for electrons in the inner zone have now been reported. Brown [145] has published a diffusion coefficient of D, $D_E \approx 7(10)^{-5}$ R_E^2/day at $L \sim 1.76$ obtained from the observed increase in the width of the ≥ 1.9-MeV electron intensity peak at $L \sim 1.76$ produced by the third U.S.S.R. high altitude nuclear detonation. At lower L values, Newkirk and Walt [145a] have empirically evaluated the radial diffusion coefficient of ≥ 1.6-MeV electrons by integrating the Fokker–Planck equation using the observed radial profile and long-term decay of electron intensities. They obtain a diffusion coefficient which *increases* from $2(10)^{-7}$ R_E^2/day to $4(10)^{-6}$ R_E^2/day as L *decreases* from 1.21 to 1.15 R_E. Such an increase at low altitudes may be due to an increase in the relative importance of ionospheric effects [141] or higher multiple moments of the earth's field [91], or both. From the above results of Brown [145], these effects are seen to become important in the L range 1.21–1.76.

Summary of Inner Zone Electrons. The inner zone electron situation can be summarized as follows:

Nuclear explosions (in particular Starfish) have been a major source of inner zone electrons.

Atmospheric loss is the major loss process for $L \lesssim 1.4$. Additional loss mechanisms are required for $L \gtrsim 1.4$.

Resonant acceleration is an important source of inner zone electrons for $L \lesssim 1.5$.

Cross-L diffusion from higher L-shells is an important source of electrons throughout the inner zone.

Electrons greater than a few hundred kilovolts in energy are not observed to appear in large numbers nor over a large spatial extent in the inner zone during naturally occurring phenomena. This may be a consequence of the characteristics of the source or of the loss mechanism or both.

Neutron decay is not an important source of inner zone electrons.

4.3. Outer Zone Protons

4.3.1. Steady State. As mentioned previously in the discussion of inner zone protons, the transport of charged particles across L shells under violation of the third adiabatic invariant, while conserving the first and second invariant, was shown some time ago to be a potentially effective method of populating the trapping regions with energetic particles [146–148]. There have been many excellent theoretical studies performed during the intervening years. Thus the problem of charged-particle diffusion across field lines has become the second large-scale source-transport problem in the trapping regions capable of quantitative comparisons with experimental data [125,149–157].

This problem and its importance as a source and transport mechanism has been recently reviewed [158]. Let us briefly summarize those results and comment further on a possible source location.

Quantitative studies have been completed using a Fokker–Planck formalism

$$(4.1) \qquad \frac{\partial n}{\partial t} = -\frac{\partial}{\partial r} D_1 n + \frac{1}{2} \frac{\partial^2}{\partial r^2} D_2 n - \frac{\partial}{\partial \mu} \left\langle \frac{\Delta \mu}{\Delta t} \right\rangle n - \frac{n}{\tau}$$

Here n = number of particles in dr, $d\mu$, and dJ, μ = first adiabatic invariant, the magnetic moment, J = second adiabatic invariant, $D_1 = \langle \Delta r/\Delta t \rangle$ = average radial displacement per unit time, $D_2 = \langle (\Delta r)^2/\Delta t \rangle$ = average radial displacement squared per unit time, and τ = e-fold lifetime for charge exchange.

Using the relation between D_1 and D_2 obtained by Fälthammar [155] for a dipole field

$$(4.2) \qquad D_1 = \frac{r^2}{2} \frac{\partial}{\partial r} \frac{D_2}{r^2} = r^2 \frac{\partial}{\partial r} \frac{D}{r^2}, \qquad D \equiv D_2/2$$

The above may be written in a "diffusion-like" form

$$(4.3) \qquad \frac{\partial n}{\partial t} = \left\{ \frac{\partial}{\partial r} \frac{D}{r^2} \frac{\partial}{\partial r} (r^2 n) \right\} - \frac{\partial}{\partial \mu} \left\langle \frac{\Delta \mu}{\Delta t} \right\rangle n - \frac{n}{\tau}$$

The last two terms on the right above represent losses from coulomb collisions and charge exchange, respectively. This equation has been solved by Davis and Chang [149], Tverskoy [150], and Nakada and Mead [125].

The latter authors evaluated the size of the diffusion coefficient D by obtaining the rate and size of sudden impulses occurring in the magnetosphere and by using a distorted magnetosphere without a tail configuration [39]. They obtained a value of

$$D = 0.3(10)^{-9} r^{10} \; \text{R}_\text{E}^2/\text{day}.$$

Using this value for D, Nakada and Mead [125] compared the predicted and observed proton steady state distribution functions. These are shown in Fig. 36.

The basic difference between the observed and calculated curves in Fig. 36 is that the peaks of the observed distributions are at a lower altitude than the theoretical curves. Nakada and Mead [125] found that an increase of a

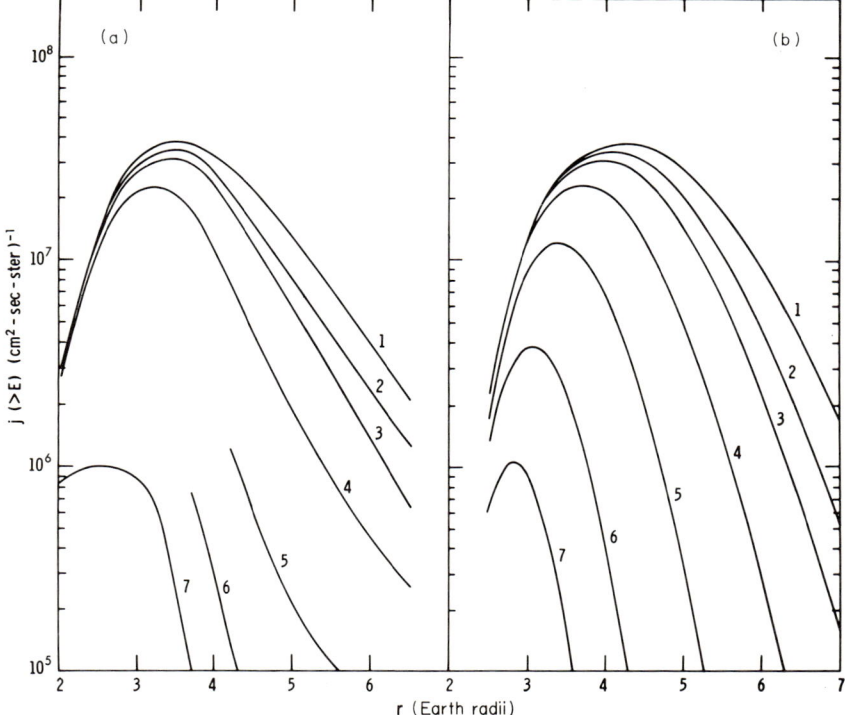

Fig. 36. Comparison of observed integral fluxes with those predicted by cross-L diffusion calculations. (a) Experimental results E.P.A. = 90°. (b) Theory Curve 1 : $E \geq$ 98 keV; 2, $E \geq 134$ keV; 3, $E \geq 168$ keV; 4, $E \geq 268$ keV; 5, $E \geq 498$ keV; 6, $E \geq 988$ keV; 7, $E > 1.69$ keV. Calculated curves are normalized to the same peak flux for the lowest energy threshold (from Nakada and Mead [125]).

factor of eight in D would bring the observed and calculated peaks into coincidence. They also argued that a larger value of D should be expected, since (1) inclusion of tail field effects would increase D by about a factor of four and (2) perturbations other than sudden impulses probably contribute to the diffusion process. The importance of other variations (including electric fields) may be seen from the fact that D can be expressed in terms of the power spectrum of the disturbance [154,156].

In addition Tverskoy [151], after an earlier low estimate has reevaluated D and obtained

$$D = (4 - 13)10^{-9} r^{10} \ R_E{}^2/\text{day}$$

a value which would produce agreement between the observed and calculated proton intensity distributions.

Therefore, with apparently reasonable values for the diffusion coefficient, the relative steady state ($\partial n/\partial t = 0$) equatorially mirroring proton distributions can be described by radial diffusion theory. Absolute flux comparisons remain to be done, whereby an observed steady source of low-energy protons at high altitudes can be used to predict observed fluxes of higher-energy protons at low altitudes. The location and characteristics of such a source has been a free parameter in diffusion calculations for many years.

4.3.2. Sources. Frank and Owens [159] have recently displayed low energy ($0.5 \lesssim E_p \lesssim 50$ keV) proton intensity contours normalized to the magnetic equator from $L \sim 3$ to $L \sim 12$ for the period 9 June–23 July 1966. While the data show spatial characteristics which depend on energy and magnetic activity, there is a tendency for the intensities to exhibit a broad quiet time nightside maximum in the $L \approx 6.5 - 8.5 \ R_E$ range. As an example, we show in Fig. 37 (from Frank and Owens [159]) the intensity contours for 30–50 keV protons observed in the local evening–midnight quadrant. As previously surmised [160,161], it has now been observed that protons in this energy range play a dominant role in the development of the storm time ring current [162,163]. Energy densities reported by Frank for these protons seem sufficient to account for quiet time and storm time magnetic effects observed at synchronous altitudes and at the earth's surface.

Figure 38 shows an attempted comparison of the low-energy proton population presented by Frank and Owens [159] with the higher-energy population at lower altitudes discussed in the survey section. Included in Fig. 38 are several of the integral flux radial profiles displayed in Fig. 13 along with omnidirectional fluxes in the 30–50 keV and 16–25 keV differential energy intervals. These values have been obtained from Fig. 37 and a corresponding figure in Frank and Owens [159] for 16–25 keV protons. The

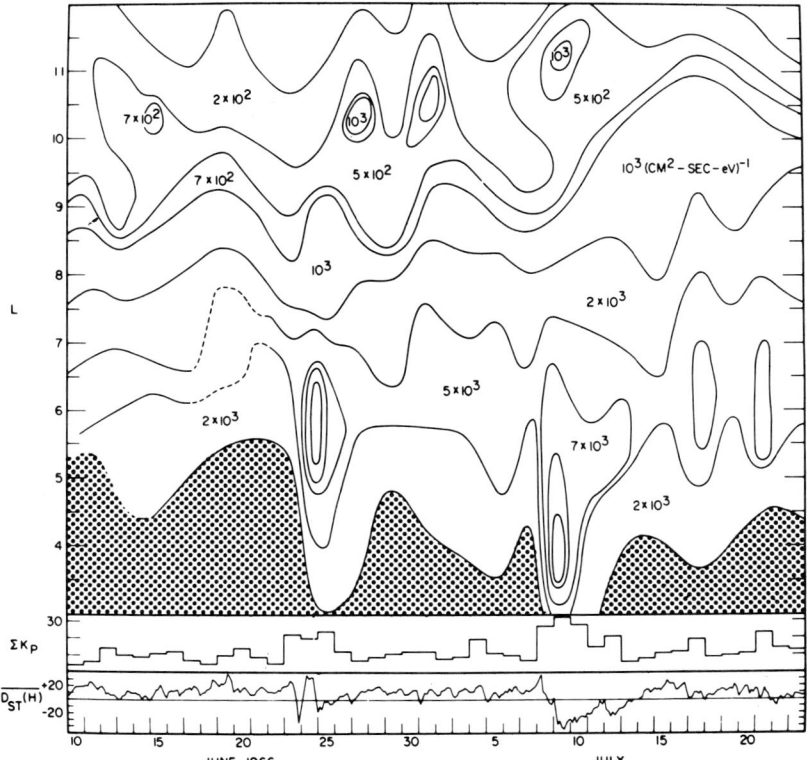

Fig. 37. Equatorial isoflux contours for 30–50 keV protons observed on the nightside hemisphere for $L = 3$–12 R_E (from Frank and Owens [159]).

spread represents the total variation seen for two "quiet" days, 22 June and 18 July. Using these data as a guide, we have also drawn in a "best-guess" curve for the omnidirectional integral flux above 30 keV.

The resulting curves suggest that the nightside hemisphere $\lesssim 50$-keV proton population is a strong source candidate for the diffusion calculations. Figure 37 shows that the inner edge of the low-energy distribution can vary a considerable distance even during relatively quiet periods. We note that it is just this type of variation operating on a time scale smaller than particle drift period which initiates the diffusion process.

4.3.3. Non-Steady State. Having seen evidence indicating the importance of cross-L diffusion in populating the trapping regions in the steady state ($\partial n/\partial t = 0$) condition, given suitable sources, it remains to be seen whether

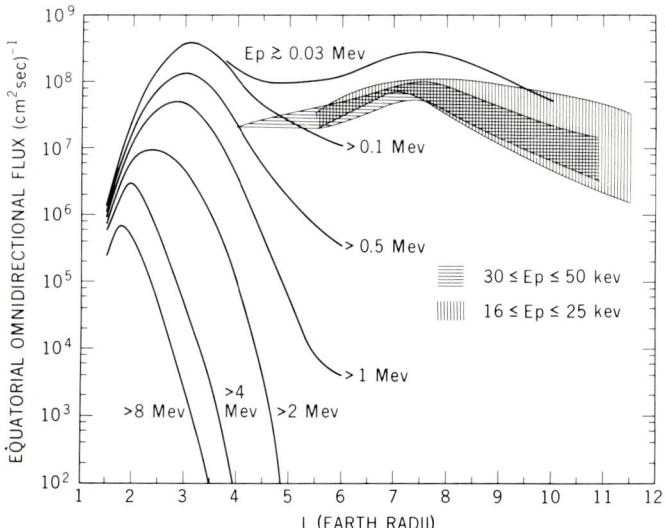

FIG. 38. Omnidirectional proton flux equatorial radial profiles. Curves for $E_p \geq 0.1$ MeV are from Fig. 13. The 30–50 keV and 16–25 keV curves are from two quiet periods in Frank and Owens [159]. The $E_p \gtrsim 0.03$ MeV curve is an estimate using the 30–50 keV and >100 keV values. It is suggested that the nightside peak of 16–50 keV protons at $6.5 \leq L \leq 8.5$ R_Ξ is an important source of outer zone protons via cross-L diffusion.

non-steady state ($\partial n/\partial t \neq 0$) effects can be explained. To identify nonadiabatic effects, Söraas and Davis [66] have removed adiabatic variations, as previously described in the survey section, during a 5-month (January–June 1965) history of outer zone proton intensities for several integral energy thresholds in the range 134–1700 keV. With the adiabatic variations removed, they find remaining

(1) rapid nonadiabatic variations occurring in the geomagnetic storm main phase during which the low-energy proton intensities are enhanced and high-energy intensities depleted, and

(2) a slow, nonadiabatic poststorm variation during which both low, and high-energy intensities recover toward pre-storm values—low-energy intensities decay and high-energy intensities increase during this period.

Using these results, Söraas [164] has subsequently considered the non-steady state ($\partial n/\partial t \neq 0$) condition by studying the slow nonadiabatic variation in (2) above. Using the same formalism as Nakada and Mead [125], Söraas [164] obtains a value for the diffusion coefficient by best fitting the time evolution of the measured distribution function.

The measured proton distribution functions representing the initial conditions for the solution of the Fokker–Planck equation are shown in

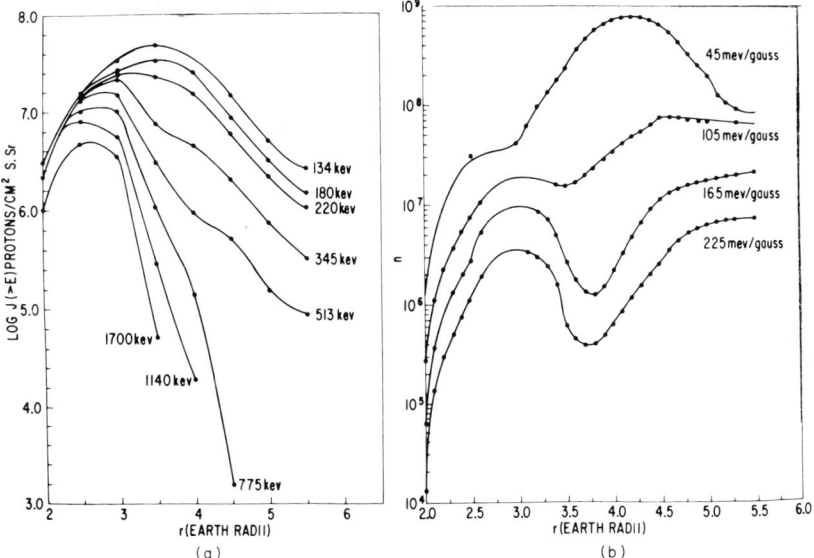

Fig. 39. Initial distribution used in solution of time-dependent Fokker–Planck equation governing cross-L diffusion of trapped protons. (a) The integral proton fluxes above various energy thresholds versus radial distance as measured after the storm of 18 April, 1965. (b) The distribution function n obtained from the flux distribution, for various values of the magnet moment (from Söraas [164]).

Fig. 39 for the poststorm period following the 18 April 1965 storm. The measured and computed time evolution of this distribution are shown in Fig. 40. Calculated curves are shown for the best fit value of D as well as for the earlier estimate of Nakada and Mead [125].

The experimental and theoretical results show good agreement at a value required by Nakada and Mead [125] to fit the equilibrium distribution and in good agreement with the estimate of Tverskoy [151].

The calculated curves for both values of D in Fig. 40 are identical for the three lowest energies shown, indicating that the proton time behavior is dominated by losses at these energies. Also the low-energy protons decay faster than the predicted curves. Therefore, in the loss-dominated region, either the coulomb and charge exchange loss terms are not properly accounted for, or additional loss mechanisms are operating.[1]

---

[1] Note that the omission of a μ^{-1} term in the conversion from flux to density by Nakada and Mead [125] and subsequently by Söraas [164] leads to errors in the coulomb loss term which can be as large as a factor of two for the cases considered by Söraas [164]. Such an error is within existing atmospheric uncertainties and will thus not affect Söraas' main results concerning the effectiveness of the diffusion mechanism.

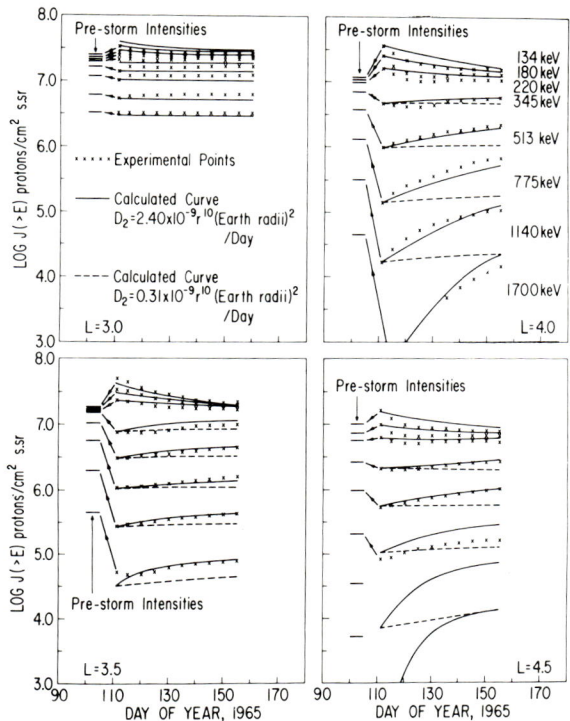

Fig. 40. Comparison of the observed time evolution of the proton distribution with the calculated time evolution using the time-dependent Fokker–Planck equation. Four L shells are shown. Solid line gives best fit value of D_2. Dashed line gives result of an earlier estimate of D_2 (from Söraas [164]).

Söraas and Davis [66] also report time variations in the outer zone proton pitch angle distributions which are both adiabatic and nonadiabatic in appearance. However, additional calculations are required in order to perform adiabatic corrections for arbitrary pitch angles. When this is done, nonadiabatic effects at various pitch angles may be identified and important information obtained concerning the motion of proton mirror points down field lines. Initial work on the coupling between radial and pitch angle diffusion has been presented by Haerendel [165].

4.3.4. Summary of Outer Zone Protons. The following conclusions concerning outer zone protons are presented:

Cross-L diffusion (driven by violation of the third adiabatic invariant while conserving the first and second invariants) is the major source and transport mechanism for outer zone protons.

Cross-L diffusion satisfactorily describes the steady state, $\partial n/\partial t = 0$, equatorial distributions with a value for the diffusion coefficient of

$$D = 2.4(10)^{-9} r^{10} \, R_E^2/\text{day}$$

Cross-L diffusion can satisfactorily describe the observed time evolution of the proton distribution in the non-steady state case, $\partial n/\partial t \neq 0$, during the slow nonadiabatic recovery toward prestorm intensities following a magnetic storm with the same value of D as in the steady state case

$$D = 2.4(10)^{-9} r^{10} \, R_E^2/\text{day}$$

Low-energy protons ($\lesssim 50$ keV) in the region $6.5 \lesssim L \lesssim 8.5$ on the nightside hemisphere are the most probable source of higher-energy outer zone protons.

Rapid nonadiabatic intensity variations during geomagnetic storm main phase are not yet understood.

Identification of nonadiabatic effects at arbitrary pitch angles is required in order to study pitch angle scattering effects within the proton population.

4.4. Outer Zone Electrons

4.4.1. Stable Trapping in the Geomagnetic–Geoelectric Field.

It has been well established that the steady state outer zone electron population displays a characteristic local time variation over a wide energy range at both high and low altitudes [19,61,71,166–168]. At low altitudes, the diurnal variation exhibited by energetic ($\gtrsim 280$ keV and $\gtrsim 1.2$ MeV) electrons has been shown to be consistent with their drift in a model distorted magnetosphere under conservation of the adiabatic invariants [20]. With respect to the accuracy of the charged particle observations, the magnetospheric model used was a realistic one, and these energetic electrons appear to be stably trapped in the earth's field.

Low-altitude $\gtrsim 40$-keV electrons, observed at much higher invariant latitudes ($\Lambda \sim 75°$) on the dayside hemisphere than $\gtrsim 280$-keV electrons ($\Lambda \sim 70°$), exhibit diurnal variations far too large to be explained by drift in a realistic distorted field. This is demonstrated in Fig. 41 where the amount of latitude shift $\Delta\Lambda$ is plotted as a function of the noon meridian invariant latitude Λ_D for both low-altitude locally mirroring 40-keV and 280-keV electrons. This observation has led to the suggestion of a local dayside source of low-altitude ≥ 40-keV electrons [20] and was qualitatively similar to the earlier suggestion [166] of a local dayside acceleration mechanism for ≥ 40-keV electrons based on low-altitude observations of a high-latitude enhancement of 40-keV electron precipitation intensities near noon.

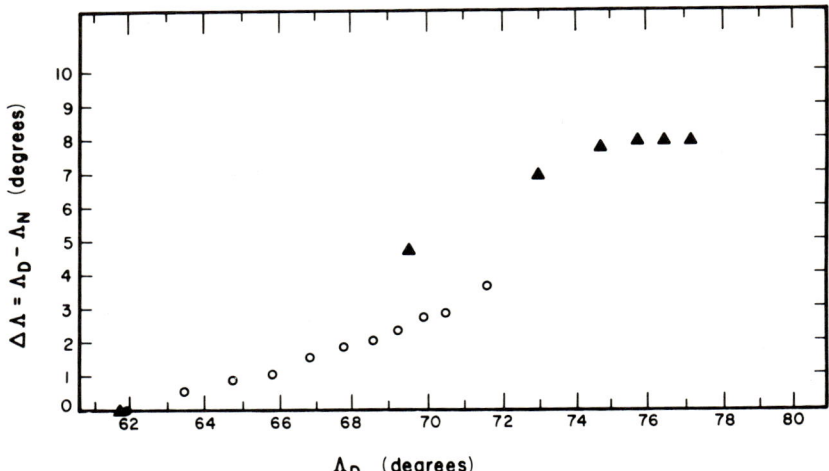

FIG. 41. Diurnal variation of outer zone electrons as a function of invariant latitude $\Lambda = \cos^{-1} L^{-1/2}$. Amount of latitude shift $\Delta\Lambda$ on the noon–midnight meridian is plotted versus the latitude of observation on the noon meridian Λ_D. The variation shown by the ≥ 280 keV and ≥ 1.2 MeV electrons (O) is that expected from adiabatic motion in the distorted geomagnetic field (see Fig. 4b). The ≥ 40 keV electrons (▲) display a diurnal variation which is far too large to be explained by simple adiabatic theory. The main manifestation of this anomalously large variation is the appearance of electrons at high dayside latitudes ($\sim 77°$) (from Williams and Mead [20]).

In lieu of a source mechanism, Taylor [169] has calculated the adiabatic motion of electrons in a combined geoelectric–geomagnetic field model (electro–magnetosphere) and found that the high-latitude appearance of ≥ 40-keV electrons on the dayside could be explained by the inclusion of an electric field. Energetic electrons, being relatively unaffected by the electric field assumed, remained controlled primarily by the magnetic field. Thus the magnitude differences in the diurnal variations exhibited by low-altitude trapped ≥ 40-keV and ≥ 280 keV electrons seemed to be explained simply by adiabatic drift in a geoelectric–geomagnetic field model.

However, this agreement is probably fortuitous. While such effects occur and are important, their magnitude at a given particle energy depends strongly on the electric field configuration assumed. The electric field used by Taylor [169] is that obtained by Taylor and Hones [38] and was deduced from an ionospheric current system which in turn was obtained from ground-level magnetic disturbances. Grave difficulties exist with this deduced field, particularly when quantitative calculations are to be compared with magnetospheric particle observations. The following have been noted by Taylor [169]:

(1) the current system itself is uncertain and has not been determined unequivocally;

(2) the current system is time-variable with hourly variations common;

(3) large uncertainties exist in the magnitude and time variation of ionospheric conductivities.

These difficulties plus the use of a current system obtained during a period of moderate magnetic activity may tend to overestimate the field intensity when applying the calculations to "average" conditions. In addition, the absence of "monoenergetic" auroral electron beams with characteristic energies of tens of kilovolts indicates field strengths much less than those obtained by Taylor and Hones [169,170].

These plus other more qualitative, although well taken, arguments have been used to further point out that calculations of particle motion in model geoelectric–geomagnetic fields should be considered primarily as useful guidelines at energies where electric field effects are important [171]. Thus it seems that the existence of $\gtrsim 40$-keV electrons at high dayside latitudes still requires explanation.

4.4.2. Stable Trapping in the Geomagnetic Field with Weak Pitch Angle Diffusion. Based on the preceding discussion the low-altitude high-latitude profiles of $\gtrsim 40$-keV electrons will be considered in more detail, with this section presenting the argument that inclusion of pitch angle diffusion within the geomagnetic trapping regions is able to explain consistently the low-altitude 40-keV electron spatial distribution.

Figure 42 shows two sample ~ 40-keV low-altitude profiles for both locally mirroring and precipitated electrons (from O'Brien [172] and Fritz, personal communication). Not only does a high-latitude region of flux isotropy exist during an auroral event, as shown in Fig. 42a, but such a region as shown in Fig. 42b is also a characteristic feature of the 40-keV low-altitude electron distribution even during magnetically quiet periods [173,174]. The ratio ϕ of the precipitated to the locally mirroring 40 keV electron fluxes is also shown in Fig. 42b. Fritz [173] has used the latitude at which ϕ breaks from the quiescent value of $\sim 2(10)^{-2}$ as an indication of the boundary between the isotropic $\phi \sim 1$ and anisotropic ($\sim 2(10)^{-2}$) flux regions. This ϕ boundary is very similar to the "smooth" boundary defined by McDiarmid and Burrows [175] as a measure of a gross character change in the 40-keV electron intensity latitude profile. Fritz [174] has studied this region in more detail and has defined several additional boundaries at various values of ϕ which more fully describe the local time variation of the isotropic flux region.

The isotropic flux region shown in Fig. 42 can be associated with the strong diffusion region within the auroral discontinuity, as discussed by Kennel

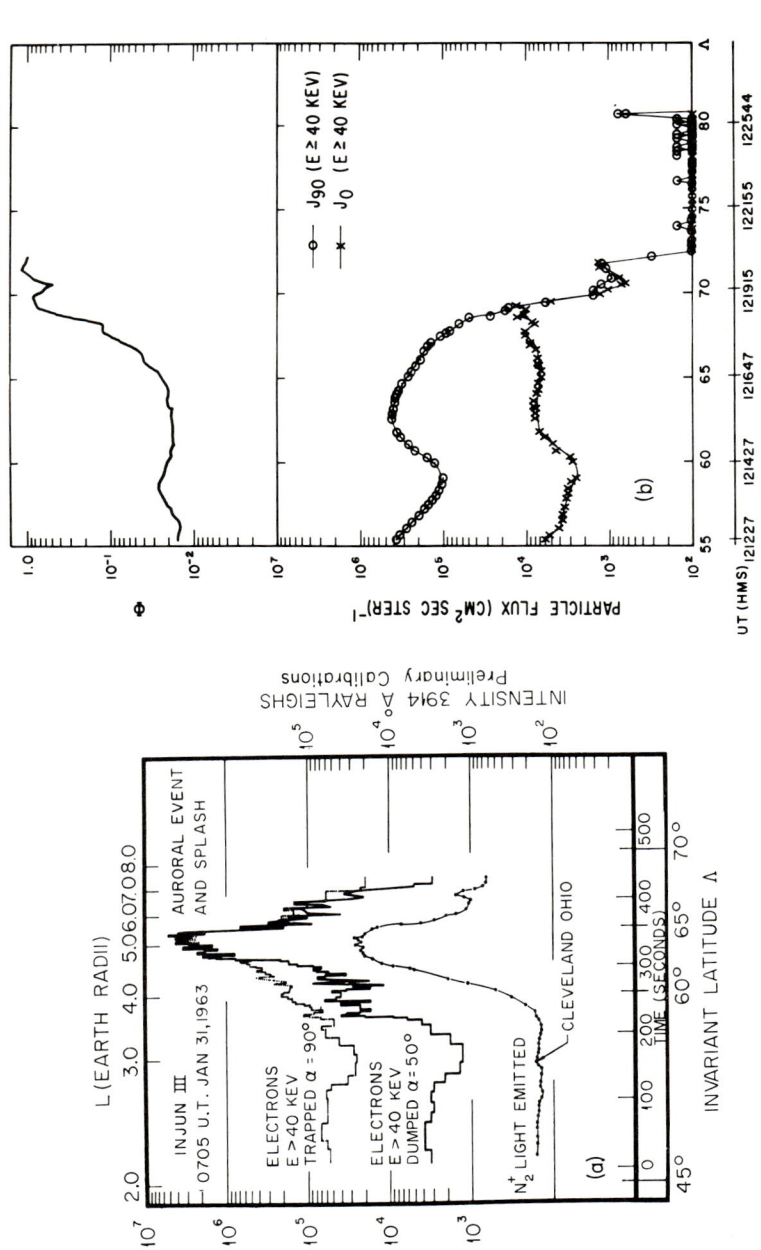

FIG. 42. Low-altitude profiles of trapped and precipitated ≥ 40 keV electrons. (a) from O'Brien [172] and (b) from Fritz (personal communication). In (b), Φ is the ratio of the precipitated to locally trapped ≥ 40 keV electron flux. Magnetic conditions during day when pass (b) was observed were very quiet and show that a region of isotropy exists at high latitudes near local midnight during quiet periods as well as during strong auroral events as in (a).

[176]. At high altitudes this region can be linked to the total pseudo-trapping region presented earlier in Fig. 12. We note again that this strong diffusion region is apparently a spatial effect existing at all times and requiring neither high magnetic activity nor large particle fluxes of $\geq$40-keV electrons needed to create the necessary field turbulence (see Fig. 42a and 42b). This strong diffusion region will be considered in more detail in a later section.

Here a look at the weak precipitation region in more detail seems in order. This region, a weak pitch angle diffusion region, is characterized by a steady and persistent ratio of precipitated to trapped electrons of $\phi \sim 2(10)^{-2}$, as shown in Fig. 42b [173,174]; note that the weak diffusion region is also evident in the earlier pass shown in Fig. 42a. Using the steady state weak pitch angle diffusion formalism of Kennel [176], Fritz [177] is able to consistently explain the value $\phi \simeq 2(10)^{-2}$ with a diffusion coefficient of the form

$$D_\alpha = D_0 \sin^{-1.5} \alpha, \qquad D_0 = (2 \pm 1)\, 10^{-6}\, \text{sec}^{-1}$$

This form of the pitch angle diffusion coefficient yields diffusion lifetimes (the time required for an electron to diffuse into the loss cone and thus, for the weak diffusion case considered, be lost) such that 40-keV electrons having equatorial pitch angles of $>29°$ at $L=6$ will not complete their drift around the earth. Thus 40 keV electrons mirroring at altitudes $\leq 4\, R_E$ on the $L=6$ shell are lost to the atmosphere before they complete their drift and are not stably trapped.

These results agree with the earlier results of McDiarmid and Burrows [167,178,179] who came to the same conclusion based on observed 40-keV electron precipitation and decay rates. Using their observations that:

(1) low-altitude trapped and precipitated $\geq$40-keV electrons exhibit similar diurnal variations;

(2) measured precipitation rates during magnetically quiet periods are of the right order to account for the observed decay at equatorial altitudes at $L \sim 6$; and

(3) trapped 40-keV electrons at low altitudes and at the equator simultaneously decay at the same rate, indicating an equilibrium situation throughout the flux tube with no long-term particle build-up at low altitudes.

McDiarmid and Burrows [179] concluded that the mirror point velocity down the field line must increase at least as fast as $|\bar{B}|$ increases along the field line. Figure 43 (from McDiarmid and Burrows [179]) shows for $L \sim 6$ the ratio of the mirror point travel times to the atmosphere ($R=1$) for 40-keV electrons starting, respectively, from an altitude R and from the equatorial plane. Also shown are the lifetimes of ~ 40 keV electrons at various altitudes using a mean observed equatorial lifetime of a few days. At 1000 km the lifetime shown in Fig. 43 (~ 55 sec) is much less than the drift period

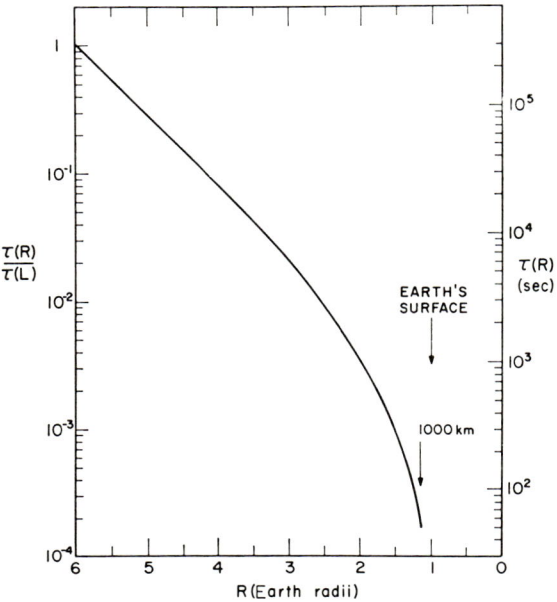

FIG. 43. Pitch angle diffusion lifetime for ≥ 40 keV electrons on $L = 6$ shell as deduced from experimental observations on ≥ 40 keV trapped and precipitated electrons. The ratio $\tau(R)/\tau(L)$ of the time required for the electron's mirror point to diffuse from an altitude R to the atmosphere, $\tau(R)$, to the time required to diffuse from the equator to the atmosphere, $\tau(L)$, is plotted versus R on the $L = 6$ shell. Assuming a few days' lifetime at the equator, $\tau(R)$ is also shown as a function of R. From these curves ≥ 40 keV electrons mirroring at $R < 3.5$ R$_E$ in the $L = 6$ shell will be unable to complete their drift around the earth before being lost to the atmosphere (from McDiarmid and Burrows [179]). Reproduced by permission of the National Research Council of Canada.

and indicates that the high-altitude low-altitude 40-keV electron population is not stably trapped at $L = 6$.

These results can be extended throughout the outer zone and *hence 40-keV electrons mirroring at low altitudes in the outer zone are not stably trapped*. Furthermore, the low-altitude spatial distributions of these electrons are projections along field lines of the high-altitude distributions to low altitudes via pitch angle diffusion (both weak and strong diffusion). Using measured high-altitude electron distributions [54,180] and the field configuration shown in Figs. 11 and 12, the above-discussed process is able to explain the anomalously large low-altitude diurnal variation of 40-keV electrons shown in Fig. 41. This may be a possible answer to the problem discussed in the previous section concerning the appearance of low-altitude locally mirroring 40-keV electrons at very high latitudes ($\sim 77°$) on the noontime meridian. The low-altitude sources required by Frank *et al.* [166] and Williams and

Mead [20] become simply the pitch angle diffusion of the high-altitude population to low altitudes. Low-altitude 40-keV electron observations should thus give a direct measure of the behavior of the high-altitude 40-keV electron population. The related question of what are the sources of the high-altitude population is considered in a later section.

The importance of the pitch angle diffusion process in stable trapping consideration diminishes as the electron energy increases. First, drift times decrease as the energy increases. Thus for a particular pitch angle diffusion lifetime, a high-energy electron might be stably trapped simply because it drifts faster than a nonstably trapped lower energy electron.

Second, pitch angle diffusion is less effective at high electron energies. This point is supported by the following observations:

(a) Fluxes of locally mirroring $\geq$1-MeV electrons at low altitude remain steady and unperturbed even during strong $\geq$40-keV precipitation events [172]

(b) Intensity decay times increase with increasing energy during magnetically quiet periods

(c) Trapped $\geq$1-MeV electrons require many days to come to an equilibrium distribution within a flux tube [143].

These observations further indicate that the magnitude of D_0 *and* the pitch angle dependence of D_α change significantly as the energy increases. In fact D_0 decreases and the exponent of $\sin^q \alpha$ moves from $q = -1.5$ at 40 keV toward zero and possibly to positive values at higher energies.

To display qualitatively the effects of weak pitch angle diffusion for various energy electrons in the outer zone, we have calculated in the manner of McDiarmid and Burrows [179] the pitch angle diffusion lifetimes for $\geq$40-keV, $\geq$300-keV, and $\geq$1-MeV electrons trapped at low altitude (1000 km). This is done by assuming (1) that electron intensity decay times at the equator and at 1000 km are the same, that (2) the equatorial decay is a result of precipitation into the atmosphere. The pitch angle diffusion coefficient at the equator D_0 is given by the inverse of the observed equatorial intensity decay time. The pitch angle diffusion lifetime at 1000 km is then obtained from

$$\tau_D = (B_{\text{eq}}/B_{1000})D_0^{-1}$$

The equatorial lifetimes used are shown in Fig. 44. Because low-altitude decay rates at 300 keV are similar to equatorial decay rates [143] the low-altitude values at 300 keV have been used to extend our L range. Inner zone decay rates are shown for reference [137]. An enhanced decay at $L \sim 3$ immediately following a storm is also indicated. Only the decay rates for magnetically quiet periods are used for the diffusion lifetime estimates.

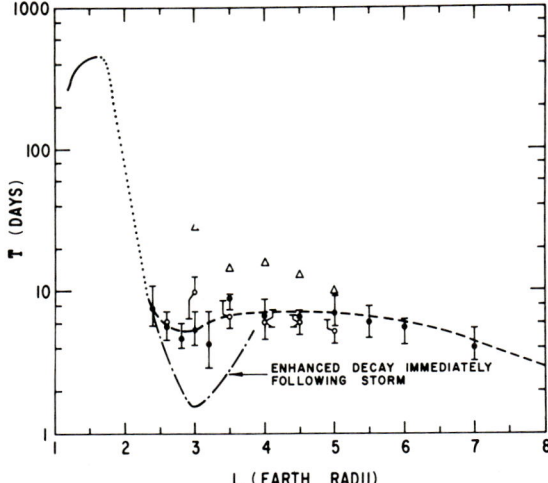

FIG. 44. Electron intensity decay rates used in estimating equatorial pitch angle diffusion coefficient D_0. Outer zone data from Williams et al. [143]. ● $\geq$280 keV, 1100 km; ○, $\geq$300 keV, equator; △, $\geq$1 MeV, equator; ——, $\geq$280 keV, 1100 km (Beall et al. [137]); ...., no available data.

Intensity decay rates for $\geq$40-keV electrons are 70–80% of the $\gtrsim$300 keV decay rates [181], and this value has been used in the calculations.

Electron pitch angle diffusion lifetimes at 1000 km, as described above, along with drift times are shown in Fig. 45, Whenever drift times are greater than the diffusion lifetime, stable trapping is not possible because electrons are unable to drift around the earth before being lost to the atmosphere. Figure 45 thus represents a qualitative guide to stable trapping at low altitudes in the presence of weak pitch angle diffusion. It can be seen that at 1000 km, 40-keV electrons are not stably trapped on any L shell in the outer zone; 300-keV electrons are stably trapped out to $L \sim 6$ (the beginning of the strong diffusion region on the nightside hemisphere); 1-MeV electrons can be stably trapped throughout the outer zone.

While useful as a guide, Fig. 45 is quantitatively inaccurate because the assumptions used in obtaining a diffusion lifetime at 1000 km are not satisfied at all energies. The $\gtrsim$40-keV electron population does reasonably satisfy these assumptions [179] and thus is not stably trapped at low altitudes in the outer zone, as is clear from Figs. 43 and 45.

Assumption 1 is satisfied by the $\gtrsim$300-keV electron population for mirror point altitudes which remain above the atmosphere at all longitudes [143]. It is not at all clear whether or not assumption 2 is justified, even for observations relating to the South Atlantic anomaly [182]. However, the build-up

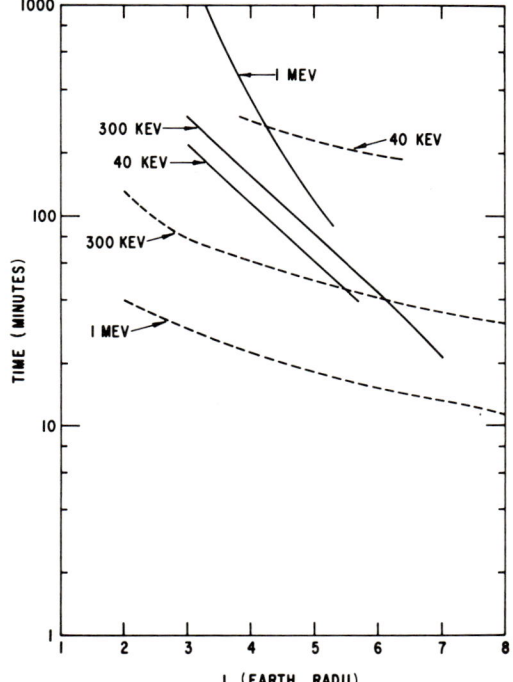

FIG. 45. Qualitative guide to stable trapping of electrons at 1000 km in geomagnetic field with weak pitch angle diffusion. Where drift times (- - -) are larger than pitch angle diffusion lifetimes (———), stable trapping is not possible in the sense that the electron is unable to complete its drift motion around the earth.

of energetic electron intensities at all longitudes away from the South Atlantic anomaly for B, L values which impact the atmosphere in the anomaly [182,183] shows that these low-altitude energetic electrons are able to complete their drift motion in the range $2 \lesssim L \lesssim 4$, consistent with Fig. 45. Estimates of D_α at these altitudes range from $\sim 2(10)^{-6}$ sec^{-1} to $9(10)^{-6}$ sec^{-1} for $2 \leq L \leq 4$ [182].

Assumption 1 is not satisfied by the $\gtrsim$1-MeV electron population, and no information is available concerning assumption 2. The persistent flux build-up of $\gtrsim$1-MeV electrons at low altitudes over periods of several days, while the equatorial $\gtrsim$1-MeV population shows a steady decay, clearly demonstrates that the pitch angle dependence of D_α is significantly different at $\sim$1-MeV than at 40 or 300 keV [143]. This result shows that the 1-MeV diffusion lifetime given in Fig. 45 is much too low and that 1-MeV electrons are even less affected by pitch angle diffusion than Fig. 45 would indicate.

4.4.3. Cross-L Disffusion. The preceding section has ignored the effects of cross-L diffusion. Although this omission does not change the general results obtained above, the coupling of cross-L diffusion and pitch angle diffusion must be done in the weak pitch angle diffusion region to obtain an accurate description of the overall behavior of outer zone electrons. The importance of this coupling in the case of outer zone protons has been stressed by Haerendel [165]. A treatment of the coupling between the processes of convection and pitch angle diffusion in the strong pitch angle diffusion region has been presented by Kennel [176].

Cross-L diffusion was shown in an earlier section to be a major source-transport mechanism for outer zone protons as well as inner zone electrons. The best evidence for the effectiveness of cross-L diffusion for outer zone electrons are the observations of Frank [184]. He reports observing an apparent inward motion of the inner edge of the equatorial ≥ 1.6-MeV electron distribution through the range $L \sim 4.8$ to $L \sim 3.2$ during extended quiet periods following magnetic storms in December 1962 and April 1963. This apparent inward motion proceeded at a velocity (earth radii per day) proportional to L^8. However, it has been also noted [181] that low-altitude $\gtrsim 3.9$-MeV electrons behave quite differently and do not simply reflect the behavior of the equatorial energetic population. Cross-L diffusion is also qualitatively seen proceeding to both higher and lower altitudes after the sudden appearance of energetic electrons deep within the trapping regions during magnetic storms [143].

The quantitative evaluation of cross-L diffusion as it affects electrons is much more difficult to determine than it is for protons because of the far stronger effects of pitch angle diffusion. However, useful insights into the outer zone situations can be obtained by assuming radial diffusive processes with the exclusion of pitch angle scattering effects. Newkirk and Walt [185] have made this assumption and, using the equatorial energetic electron data of Brown [145] and Frank [184], have obtained from the Fokker–Planck equation the magnitude and radial dependence required for the diffusion coefficient. Their results are shown in Fig. 46 along with their previous inner zone results.

Using the diffusion coefficients in Fig. 46. Newkirk and Walt [185] find that they are able to fit the outer zone equatorial radial distribution of ≥ 1.6-MeV electrons, provided a constant source exists at $L \sim 6$.

4.4.4. Strong Pitch Angle Diffusion Region and Sources. We have associated the region of loss cone isotropy for 40-keV electron intensities observed at low altitudes (Fig. 42) with a strong pitch angle diffusion region. In addition, this region was found to be a characteristic feature of the low-altitude, high-latitude profiles of 40-keV electrons, indicating that the strong pitch angle diffusion region is a characteristic spatial feature of the magnetosphere.

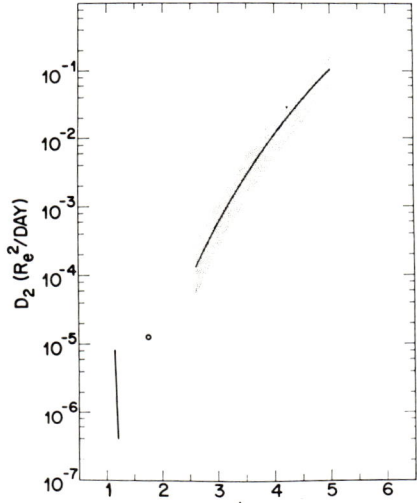

FIG. 46. Summary of evaluations of the cross-L diffusion coefficient for energetic electrons throughout the trapping regions (from Newkirk and Walt [185]).

We thus suggest that even in a steady state situation (here considered to be the absence of large events or fluxes of 40-keV electrons in the strong diffusion region at low altitudes) the convection-pitch angle diffusion coupling discussed by Kennel [176] provides the major input of particles into the strong diffusion region and consequently is an important source of electrons in the outer zone.

In this model, 1–10 keV plasma sheet electrons are continually convected toward the earth in the nightside hemisphere. As they are convected inward (see [171] for a review of magnetospheric convection) they are accelerated to some tens of kilovolts due to flux tube compression (or equivalently by the combination of betatron acceleration and Fermi acceleration). When the precipitation time scale becomes comparable to the convective flow time scale, flux tubes are depleted along the flow fast enough to produce the high-latitude edge of the strong diffusion region shown in Fig. 42. In the strong diffusion case, this high-latitude edge is mainly determined by the size of the loss cone which determines the minimum precipitation lifetime. Thus, as the electrons approach the earth, the loss cone increases and eventually becomes large enough so that precipitation can effectively compete with the convective flow. The flux tube is then either depleted as it convects through the strong diffusion region, as suggested by Kennel [176], or it continues inward carrying the remnants of its previous population into the more stable, weak diffusion region, characterized in Fig. 42b by $\phi \sim 2(10)^{-2}$.

This process is illustrated in Fig. 47 where the nightside magnetosphere is shown along with the plasma sheet, neutral sheet, and pseudo-trapping regions in a manner consistent with Figs. 11 and 12. Convective flow from the tail and pitch angle diffusion down field lines are shown as solid arrows whose size indicates relative intensity. Electrons participating in this flow

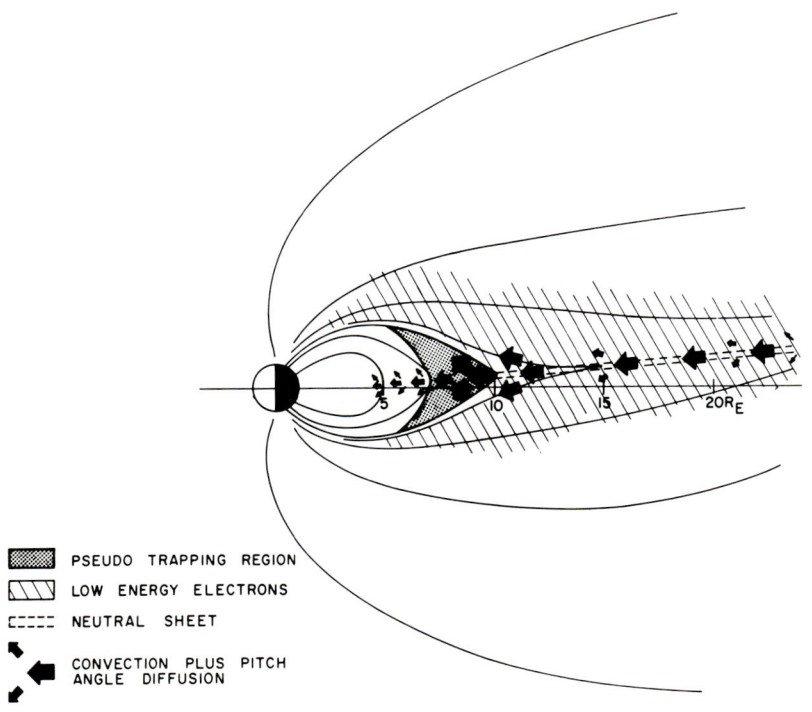

FIG. 47. Illustration of the coupling between convection and pitch angle diffusion as discussed by Kennel [176]. Size of arrows representing convection and pitch angle diffusion are meant to give qualitative relative magnitudes. Definition of low altitude region of ≥ 40 keV electron flux isotropy can be seen. In weak diffusion region, $R < 7\,R_E$, cross-L diffusion occurs and contributes to the outer zone energetic electron population. The magnetic field configuration, pseudo-trapping region, low-energy electron distribution (plasma sheet), and neutral sheet are as shown in Fig. 11.

are ~ 1–10 keV in energy in the tail sheet and are energized to tens of kilovolts by the time they reach the strong diffusion region. The spatial extent of the strong diffusion region is difficult to determine, but it extends beyond the high-altitude limit of pseudo-trapping region and probably extends below the lower limit of the pseudo-trapping region. Electrons in this region can be further diffused inward across field lines via third invariant violation and contribute to the energetic outer zone electron population forming the source required by Newkirk and Walt [185] in their diffusion calculations, as discussed in the previous section. Electrons convected into the strong diffusion region at the equator on the nightside will drift eastward, remain in a strong diffusion region, and produce isotropic or near isotropic fluxes at low altitudes for all local times for which significant intensities still remain at the equator. The observations of Fritz [174], wherein the probability of occurrence of low-altitude isotropic 40-keV electron fluxes decrease as local time progresses eastward from midnight, indicates the drifting equatorial population is depleted by the pitch angle diffusion process.

Observations of 40 keV electrons at synchronous altitudes have shown that the nightside plasma sheet is a major source of electrons during substorms ([186–188], and for a review of particle and field observations at synchronous altitudes, see [189]). Electrons appearing at 6.6 R_E, subsequently drift eastward, and are lost to the atmosphere via pitch angle diffusion. The field geometry of Figs. 10, 11, and 12 indicates that equatorial electrons appearing at 6.6. R_E near midnight are within the strong diffusion region discussed above.

Observational evidence, both direct and indirect, has been reported, supporting the existence of sources other than the nightside plasma sheet which could supply tens of kilovolt particles to the distant trapping regions. Briefly this evidence consists of

(1) observation of plasma sheet electrons at almost all local times, particularly the observation of ∼1-keV electrons extending to plasmapause altitudes near the dawn meridian [57];

(2) observation of low-energy electron fluxes at latitudes associated with the dayside neutral point or line (see Gringauz [58]);

(3) identification of a "diffuse" precipitation region located mainly in the midnight-to-noon hemisphere [190]. This region, associated with mantle aurora [191], may be also associated with observation (1) above;

(4) observation of a high-latitude, noon-time region of enhanced precipitation [166]. This effect may be associated with (2) above;

(5) marked change in character of the low altitude isotropic flux region as local time varies from midnight to dawn and on through to late evening [174]. This region, well defined and localized at midnight, becomes much less well defined and more diffuse toward dawn and noon, and becomes very difficult to identify clearly in the afternoon to early evening sector.

4.4.5. Pitch Angle Diffusion Processes. The continual existence of a precipitated flux of 40 keV electrons [172], the fact that "average" flux contours of precipitated 40-keV electrons exist [166–168], and the observation of a continual flux build-up with longitude at B, L values impacting the atmosphere in the South Atlantic anomaly [182,183] support the suggestion of an ever present mechanism producing a steady pitch angle diffusion on field lines throughout the outer zone.

Roberts [192] has demonstrated the effectiveness of pitch angle scattering mechanisms determining both the lifetimes and maximum intensities for outer zone trapped electrons (see review by Roberts [193]). He has shown that rms whistler-mode noise B vector of ~ 1 milligamma distributed over a 10^4-Hz bandwidth can account for observed electron "steady state" lifetimes of ~ 4–7 days in the outer zone and could thus account for the steady "drizzle" of electrons precipitated into the atmosphere throughout the outer zone. Noise of this intensity has been observed to be a common feature in the 50°–70° invariant latitude range at altitudes at least up to 2700 km [194,195].

Kennel and Petschek [196] have shown that, if the equatorial anisotropic fluxes observed in the outer zone exceed a critical limit, an instability is triggered which generates additional whistler-mode turbulence and enhances the scattering of electrons to lower pitch angles and into the atmosphere. A brief description of this effect is as follows. Ever-present whistler-mode noise in the outer magnetosphere (as described above) interacts with electrons of energies greater than $B^2/8\pi N$, resulting in the pitch angle diffusion of these electrons (B = local magnetic field strength and N = total electron number density). This steady weak diffusion process yields an anisotropic equatorial pitch angle distribution which in turn leads to a growth of the waves causing the diffusion. The growth rate of the waves is proportional to the number of interacting particles and thus places a limit on the number of stably trapped particles of energy $\gtrsim B^2/8\pi N$. If the number of particles above the threshold energy E_T increased above its limit, a rapid wave growth would result which in turn would cause a rapid loss of particles via pitch angle diffusion until the intensities were once again at or below stable limit. Such strong diffusion is supported by the observation that equatorial 40-keV electron intensities are generally near, but rarely exceed, the critical limit [196] and by the observation of 40-keV electron isotropic loss cone distributions during magnetically quiet times. However, the observation of a characteristic isotropic loss cone distribution, even at low electron intensities, indicates the existence of other strong pitch angle scattering mechanisms. In fact, if the Kennel–Petschek mechanism operates *only* when the electron flux exceeds a critical limit, it may only be effective during the very short time interval required to attenuate the fluxes to below the critical limit.

4.4.6. Summary of Outer Zone Electrons. The outer zone electron trapping region is characterized by the continual presence of two pitch angle diffusion regions, one strong and one weak. Near midnight, the strong diffusion region extends slightly below the equatorial stable trapping boundary and extends outward to some unknown distance. This region is easily identified at low altitudes by the observation of isotropic 40-keV electron fluxes. The weak diffusion region begins at the lower boundary of the strong diffusion region, comprises most of the outer zone, and is characterized by a steady and persistent ratio of precipitated to locally mirroring 40-keV electron intensities of $\phi \sim 2(10)^{-2}$.

The weak pitch angle diffusion process has a significant effect on the stable trapping limits for electrons. Comparisons of drift times with rough estimates of pitch angle diffusion lifetimes indicate that at 1000 km, 40-keV electrons are nowhere stably trapped in the outer zone, 300-keV electrons are stably trapped out to the vicinity of the strong diffusion region, and 1-MeV electrons are stably trapped throughout the outer zone.

The above results explain in a consistent way the anomalously large diurnal variations exhibited by 40-keV electrons observed at low altitudes. The inclusion of pitch angle diffusion (both strong and weak) into the geomagnetic trapping regions predicts that the observed low-altitude spatial distributions should be the result of mapping a given high-altitude distribution to low altitudes via the diffusion process. Observed high- and low-altitude 40-keV electron distributions within the observed geomagnetic field support this result.

Cross-L diffusion is a major source and transport mechanism for energetic outer zone electrons during magnetically quiet periods. During such quiet periods, the tens of kilovolt electrons in the nightside strong-diffusion region are the most probable source for this transport process. Furthermore electrons of all energies appearing at various locations in the outer zone during magnetic storms and substorms are also affected by cross-L diffusion and are transported both to lower and higher L shells.

During magnetically quiet periods, the major source of several tens of kilovolt electrons in the strong diffusion region is the convection and consequent acceleration via flux tube compression of plasma sheet electrons down the tail toward the earth. As electrons are convected toward the earth, the loss cone becomes larger and eventually reaches a value allowing precipitation to compete with the convective flow [176]. At this stage the high-latitude edge of the strong diffusion region (isotropic 40-keV fluxes) is seen at low altitude. These electrons enter the weak diffusion region at the low-altitude edge of the strong diffusion region by continued convection and by mechanisms violating the third invariant and thereby driving the cross-L diffusion process.

It is highly probable that other quiet-time sources feed electrons into the strong diffusion region at various local times, particularly near dawn and near noon.

Quantitative treatments coupling cross-L diffusion with weak pitch angle diffusion are required to consider realistically the transport of electrons throughout the outer zone even during magnetically quite periods.

5. Concluding Remarks and Future Directions

It is clear that much is now known, not only of the morphology of the magnetosphere and its charged particle population, but also of the sources, losses, and transport of magnetospherically trapped particles. There remain storm-time effects which are not yet understood and which may require new mechanisms for explanation. Examples of such effects are:

(1) short term nonadiabatic time variations in proton intensities [66,127, 197] which may be related to substorms [198,199];

(2) substorm related electron intensity enhancements [186,200,201];

(3) sudden appearance of energetic electrons deep within the trapping regions and their correlation with ionospheric depletion [143];

(4) the apparent rapid movement of low-energy protons to lower altitudes resulting in the magnetic storm effects observed at ground level [162,163].

However, with the knowledge now available concerning the magnetosphere and its environs, it is possible to design experiments, both passive and active, which can directly test and observe specific mechanisms thought to be responsible for various magnetospheric phenomena. Passive experiments generally observe parameters over which they have no control (i.e., naturally occurring phenomena) whereas active experiments directly stimulate the environment and observe the results of such stimulation. The Argus and Starfish high-altitude nuclear explosions [16,132] and the injection of accelerator beams into the magnetosphere at altitudes of a few hundred kilometers [202,203] are examples of active experiments conducted in the past. In addition, the technique of barium release into the magnetosphere has been successfully used to study magnetospheric electric fields [204–208].

In the future controlled active experimentation with the earth's space environment (i.e., space weather modification) will afford one of the best possible methods of gaining an understanding of that environment. The possibility of such space weather modification has recently been discussed by Brice [209]. He has noted that recent results [210] show a lack of cold plasma at synchronous altitudes near local midnight. Further noting that the threshold energy involved in the whistler-mode instability, E_T in Section 4.4.5, is inversely proportional to the total number density N, he then suggests that the injection of cold plasma near local midnight at 6.6 R_E will have the effect of lowering the threshold energy corresponding to the stable trapping limit. Electrons of energy greater than the new threshold energy then become subject to the trapping limit and the initially higher intensities (corresponding to the lower energy) will be rapidly diffused into the atmosphere until the new trapping limit is satisfied.

Thus by injecting cold plasma into the ambient environment at nighttime altitudes of 6.6 R_E, it may be possible to create artificially ionospheric disturbances or auroral events, or both. In so doing, valuable information will be gained concerning the operation of plasma instabilities in the magnetosphere which are responsible for much of the particle behavior throughout the trapping regions. It should be noted that the above effects are self-quenching, i.e., as soon as the scattering brings the appropriate particle intensities to the new trapping limit, the wave growth rate is severely attenuated and a stable situation is restored.

The Space Environment Laboratory of NOAA (National Oceanographic and Atmospheric Agency) has proposed to conduct this space environment modification experiment. The technique envisaged is to fly a small satellite to synchronous orbit carrying the plasma injection device and the necessary instrumentation to monitor the effects of the injection (e.g., particle distributions, magnetic and electric fields). Repeated injections appear feasible with direct readout of the onboard instrumentation being used to determine the optimum ambient conditions for each injection.

In conclusion, pursuit of similar active and controlled experiments with the earth's environment in the future is urged. In this way man not only will gain an understanding of his environment, but will also discover ways to both protect and make use of it.

Acknowledgments

I wish to acknowlege the generous assistance of Dr. G. D. Mead in the final preparation of this paper. I also thank Drs. C. O. Bostrom, C. E. McIlwain, and J. I. Vette for various figures used in the paper. This work was done both at Goddard Space Flight Center, NASA, and at the Space Environment Laboratory of NOAA.

List of Symbols

B	Magnetic field strength	E_p	Proton energy
B	Magnetic field vector	E_T, E	Energy threshold
B_M	Mirror point B value	F	Omnidirectional flux, particles per square centimeter per second
c	Velocity of light		
D	Diffusion coefficient	g	Acceleration of gravity
e	Electronic charge	I	Integral invariant
E	Electric field strength	j	Directional flux, particles per square centimeter per second per steradian
E	Electric field vector		
E_e	Electron energy		
E_N	Energy per nucleon	J	Second adiabatic invariant

K_p	Geomagnetic activity index	Z	Nuclear charge
L	L coordinate used in ordering charged particle data in the geomagnetic field	α	Pitch angle: angle between the magnetic field line and the particle velocity vector
m	Particle mass	α_e	Equatorial pitch angle
M, M^*	Mirror point, conjugate mirror point	τ	Charge exchange lifetime
		τ_c	Gyroperiod
n	Number density in dr, $d\mu$, and dJ	τ_D	Pitch angle diffusion lifetime
		ρ	Gyroradius
P	Momentum	ρ	Position on helix measured from guiding center
r	Instantaneous position of a particle	μ	Particle's magnetic moment due to gyration
$\mathbf{R}$	Position of guiding center		
R_E	Geocentric distance in earth radii	Φ_M	Third adiabatic invariant, the magnetic flux
s	Distance along field line		
v	Speed	Λ	Invariant latitude
$\mathbf{v}$	Velocity vector	γ	10^{-5} gauss
V	Velocity component perpendicular to field line	ϕ	Ratio of precipitated to trapped electron fluxes at low altitudes
$V_{\parallel}$	Velocity component parallel to field line		

References

1. Störmer, C. (1907). Sur les trajectoires des corpuscules electrisiés dans l'éspace sous l'action du magnetism terrestre. *Arch. Sci. Phys.* **24**, No. 5, 113, 221, 317.
2. Störmer, C. (1933). On the trajectories of electrical particles in the field of a magnetic dipole with applications to the theory of cosmic radiation. *Avh. Norske Vid. Akad. Oslo Mat.-Nat. Kl.* **11**.
3. Störmer, C. (1955). "The Polar Aurora," 403 pp. Oxford Univ. Press, London and New York.
4. Chapman, S., and Ferraro, V. C. A. (1931). A new theory of magnetic storms (sect. 6 and 7). *Terr. Magn. Atmos. Elec.* **36**, 77, 171.
5. Chapman, S., and Ferraro, V. C. A. (1932). A new theory of magnetic storms (sect. 8). *Terr. Magn. Atmos. Elec.* **37**, 147.
6. Alfvén, H. (1939). Theory of magnetic storms. 1. *Kgl. Sv. Vetenskapsakad. Handl.* [3] **18**, No. 3.
7. Alfvén, H. (1940). Theory of magnetic storms. 2, 3. *Kgl. Sv. Vetenskapsakad. Handl.* [3] **18**, No. 9.
8. Alfvén, H. (1950). "Cosmical Electrodynamics," 1st ed. Oxford Univ. Press, London and New York.
9. Singer, S. F. (1956). Trapped orbits in the earth's dipole field. (Abstr.) *Bull. Amer. Phys. Soc.* [2] **1**, 229.
10. Singer, S. F. (1957a). A new model of magnetic storms and aurorae. *Trans. Amer. Geophys. Union* **38**, No. 2, 175–190.
11. Singer, S. F. (1957b). Project "Far Side." *Missiles Rockets* **2**, Part 2, 120–128.
12. Van Allen, J. A., Ludwig, G. H., Ray, E. C., and McIlwain, C. E. (1958). Observation of high intensity radiation by satellites 1958 α and γ. *Jet Propul.* **28**, 588–592.
13. Yoshida, S., Ludwig, G. H., and Van Allen, J. A. (1960). Distribution of trapped radiation in the geomagnetic field. *J. Geophys. Res.* **65**, 807–813.

14. Spitzer, L. (1956). "Physics of Fully Ionized Gases," 105 pp. Wiley (Interscience), New York,
15. Northrup, T. G. (1963). "The Adiabatic Motion of Charged Particles," 109 pp. Wiley (Interscience), New York.
16. Christofilos, N. C. (1959). The Argus experiment. *J. Geophys. Res.* **64**, 869–876.
17. Van Allen, J. A., McIlwain, C. E., and Ludwig, G. H. (1959). Satellite observations of electrons artificially injected into the geomagnetic field. *J. Geophys. Res.* **64**, 877–891.
18. McIlwain, C. E. (1961). Coordinates for mapping the distribution of magnetically trapped particles. *J. Geophys. Res.* **66**, 3681–3691.
19. O'Brien, B. J. (1963). A large diurnal variation of the geomagnetically trapped radiation. *J. Geophys. Res.* **68**, 989–995.
20. Williams, D. J., and Mead, G. D. (1965). Nightside magnetosphere configuration as obtained from trapped electrons at 1100 kilometers. *J. Geophys. Res.* **70**, 3017–3029.
21. Vernov, S. N., and Chudakov, A. E. (1960). Terrestrial corpuscular radiation and cosmic rays. *Space Res.* **1**, 751–796.
22. Dessler, A. J. (1961). Penetrating particle radiation. *In* "Satellite Environment Handbook" (F. S. Johnson, ed.), 1st ed., pp. 49–74. Standford Univ. Press, Stanford, California. (2nd ed., 1965.)
23. Van Allen, J. A. (1961). Dynamics, composition and origin of the geomagnetically trapped corpuscular radiation. Invited paper given at 11th General Assembly of the International Astronomical Union 16 August. [Also available *in* "Space Science" (Le Galley, ed.) pp. 226–274. Wiley, New York, 1963.]
24. Singer, S. F., and Lencheck, A. M. (1962). Geomagnetically trapped radiation. *Progr. Elem. Particle Cosmic Ray Phys.* **11**, 247–344.
25. Elliott, H. (1963). The Van Allen particles. *Rep. Progr. Phys.* **26**, 145–180.
26. Farley, T. A. (1963). The growth of our knowledge of the earth's outer radiation belt. *Rev. Geophys.* **1**, 3–34.
27. Rosser, W. G. V. (1964). The Van Allen radiation zones. 1. The theory of trapping and the inner zone. *Contemp. Phys.* **5**, 198–211; 2. The outer zone, the geomagnetic cavity, and magnetic storms. *Ibid.* **5**, 255–269.
28. Shabansky, V. P. (1965). The radiation belts. *Geomagn. Aeron.* **5**, 765–789.
29. Anderson, K. A. (1966). Energetic particles in the earth's magnetic field. *Ann. Rev. Nucl. Sci.* **16**, 291–344.
30. O'Brien, B. J. (1967). Interrelations of energetic charged particles in the magnetosphere. "Solar Terrestrial Physics" (J. W. King and W. S. Newman, eds.), Chapter 7, pp. 169–211. Academic Press, New York.
31. Hess, W. N. (1968). "The Radiation Belt and Magnetosphere," 548 pp., Ginn (Blaisdell), Boston, Massachusetts.
32. Williams, D. J., and Mead, G. D. eds. (1969). "Magnetospheric Physics," 459 pp. Am. Geophys. Union, Washington, D.C. [also in *Rev. Geophys.* **7** Nos. 1, 2).]
33. McCormac, Billy M., ed. (1966). "Radiation Trapped in the Earth's Magnetic Field," 901 pp. Reidel, Dordrecht, Holland.
34. McCormac, Billy M., ed. (1968). Earth's Particles and Fields," 464 pp. Van Nostrand-Reinhold, Princeton, New Jersey.
35. McCormac, Billy M., ed. (1970). "Particles and Fields in the Magnetosphere," 453 pp. Reidel, Dordrecht, Holland.
36. Vette, J. I., Lucero, A. B., and Wright, J. A. (1966). Model of the trapped radiation environment. Vol. 2: Inner and outer zone electrons. *NASA Rep.* SP-3024.
37. King, J. H. (1967), Models of the trapped radiation environment. Vol. 4: Low energy protons. *NASA Rep.* SP-3024.

38. Taylor, H. E., and Hones, E. W. Jr. (1965). Adiabatic motion of auroral particles in a model of the electric and magnetic fields surrounding the earth. *J. Geophys. Res.* **70**, 3605–3628.
39. Mead, G. D. (1964). Deformation of the geomagnetic field by the solar wind. *J. Geophys. Res.* **69**, 1181–1195.
40. Roederer, J. G. (1967). On the adiabatic motion of energetic particles in a model magnetosphere. *J. Geophys. Res.* **72**, 981–992; See also," Dynamics of Geomagnetically Trapped Radiation." Springer, Heidelberg, 1970.
41. Pfitzer, K. A., Lezniak, T. W., and Winckler, J. R. (1969). Experimental verification of drift-shell splitting in the distorted magnetosphere. *J. Geophys. Res.* **74**, 4687–4693.
42. Lanzerotti, L. J., Roberts, C. S., and Brown, W. L. (1967). Temporal variations in the electron flux at synchronous altitudes. *J. Geophys. Res.* **72**, 5893–5902.
43. Vernov, S. N., Vakulov, P. V., Kuznetsov, S. N., Logachev, Yu. I., Sosnovets, E. N., and Stolpovsky, V. G. (1967). Boundary of the outer radiation belt and zone of unstable radiation. *Geomagn. Aeron.* **7**, 335–339.
44. Williams, D. J., and Ness, N. F. (1966). Simultaneous trapped electron and magnetic tail field observations. *J. Geophys. Res.* **71**, 5117–5128.
45. Roederer, J. G. (1969). Quantitative models of the magnetosphere. In "Magnetospheric Physics" (D. J. Williams and G. D. Mead, eds.), pp. 77–96. Am. Geophys. Union, Washington, D.C. [also in *Rev. Geophys.* **7**, Nos. 1, 2.]
46. Williams, D. J. (1966). Outer zone electrons. In " Radiation Trapped in the Earth's Magnetic Field " (B. M. McCormac, ed.), pp. 263–280. Reidel, Dordrecht, Holland.
47. Ness, N. F., and Williams, D. J. (1966). Correlated magnetic tail and radiation belt observations. *J. Geophys. Res.* **71**, 322–325.
48. Mead, G. D. (1966). The motion of trapped particles in a distorted field. In "Radiation Trapped in the Earth's Magnetic Field " (B. M. McCormac, ed.), pp. 481–490. Reidel, Dordrecht, Holland.
49. Fairfield, D. H. (1968). Average magnetic field configuration of the outer magnetosphere. *J. Geophys. Res.* **73**, 7329–7338.
50. Serlemitsos, P. (1966). Low energy electrons in the dark magnetosphere. *J. Geophys. Res.* **71**, 61–77.
51. Haskell, G. P. (1969). Anisotropic fluxes of energetic particles in the outer magnetosphere. *J. Geophys. Res.* **74**, 1740–1748.
52. Gringauz, K. I., Bezrukikh, V. V., Ozerov, V. D., and Rybchinsky, R. Ye. (1960). A study of interplanetary ionized gas, energetic electrons, and corpuscular solar emission, using three-electrode charged-particle traps set up on the second Soviet cosmic rocket Luna 2. *Dokl. Akad. Nauk USSR* **131**, 1301–1304. (Engl. transl.: *Sov. Phys.—Dokl.* **5**, 361–364, 1960.)
53. Gringauz, K. I., Kurt, V. G., Moroz, V. I., and Shklovsky, I. S. (1960). Results of observations of charged particles observed out to 100,000 km with the aid of charged particles traps on Soviet space probes. *Astron. Zh.* **37**, 716–735. (Engl. transl.: *Sov. Astron.—AJ* **4**, 680–695, 1961.)
54. Freeman, J. W., Jr. (1964). The morphology of the electron distribution in the outer radiation zone and near the magnetospheric boundary as observed by Explorer 12. *J. Geophys. Res.* **69**, 1691–1723.
55. Bame, S. J., Asbridge, J. R., Felthauser, H. E., Hones, E. W., and Strong, I. B. (1967). Characteristics of the plasma sheet in the earth's magnetotail. *J. Geophys. Res.* **72**, 113–129.
56. Vasyliunas, V. M. (1968). A survey of low-energy electrons in the evening sector of the magnetosphere with OGO 1 and OGO 3. *J. Geophys. Res.* **73**, 2839–2884.

57. Vasyliunas, V. M. (1968). Low-energy electrons in the day side of the magnetosphere. *J. Geophys. Res.* **73**, 7519–7523.
58. Gringauz, K. I. (1969). Low energy plasma in the earth's magnetosphere. In "Magnetospheric Physics" (D. J. Williams and G. D. Mead, eds.), pp. 339–378. Am. Geophys. Union, Washington, D.C. [also in *Rev. Geophys.* **7**, Nos. 1, 2.]
59. Carpenter, D. L. (1966). Whistler studies of the plasmapause in the magnetosphere. 1. Temporal variations in the position of the knee and some evidence on plasma motions near the knee. *J. Geophys. Res.* **71**, 693–709.
60. Anderson, K. A. (1965). Energetic electron fluxes in the tail of the geomagnetic field. *J. Geophys. Res.* **70**, 4741–4763.
61. Frank, L. A., Van Allen, J. A., and Macagno, E. (1963). Charged-particle observations in the earth's outer magnetosphere. *J. Geophys. Res.* **68**, 3543–3554.
62. Frank, L. A. (1965). A survey of electrons E > 40 kev beyond 5 earth radii with Explorer 14. *J. Geophys. Res.* **70**, 1593–1626.
63. Anderson, K. A., Harris, H. K., and Paoli, R. J. (1965). Energetic electron fluxes in and beyond the earth's outer magnetosphere. *J. Geophys. Res.* **70**, 1039–1050.
64. Vette, J. I. (1970). Summary of particle populations in the magnetosphere. In "Particles and Fields in the Magnetosphere" (B. M. McCormac, ed.), pp. 305–318. Reidel, Dordrecht, Holland.
65. McIlwain, C. E. (1966). Ring current effects on trapped particles. *J. Geophys. Res.* **71**, 3623–3628.
66. Söraas, F., and Davis, L. R. (1968). Temporal variations of the 100 kev to 1700 kev trapped protons observed on satellite Explorer 26 during the first half of 1965. NASA TM X-63320.
67. Frank, L. A., and Van Allen, J. A. (1966). Correlation of outer radiation zone electrons ($E_e \sim 1$ Mev) with the solar activity cycle. *J. Geophys. Res.* **71**, 2697–2700.
68. Vernov, S. N., Gorchakov, E. V., Kuznetsov, S. N., Logachev, Yu, I., Sosnovets, E. N., and Stolpovsky, V. G. (1969). Particle fluxes in the other geomagnetic field. In "Magnetospheric Physics" (D. J. Williams and G. D. Mead, eds.), pp. 257–280. Am. Geophys. Union, Washington, D.C. [also in *Rev. Geophys.* **7**, Nos. 1, 2].
69. Vette, J. I., and Lucero, A. B. (1967), Models of the trapped radiation environment. Vol. 3: Electrons at synchronous altitudes. NASA Rep. SP-3024.
70. Paulikas, G. A., Blake, J. B., and Palmer, J. A. (1969). Energetic electrons at the synchronous altitude; a compilation of data. *Aerospace Corporation Rep.*, TR-0066 (5260-20)-4, November.
71. Williams, D. J., and Palmer, W. F. (1965). Distortions in the radiation cavity as measured by an 1100-kilometer polar orbiting satellite, *J. Geophys. Res.* **70**, 557–567
72. McIlwain, C. E. (1966). Processes acting upon outer zone electrons. *Inter-Union Symp. Solar-Terrestrial Phys., Belgrade, Yugoslavia*, Univ. of California, San Diego, Rep. SP-66-5.
73. Arens, J. F., and Williams, D. J. (1967). Examination of storm time outer zone electron intensity changes at 1100 kilometers. *Proc. Conjugate Point Symp. Boulder Colorado* **IV**, 22/1-22/4.
74. Williams, D. J. (1967). On the low altitude trapped electron boundary collapse during magnetic storms. *J. Geophys. Res.* **72**, 1644–1646.
75. Krimigis, S. M., and Van Allen, J. A. (1967). Geomagnetically trapped alpha particles. *J. Geophys. Res.* **72**, 5779–5797.
76. Heckman, H. A., and Armstrong, A. H. (1962). Energy spectrum of geomagnetically trapped protons. *J. Geophys. Res.* **67**, 1255–1262.
77. Naugle, J. E., and Kniffen, D. A. (1963). Variations of the proton energy spectrum with position in the inner Van Allen belt. *J. Geophys. Res.* **68**, 4065–4078.

78. Fenton, K. B. (1967). A search for α particles trapped in the geomagnetic field. *J. Geophys. Res.* **72**, 3889–3894.
79. Blake, J. B., and Paulikas, G. A. (1970). Measurements of trapped alpha particles, $2 \leq L \leq 4.5$. In "Particles and Fields in the Magnetosphere" (B. M. McCormac, ed.), pp. 380–384. Reidel, Dordrecht, Holland.
80. Fritz, T. A., and Krimigis, S. M. (1969). Initial observations of geomagnetically trapped protons and alpha particles with OGO 4. *J. Geophys. Res.* **74**, 5132–5138.
81. Krimigis, S. M. (1970). Alpha particles trapped in the earth's magnetic field. In "Particles and Fields in the Magnetosphere" (B. M. McCormac, ed.), pp. 364–379. Reidel, Dordrecht, Holland.
82. Vette, J. I. (1968). New proton environment Ap. 6, $4 < E_p < 30$ Mev. National Space Science Data Center, Letter to Distribution, October 4.
83. Vette, J. I. (1969). New proton environment Ap. 7, $E_p > 50$ Mev. National Space Science Data Center, Letter to Distribution, December 1.
84. Krimigis, S. M., Verzariu, P., Van Allen, J. A., Armstrong, T. P., Fritz, T. A., and Randall, B. A. (1970). Trapped energetic nuclei $Z \geq 3$ in the earth's outer radiation zone. *J. Geophys. Res.* **75** 4210–4215.
85. Van Allen, J. A., Randall, B. A., and Krimigis, S. M. (1970). Energetic carbon, nitrogen, and oxygen nuclei in the earth's outer radiation zone. *J. Geophys. Res.* **75**, 6085–6091.
86. Axford, W. I. (1970). On the origin of radiation belt and auroral primary ions. In "Particles and Fields in the Magnetosphere" (B. M. McCormac, ed.), pp. 46–59. Reidel, Dordrecht, Holland.
87. Singer, S. F. (1958). Radiation belt and trapped cosmic ray albedo. *Phys. Rev. Lett.* **1**, 171–173.
88. Singer, S. F. (1958). Trapped albedo theory of the radiation belt. *Phys. Rev. Lett.* **1**, 181–183.
89. Vernov, S. N., Grigorov, N. L., Ivanenko, I. P., Lebedinskii, A. I., Murzin, V. S., and Chudakov, A. E. (1959). Possible mechanism of production terrestrial corpuscular radiation under the action of cosmic rays. *Sov. Phys.—Dokl.* **4**, 154–157.
90. Freden, S. C. (1966). Energy spectrum of inner zone protons. In "Radiation Trapped in the Earth's Magnetic Field" (B. M. McCormac, ed.), pp. 116–128. Reidel, Dordrecht, Holland.
91. Freden, S. C. (1969). Inner-belt Van Allen radiation. *Space Sci. Rev.* **9**, 198–242.
92. Lenchek, A. M. (1966). Origin and loss of inner zone protons. In "Radiation Trapped in the Earth's Magnetic Field" (B. M. McCormac, ed.), pp. 287–301. Reidel, Dordrecht, Holland.
93. Hess, W. N., Canfield, E. H., and Lingenfelter, R. E. (1961). Cosmic ray neutron demography. *J. Geophys, Res.* **66**, 665–677.
94. Newkirk, L. L. (1963). Calculation of low-energy neutron flux in the atmosphere by the S_n method. *J. Geophys. Res.* **68**, 1825–1833.
95. Lingenfelter, R. E. (1963). The cosmic-ray neutron leakage flux. *J. Geophys. Res,* **68**, 5633–5640.
96. Williams, D. J., and Bostrom, C. O. (1963). *Proc. Conf. Earth's Albedo Neutron Flux*, held at *Johns Hopkins Univ. Appl. Phys. Lab.*, APL Rep. TG-543, 221 pp.
97. Williams, D. J., and Bostrom, C. O. (1964). Albedo neutrons in space. *J. Geophys. Res.* **69**, 377–391.
98. Haymes, R. C. (1965). Terrestrial and solar neutrons. *Rev. Geophys.* **3**, 345–364.
99. Jenkins, R. W., Lockwood, J. A., Ifedili, S. O., and Cupp, E. L. (1970). Latitude and altitude dependence of the cosmic ray albedo neutron flux. *J. Geophys. Res.* **75**, 4197–4204.

100. Jenkins, R. W., Lockwood, J. A., Razdan, H., and Ifedili, S. O. (1970). Energy dependence of the cosmic ray neutron leakage flux. Preprint, Univ. of New Hampshire.
101. Filz, R. C., and White, R. S. (1969). High energy protons in the earth's radiation belt. Preprint Univ. of Calif., Riverside, October 1969. Submitted to *Rev. Geophys.*
102. Heckman, H. H., Lindstrom, P. J., and Nakano, G. H. (1969). Long-term behavior of energetic inner-belt protons. Preprint, Univ. of Calif., Berkeley, UCRL-19309, October, 1969.
103. Macy, W. W., White, R. S., Filz, R. C., and Holeman, E. (1970). Time variations of radiation belt protons. *J. Geophys. Res.* **75**, 4322–4328.
104. Haerendel, G. (1964). Protonen in inneren Strahlungsgürtel. *Fortschr. Phys.* **12**, 271–346.
105. Dragt, A. J. (1965). Trapped orbits in a magnetic dipole field. *Rev. Geophys.* **3**, 255–298.
106. McIlwain, C. E., and Pizzella, G. (1963). On the energy spectrum of protons trapped in the earth's inner Van Allen Zone. *J. Geophys. Res.* **68**, 1811–1823.
107. Fillius, R. W. (1966). Trapped protons of the inner radiation belt. *J. Geophys. Res.* **71**, 97–123.
108. Blanchard, R. C., and Hess, W. N. (1964). Solar cycle changes in inner zone protons. *J. Geophys Res.* **69**, 3927–3928.
109. Dragt, A. J. (1969). Solar cycle modulation of the radiation belt proton flux. *Univ. of Md. Tech. Rep.* **938** (Feb.).
110. Filz, R. C., and Holeman, E. (1965). Time and altitude dependence of 55-Mev trapped protons, August 1961 to June 1964. *J. Geophys. Res.* **70**, 5807–5822.
111. Fillius, R. W., and McIlwain, C. E. (1964). Anomalous energy spectrum of protons in the earth's radiation belt. *Phys. Rev. Lett.* **12**, 609–612.
112. Bostrom, C. O., Zmuda, A. J., and Pieper, G. F. (1965). Trapped protons in the South Atlantic magnetic anomaly, July through December 1961, 2; Comparisons with Nerv and Relay 1 and discussion of the energy spectrum. *J. Geophys. Res.* **70**, 2035–2043.
113. Freden, S. C., Blake, J. B., and Paulikas, G. A. (1965). Spatial variation of the inner zone trapped proton spectrum. *J. Geophys. Res.* **70**, 3113–3116.
114. Gabbe, J. D., and Brown, W. L. (1966). Some observations of the distribution of energetic protons in the earth's radiation belts between 1962 and 1964. *In* "Radiation Trapped in the Earth's Magnetic Field" (B. M. McCormac, ed.), pp. 165–184. Reidel, Dordrecht, Holland.
115. Naugle, J. E., and Kniffen, D. A. (1961). Flux and energy spectra of the protons in the inner Van Allen belt. *Phys. Rev. Lett.* **7**, 3–6.
116. Armstrong, A. H., Harrison, F. B., Heckman, H. H., and Rosen, L. (1961). Charged particles in the inner Van Allen radiation belt. *J. Geophys. Res.* **66**, 351–357.
117. Lenchek, A. M. (1962). On the anomalous component of low-energy geomagnetically trapped protons. *J. Geophys. Res.* **67**, 2145–2158.
118. Lenchek, A. M., and Singer, S. F. (1963). The albedo neutron theory of geomagnetically trapped protons. *Planet. Space Sci.* **11**, 1151–1208.
119. Dragt, A. J., Austin, M. M., and White, R. S. (1966). Cosmic ray and solar proton albedo neutron decay injection. *J. Geophys. Res.* **71**, 1293–1304.
120. Lingenfelter, R. E., Flamm, E. J., Canfield, E. H., and Kellman, S. (1965). High energy solar neutrons. 1. Production in flares. *J. Geophys. Res.* **70**, 4077–4086.
121. Lingenfelter, R. E., Flamm, E. J., Canfield, E. H., and Kellman, S. (1965). High energy solar neutrons. 2. Flux at the earth. *J. Geophys. Res.* **70**, 4087–4095.

122. Claflin, E. S., and White, R. S. (1970). Injection of protons into the radiation belt by solar neutron decay. *J. Geophys. Res.* **75**, 1257–1262.
123. McIlwain, C. E. (1963). The radiation belts, natural and artificial. *Science* **142**, 355–361.
124. Farley, T. A., Tomassian, A. D., and Walt, M. (1970). Source of high energy protons in the Van Allen radiation belt. *Phys. Rev. Lett.* **25**, 47–49.
125. Nakada, M. P., and Mead, G. D. (1965). Diffusion of protons in the outer radiation belt. *J. Geophys. Res.* **70**, 4777–4791.
126. Bostrom, C. O., Beall, D. S., and Armstrong, J. C. (1970). Time history of the inner radiation zone, October 1963 to December 1968. *J. Geophys. Res.* **75**, 1246–1256.
127. McIlwain, C. E. (1965). Redistribution of trapped protons during a magnetic storm. *Space Res.* **5**, 374–391.
128. O'Brien, B. J. (1962). Review of studies of trapped radiation with satellite-borne apparatus. *Space Sci. Rev.* **1**, 415–484.
129. Kellogg, P. J. (1959a). Possible explanation of the radiation observed by Van Allen at high altitudes in satellites. *Nuovo Cimento* **11**, 48–66.
130. Kellogg, P. J. (1960). Electrons of the Van Allen radiation. *J. Geophys. Res.* **65**, 2705–2713.
131. Lencheck, A. M., Singer, S. F., and Wentworth, R. C. (1961). Geomagnetically trapped electrons from cosmic ray albedo neutrons. *J. Geophys. Res.* **66**, 4027–4046.
132. Hess, W. N., ed. (1963). Collected papers on the artificial radiation belt from the July 9, 1962 nuclear detonation. *J. Geophys. Res.* **68**, 605–758.
133. McIlwain, C. E. (1966). Measurements of trapped electron intensities made by the Explorer 15 satellite. *In* "Radiation Trapped in the Earth's Magnetic Field" (B. M. McCormac, ed.), pp. 593–609. Reidel, Dordrecht, Holland.
134. Bostrom, C. O., and Williams, D. J. (1965). Time decay of the artificial radiation belt. *J. Geophys. Res.* **70**, 240–242.
135. Paulikas, G. A., Blake, J. B., and Freden, S. C. (1967). Inner zone electrons in 1964 and 1965. *J. Geophys. Res.* **72**, 2011–2020.
136. Walt, M. (1966). Loss rates of trapped electrons by atmospheric collisions. *In* "Radiation Trapped in the Earth's Magnetic Field," (B. M. McCormac, ed.), pp. 337–351. Reidel, Dordrecht, Holland.
137. Beall, D. S., Bostrom, C. O., and Williams, D. J. (1967). Structure and decay of the Starfish radiation belt October 1963 to December 1965. *J. Geophys. Res.* **72**, 3403–3424.
138. Walt, M. (1964). The effects of atmospheric collisions on geomagnetically trapped electrons. *J. Geophys. Res.* **69**, 3947–3958.
139. Imhof, W. L., and Smith, R. V. (1965). Longitudinal variations of high energy electrons at low altitudes. *J. Geophys. Res.* **70**, 569–577.
140. Imhof, W. L., and Smith, R. V. (1966). Low altitude measurements of trapped electrons. *In* "Radiation Trapped in the Earth's Magnetic Field" (B. M. McCormac, ed.), pp. 100–111. Reidel, Dordrecht, Holland.
141. Cladis, J. B. (1966). Resonance acceleration of particles in the inner radiation belt. *In* "Radiation Trapped in the Earth's Magnetic Field" (B. M. McCormac, ed.), pp. 112–115. Reidel, Dordrecht, Holland.
142. Pfitzer, K. A., and Winckler, J. R. (1968). Experimental observation of a large addition to the electron inner radiation belt after a solar flare event. *J. Geophys. Res.* **73**, 5792–5797.
143. Williams, D. J., Arens, J. F., and Lanzerotti, L. J. (1968). Observations of trapped electrons at low and high altitudes. *J. Geophys. Res.* **73**, 5673–5696.

144. Imhof, W. L., Reagan, J. B., and Smith, R. V. (1967). Long-term study of electrons trapped in low L shells. *J. Geophys. Res.* **72**, 2371–2377.
145. Brown, W. L. (1966). Observations of the transient behavior of electrons in the artificial radiation belts. *In* "Radiation Trapped in the Earth's Magnetic Field" (B. M. McCormac, ed.), pp. 610–633. Reidel, Dordrecht, Holland.
145a. Newkirk, L. L., and Walt, M. (1968). Radial diffusion coefficient for electrons at low L values. *J. Geophys. Res.* **73**, 1013–1018.
146. Kellogg, P. J. (1959b). Van Allen radiation of solar origin. *Nature* **183**, 1295–1297.
147. Parker, E. N. (1960). Geomagnetic fluctuations and the form of the outer zone of the Van Allen radiation belt. *J. Geophys. Res.* **65**, 3117–3130.
148. Herlofson, N. (1960). Diffusion of particles in the earth's radiation belt. *Phys. Rev. Lett.* **5**, 414–416.
149. Davis, L., Jr., and Chang, D. B. (1962). On the effect of geomagnetic fluctuations on trapped particles. *J. Geophys. Res.* **67**, 2169–2179.
150. Tverskoy, B. A. (1964). Dynamics of the radiation belts of the earth. 2. *Geomagn. Aeron.* **4**, 351–366.
151. Tverskoy, B. A. (1965). Transport and acceleration of charged particles in the earth's magnetosphere. *Geomagn. Aeron.* **5**, 617–628.
152. Dungey, J. W., Hess, W. N., and Nakada, M. P. (1965). Theoretical studies of protons in the outer radiation belt. *Space Res.* **5**, 399–403. (Also in *J. Geophys Res.* **70**, 3529–3532.)
153. Hess, W. N. (1966). Source of outer zone protons. *In* "Radiation Trapped in the Earth's Magnetic Field" (B. M. McCormac, ed.), pp. 352–368. Reidel, Dordrecht, Holland.
154. Fälthammar, C. G. (1965). Effects of time dependent electric fields on geomagnetically trapped radiation. *J. Geophys. Res.* **70**, 2503–2516.
155. Fälthammar, C. G. (1966). On the transport of trapped particles in the outer magnetosphere. *J. Geophys. Res.* **71**, 1487–1491.
156. Fälthammar, C. G. (1968). Radial diffusion by violation of the third adiabatic invariant. *In* "Earth's Particles and Fields" (B. M. McCormac, ed.), pp. 157–170. Van Nostrand-Reinhold, Princeton, New Jersey.
157. Haerendel, G. (1968). Diffusion theory of trapped particles and the observed proton distribution. *In* "Earth's Particles and Fields," (B. M. McCormac, ed.), pp. 171–192. Van Nostrand-Reinhold, Princeton, New Jersey.
158. Williams, D. J. (1970). Trapped protons ≥ 100 kev and possible sources. *In* "Particles and Fields in the Magnetosphere" (B. M. McCormac, ed.), pp. 396–409. Reidel, Dordrecht, Holland.
159. Frank, L. A., and Owens, H. D. (1970). Omnidirectional intensity contours of low-energy protons $(0.5 \leq E \leq 50$ kev) in the earth's outer radiation zone at the magnetic equator. *J. Geophys. Res.* **75**, 1269–1278.
160. Cahill, L. J. (1966). Inflation of the magnetosphere near 8 earth radii in the dark hemisphere. *Space Res.* **6**, 662–678.
161. Hoffman, R. A., and Bracken, P. A. (1965). Magnetic effects of the quiet-time proton belt. *J. Geophys. Res.* **70**, 3541–3556.
162. Frank, L. A. (1967). On the extraterrestrial ring current during geomagnetic storms. *J. Geophys. Res.* **72**, 3753–3767.
163. Frank, L. A. (1970). Direct detection of asymmetric increases of extraterrestrial ring current proton intensities in the outer radiation zone. *J. Geophys. Res.* **75**, 1263–1268.
164. Söraas, F. (1969). Comparison of post-storm non-adiabatic recovery of trapped protons with radial diffusion. NASA TM X-63625.

165. Haerendel, G. (1970). On the balance between radial and pitch angle diffusion. *In* "Particles and Fields in the Magnetosphere" (B. M. McCormac, ed.), pp. 416–428. Reidel, Dordrecht, Holland.
166. Frank, L. A. Van Allen, J. A., and Craven, J. D. (1964). Large diurnal variations of geomagnetically trapped and of precipitated electrons observed at low altitudes. *J. Geophys. Res.* **69**, 3155–3167.
167. McDiarmid, I. B., and Burrows, J. R. (1964). Diurnal intensity variations in the outer radiation zone at 1000 km. *Can. J. Phys.* **42**, 1135–1148.
168. McDiarmid, I. B., and Burrows, J. R. (1964). High-latitude boundary of the outer radiation zone at 1000 km. *Can. J. Phys.* **42**, 616–626.
169. Taylor, H. E. (1966). Adiabatic motion of outer zone particles in a model of the geoelectric and geomagnetic fields. *J. Geophys. Res.* **71**, 5135–5147.
170. Evans, D. S. (1968). The observations of a near nonenergetic flux of auroral electrons. *J. Geophys. Res.* **73**, 2315–2323.
171. Axford, W. I. (1969). Magnetospheric convection. *In* "Magnetospheric Physics" (D. J. Williams and G. D. Mead, eds.), pp. 421–459. Am. Geophys. Union, Washington, D.C., *Rev. Geophys*, **7**, Nos. 1, 2.
172. O'Brien, B. J. (1964). High-latitude geophysical studies with satellite Injun 3. 3. Precipitation of electrons into the atmosphere. *J. Geophys. Res.* **69**, 13–43.
173. Fritz, T. A. (1968). High-latitude outer-zone boundary region for ≥ 40 kev electrons during geomagnetically quiet periods. *J. Geophys. Res.* **73**, 7245–7255.
174. Fritz, T. A. (1970). Study of high latitude outer-zone boundary region for ≥ 40 kev electrons with Injun 3, *J. Geophys. Res.* **75**, 5387–5400.
175. McDiarmid, I. B., and Burrows, J. R. (1968). Local time asymmetries in the high latitude boundary of the outer radiation zone for the different electron energies. *Can. J. Phys.* **46**, 49–57.
176. Kennel, C. F. (1969). Consequences of a magnetospheric plasma. *In* "Magnetospheric Physics" (D. J. Williams and G. D. Mead, eds.), pp. 379–419. Am. Geophys Union, Washington, D.C., *Rev. Geophys*. **7**, Nos. 1, 2.
177. Fritz, T. A. (1970). On the pitch angle distribution of ≥ 40 kev outer zone electrons. (Abstr.) *Trans. Amer. Geophys. Union* **51**, No. 4, 387. Paper in preparation.
178. McDiarmid, I. B., and Burrows, J. R. (1965). On an electron source for the outer Van Allen radiation zone. *Can. J. Phys.* **43**, 1161–1164.
179. McDiarmid, I. B., and Burrows, J. R. (1966). Lifetimes of low-energy electrons. ($E > 40$ kev) in the outer radiation zone at magnetically quiet times. *Can. J. Phys.* **44**, 669–673.
180. Frank, L. A., Van Allen, J. A., and Hills, H. K. (1964). A study of charged particles in the earth's outer radiation zone with Explorer 14. *J. Geophys. Res.* **69**, 2171–2191.
181. McDiarmid, I. B., and Burrows, J. R. (1966), Temporal variations of outer radiation zone electron intensities at 1000 km. *Can. J. Phys.* **44**, 1361–1379.
182. Imhof, W. L. (1968). Electron precipitation in the radiation belts. *J. Geophys. Res.* **73**, 4167–4184.
183. Williams, D. J., and Kohl, J. W. (1965). Loss and replenishment of electrons at middle latitudes and high B values. *J. Geophys. Res.* **70**, 4139–4150.
184. Frank, L. A. (1965). Inward radial diffusion of electrons of greater than 1.6 Mev in the outer radiation zone. *J. Geophys. Res.* **70**, 3533–3540.
185. Newkirk, L. L., and Walt, M. (1968). Radial diffusion coefficient for electrons at $1.76 < L.5$. *J. Geophys. Res.* **73**, 7231-7236.
186. Arnoldy, R. L., and Chan, K. W. (1969). Particle substorms observed at the geostationary orbit. *J. Geophys. Res.* **74**, 5019–5028

187. Pfitzer, K. A., and Winckler, J. R. (1969). Intensity correlations and substorm electron drift effects in the outer radiation belt measured with the OGO-3 and ATS-1 satellites. *J. Geophys. Res.* **74**, 5005–5018.
188. Winckler, J. R. (1970). The origin and distribution of energetic electrons in the Van Allen radiation belts. *In* "Particles and Fields in the Magnetosphere" (B. M. McCormac, ed.), pp. 332–352. Reidel, Dordrecht, Holland.
189. Axford, W. I. (1969). A review of fields and particle observations made in the vicinity of the synchronous orbit, preprint, University of California, San Diego, October, 1969. Review presented at Colloquium on ESRO Geostationary Magnetospheric Satellite, Denmark, October 1969, ESRO Spec. Rep. (to be published).
190. Hartz, T. R., and Brice, N. M .1967). The general pattern of auroral particle precipitation. *Planet. Space Sci.* **15**, 301–329.
191. Brice, N. M. (1967). Bulk motion of the magnetosphere. *J. Geophys. Res.* **72**, 5193–5211.
192. Roberts, C. S. (1968). Cyclotron-resonance and bounce-resonance scattering of electrons trapped in the earth's magnetic field. *In* "Earth's Particles and Fields" (B. M. McCormac, ed.), pp. 317–336. Van Nostrand-Reinhold, Princeton, New Jersey.
193. Roberts, C. S. (1969). Pitch-angle diffusion of electrons in the magnetosphere. *In* "Magnetospheric Physics" (D. J. Williams and G. D. Mead, eds.), pp. 305–337. Am. Geophys. Union, Washington, D.C. [also in *Rev. Geophys.* **7**, Nos. 1, 2.]
194. Gurnett, D. A. (1968). Satellite observations of VLF emissions and their association with energetic charged particles. *In* "Earth's Particles and Fields" (B. M. McCormac, ed.), pp. 337–349. Van Nostrand-Reinhold, Princeton, New Jersey.
195. Taylor, W. W. L., and Gurnett, D. A. (1968). Morphology of VLF emissions observed with the Injun 3 satellite. *J. Geophys. Res.* **73**, 5616–5626.
196. Kennel, C. F., and Petschek, H. E. (1966). Limit on stably trapped particle fluxes. *J. Geophys. Res.* **71**, 1–28.
197. Katz. L. (1966). Electron and proton observations. *In* "Radiation Trapped in the Earth's Magnetic Field" (B. M. McCormac, ed.), pp. 129–154. Reidel, Dordrecht, Holland.
198. Davis, L. R., and Williamson, J. M. (1966). Outer zone protons. *In* "Radiation Trapped in the Earth's Magnetic Field" (B. M. McCormac, ed.), pp. 215–229. Reidel, Dordrecht, Holland.
199. Konradi, A. (1967). Proton events in the magnetosphere associated with magnetic rays. *J. Geophys. Res.* **72**, 3829–3941.
200. Brown, W. L., Cahill, L. J., Davis, L. R., McIlwain, C. E., and Roberts, C. S. (1968). Acceleration of trapped particles during a magnetic storm on April 18, 1965. *J. Geophys. Res.* **73**, 153–161.
201. McDiarmid, I. B., Burrows, J. R., and Wilson, M. D. (1969). Morphology of outer radiation zone electron ($E > 35$ kev) acceleration mechanisms. *J. Geophys. Res.* **74**, 1749–1758.
202. Hess. W. N., Davis, T. N., Trichel, M. C., Mead, G. D., Mead, J. M., Beggs, W. C., Maier, E. J. R., Parker, L. W., Murphy, B. L. (1970). A series of six papers presented at the April 1970 AGU meeting in Washington D.C. describe this rocket-borne beam injection experiment. The papers are SPM 61, 62, 63, 64, 65, and 66 with abstracts in EÓS. *Trans. Amer. Geophys. Union* **51**, 394–395.
203. Hendrickson, R. A., McEntire, R. W., and Winckler, J. R. (1970). The electron echo experiment: General summary and preliminary analysis of particle measurements. Preprint #CR-153, University of Minnesota Cosmic Ray Group. (December).

204. Haerendel, G., Lust, R., and Rieger, E. (1967). Motion of artificial ion clouds in the upper atmosphere. *Planet. Space Sci.* **15**, 1–18.
205. Foppl, H., Haerendel, G., Haser, L., Lust, R., Melzner, F., Meyer, B., Neuss, H., Rabben, H., Rieger, E., Stocker, J., and Stoffregen, W. (1968). Preliminary results of electric field measurements in the auroral zone. *J. Geophys. Res.* **73**, 21–26.
206. Wescott, E. M., Stolarik, J. D., and Heppner, J. P. (1969). Electric fields in the vicinity of auroral forms from motions of barium vapor releases. *J. Geophys. Res.* **74**, 3469–3487.
207. Haerendel, G., and Lust, R. (1970). Electric fields in the ionosphere and magnetosphere. *In* "Particles and Fields in the Magnetosphere" (B. M. McCormac, ed.), pp. 213–228. Reidel, Dordrecht, Holland.
208. Wescott, E. M., Stolarik, J. D., and Heppner, J. P. (1970). Auroral and polar cap electric fields from barium releases. *In* "Particles and Fields in the Magnetosphere" (B. M. McCormac, ed.), pp. 229–246. Reidel, Dordrecht, Holland.
209. Brice, N. (1970). Artificial enhancement of energetic particle precipitation through cold plasma injection—A technique for seeding substorms? *J. Geophys. Res.* **75**, 4890–4892.
210. McIlwain, C. E. (1970). Equatorial observations of auroral plasma. International Symposium on Solar-Terrestrial Physics, Leningrad, USSR, May.

PHOTOCHEMISTRY OF ATMOSPHERIC OZONE

H. U. Dütsch

Federal Institute of Technology, Laboratory of Atmospheric Physics, Zurich, Switzerland

	Page
1. Observed Distribution of Total Ozone	219
1.1. Total Amount	219
1.2. Vertical Distribution	222
2. The Classical Photochemical Theory: "Pure Oxygen Atmosphere"	244
2.1. Introduction	244
2.2. Equilibrium Theory	245
2.3. Nonequilibrium Theory	253
2.4. Influence of Motion on Photochemistry	256
2.5. Mesospheric Ozone	258
2.6. Uncertainties in the Theory	264
3. Ozone Photochemistry in a "Moist" Atmosphere	271
3.1. Introduction	271
3.2. Semiquantitative Theory of Stratospheric Ozone	274
3.3. Nighttime Processes in a "Moist" Stratosphere	288
3.4. Mesospheric Photochemistry in the Presence of Water Vapor	289
3.5. Influence of Vertical Mixing on Mesospheric Photochemistry	287
4. Possible Importance of Nitrogen Oxides to Ozone Photochemistry	303
Nighttime Destruction of Ozone through the $\widetilde{NO}$-Mechanism	305
5. Ozone as a Tracer	306
5.1. Importance of the Photochemical Theory in General Circulation Studies Using Ozone as a Tracer	306
5.2. Ozone as a Tracer in General Circulation Models	310
6. Interrelation between Ozone Photochemistry and Stratospheric Dynamics	312
List of Symbols	315
References	315

1. Observed Distribution of Total Ozone

1.1. Total Amount

The total ozone content of the atmosphere over a certain point can be relatively easily measured because of the strong absorption of the gas in the near UV: A special spectrophotometer measuring the intensity ratio of two wavelengths in the so-called Huggins bands (above 3000 Å) was developed by Dobson [1] for this purpose. A single reading (in clear air) or the combination of such measurements on two wavelength pairs (also usable under hazy conditions) yields after some reduction for scattering influence the amount of ozone overhead [2,3]. It is given as the thickness of the pure ozone layer which would be obtained if all ozone in the vertical column were concentrated at NTP, and it is normally expressed in units of 10^{-3} cm (matmcm), also called Dobson units (D.U.). According to the type of measurement by which the ozone amount is obtained, the scale depends on the (laboratory determined) absorption coefficients of ozone and has thus been changed several times, a fact which has to be kept in mind if recent data are to be compared with those from older publications. It is felt that this scale which is internationally fixed (at present mainly on the basis of measurements by Vigroux [4, 5]) is now reasonably accurate. The correctness of the scale is not so important so long as only the distribution in space and the variation with time of total ozone is studied. It becomes critical, however, in any comparison of observations with theoretical calculations and also if the observations are compared with the chemically determined vertical distribution of ozone in the atmosphere. The relative accuracy of the measurements, when obtained with a well-kept Dobson spectrophotometer, is of the order of 2 %. If, however, filter type instruments are used (working with relatively broad spectral bands), the uncertainty is quite considerably increased [6,7].

The total ozone content of the atmosphere varies from less than 200 to more than 600 D.U. On the average it depends on latitude and season (see Fig. 1), with a minimum amount in the tropics and a maximum near the polar circle and with a variation from high values in early spring to a minimum in fall. Superimposed are (mainly in the middle and high latitudes) rapid fluctuations from day to day, which may at time of maximum variability (winter and spring) easily exceed the range of the seasonal variation and which are obviously connected with weather [8]. Changes of 100 D.U. within a few hours have been observed.

With the increase of the observational network during the last 15 years also a dependence on longitude has been detected [9–13], which is related to the rapid fluctuations of weather type and which is clearly tied to the semi-permanent long waves observed in the lower stratosphere (high ozone in the upper atmospheric troughs and low values in the ridges) (Fig. 2). Local monthly mean values may thus considerably differ from year to year in connection with the well-known variations in the general circulation [14].

Figure 1 shows an obvious asymmetry in ozone distribution between the two hemispheres [15] (with respect to latitudinal as well as to seasonal variation). This fact is established beyond doubt although some details in Fig. 1 are not well enough documented owing to the sparseness of the network in the Southern Hemisphere. The mean total ozone content of the two hemispheres differs little (it is probably a bit lower in the south),[1] however,

---

[1] Gebhard et al. [17] show less ozone in the Southern Hemisphere than our Fig. 1 which is constructed after Sticksel [15].

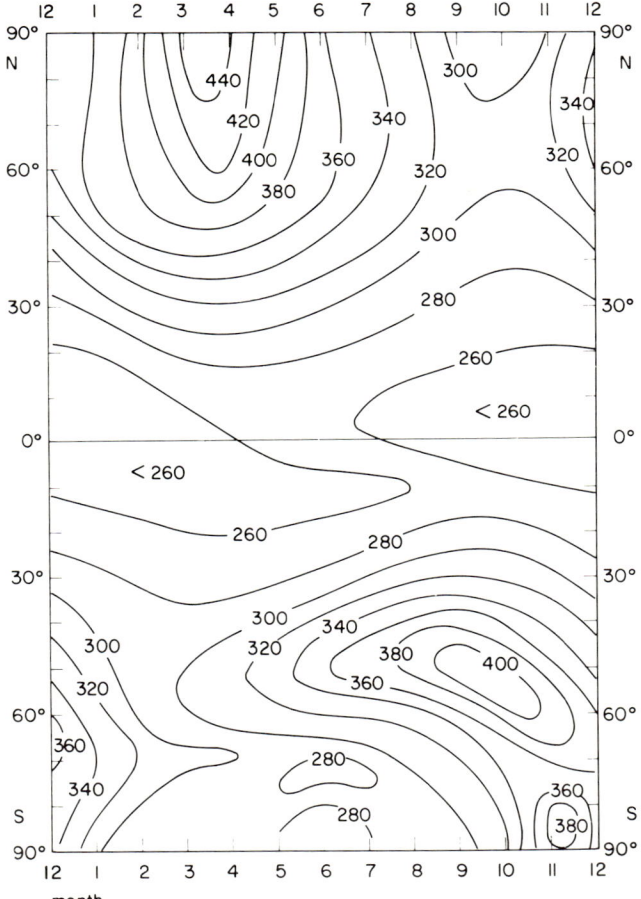

FIG. 1. Total ozone as a function of season and latitude in both hemispheres. (Northern Hemisphere [9]; Southern Hemisphere from [15]).

the seasonal variation in the Southern hemisphere [16] is only two-thirds of its northern counterpart! The belt of maximum ozone is closer to the midlatitudes in the south than in the north and the Antarctic spring maximum occurs comparatively later (by about two months) than in the Arctic. A longitude dependence has also been claimed for the southern hemisphere [17,18] but needs yet to be better documented by an improved observational network.

While climatological mean values are fairly well established at least for the Northern Hemisphere, the present network is far from adequate to

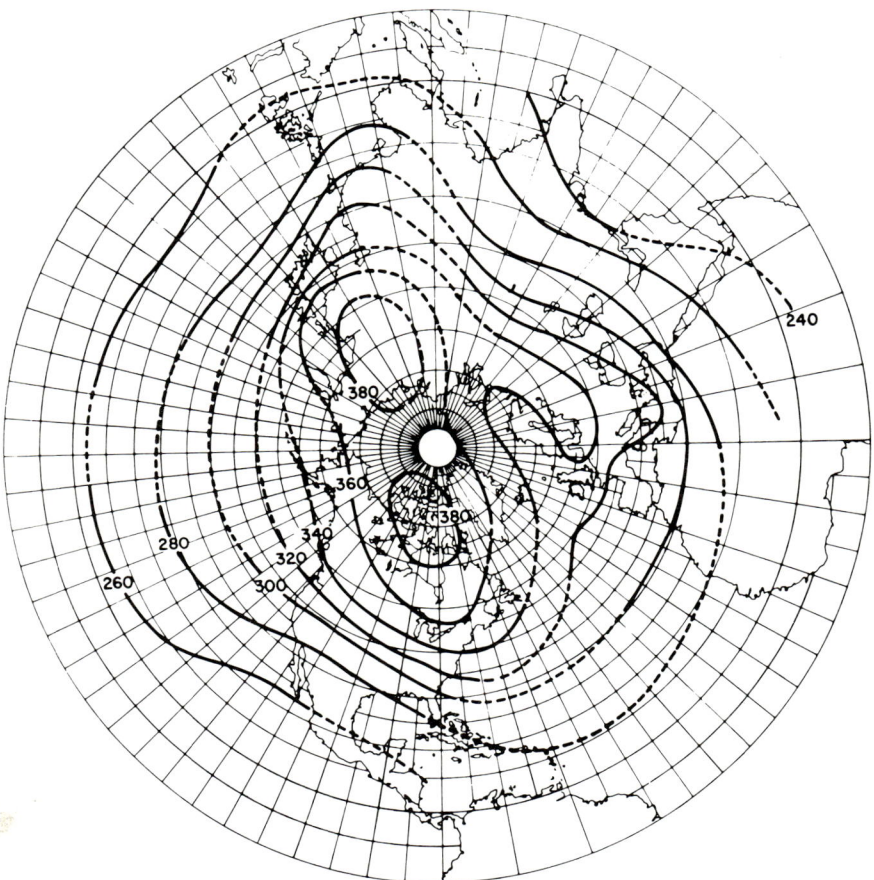

Fig. 2. Longitudinal influence on total ozone distribution in the Northern Hemisphere (after London [9]).

discuss total ozone distribution synoptically with much confidence. It is hoped, that satellite observations (in the uv and ir) will improve this situation in the near future [19–21]; it has, however, yet to be shown beyond doubt whether their accuracy is adequate.[2]

1.2. Vertical Distribution

1.2.1. Methods of Observation. Measurements of the vertical distribution of ozone are considerably more difficult than those of the total amount. The most extensive material available at present is that obtained by the indirect

---

[2] First results look very convincing (Mateer, personal communication).

"Umkehr" method [17,22–25] (measuring the intensity ratio of the two uv wavelengths of the Dobson spectrophotometer in the zenith sky-light during sunset or sunrise). It has the disadvantage of smearing out details (Fig. 3), (as almost any indirect method does). Even the main maximum is underestimated (Fig. 4). Further, there is a considerable bias toward anticyclonic ozone distributions because of the clear sky conditions for the observations. Most of our knowledge about ozone distribution in the upper stratosphere is, however, still based on Umkehr observations, because apart from a few scattered rocket experiments [26–29], direct soundings do not exceed 30–35 km (and their standard error at that level is probably of the same order as that of the indirect Umkehr method).

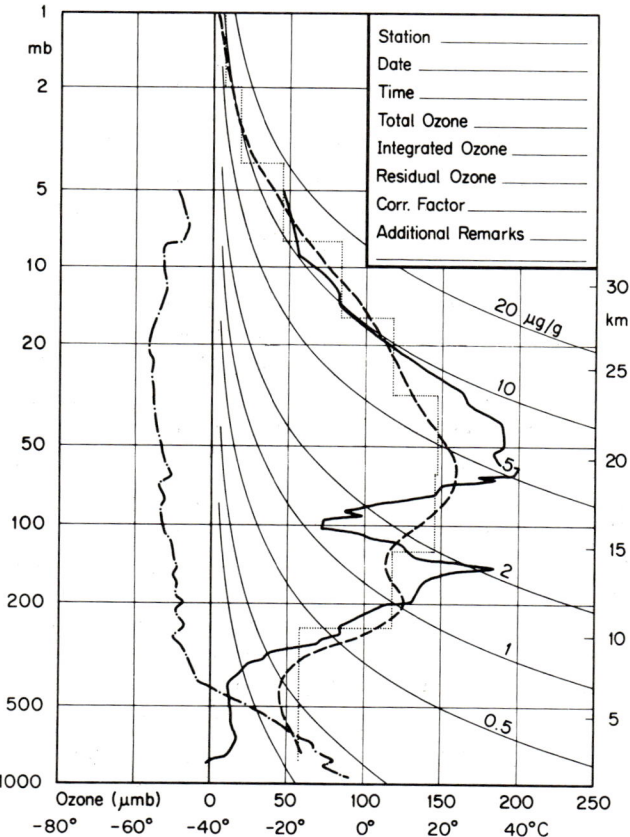

Fig. 3. Comparison between vertical ozone distribution obtained with a chemical ozone sonde (full curve) at Thalwil, Switzerland, and from simultaneous Umkehr observations at Arosa, Switzerland (mean of morning and evening); dotted curve: block distribution, dashed curve: smooth distribution.

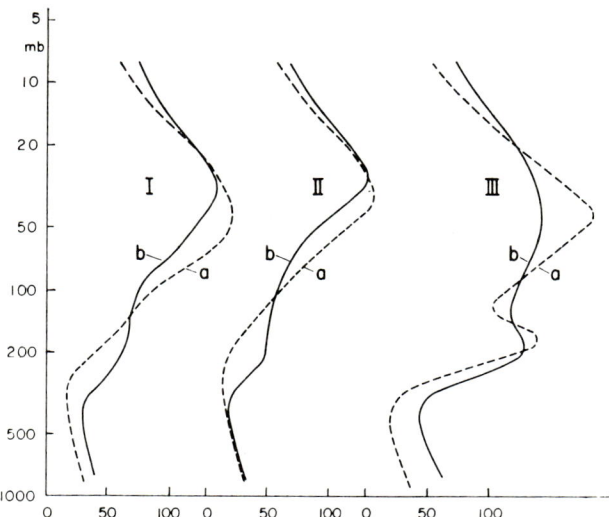

FIG. 4. Comparison between the mean vertical distributions obtained simultaneously from direct soundings (electrochemical sondes) and from the Umkehr method [subdivided in cases with low (II), medium (I), and high (III) total ozone] at 47°N.

The first direct soundings were made with optical instruments (differentiation of the total amount observed overhead) [30–32]; however, during the last decade these instruments were almost completely superseded by chemical sondes. The first large-scale network (over North America during the IQSY) used the dry chemical, chemiluminescent instrument developed by Regener [33,34]; more recently wet chemical sondes using different applications of the KI–O_3 reaction are predominant [35,36,37]. The wet chemical instruments should, in principle, be absolute (two electrons per O_3 molecule are pumped through an electronic circuit); however, experience shows, that the Brewer-type normally measures about 15–20 % low, probably mainly through ozone destruction in the pump. A recent intercomparison proved that the differences between wet chemical instruments are minor if the results are adjusted by a single factor multiplication to the simultaneously observed total amount [38]. The introduction of a certain distortion of the measured ozone distribution by the chemiluminescent type instruments by varying disk sensitivity and because of calibration problems has been claimed by Komhyr [39].

The network for routine ozone sounding is still rather scanty. It is on the basis of the available material reasonably well possible to draw meridional cross sections of the vertical ozone distribution for single months or time cross sections for a number of stations; a full three-dimensional analysis as a function of season (e.g., including longitude dependence), cannot yet been given without using much statistical inference. The meridional cross sections for the Southern Hemisphere presented in the following section are still shaky and may yet readily undergo considerable changes when additional material is obtained.

Synoptic studies of the three-dimensional ozone variation (using the gas as a tracer) are still rather difficult, not only because the network is loose but also because of the rather wide spacing in time of ascents at single stations.

Satellite measurements of the vertical ozone distribution [40] are expected to fill gaps in the near future; however, they will hardly provide much more details than Umkehr measurements and their reliability has yet to be proved.

1.2.2. Vertical ozone distribution at different latitudes. Figures 5a and 5b show mean vertical distributions at different latitudes in spring (March–April) and fall (October–November) (bimonthly means have been chosen because, except for midlatitudes, the number of soundings per single month is still too small to give representative results). The poleward increase of the total amount is clearly produced by the very pronounced positive gradient of ozone concentration with latitude in the lower stratosphere. In the middle stratosphere (above 22–24 km), however, the ozone partial pressure decreases poleward (in agreement with the photochemical theory, see Section 2.2). The described trend is in existence all year round, but while the negative gradient in the middle stratosphere is more pronounced in fall than in spring, the increase of ozone concentration with latitude in the lower stratosphere is much smaller in the second half of the year than in the first. This explains the decreased meridional gradient of total ozone in fall compared to spring. A high percentage of the seasonal variation of total ozone (spring maximum, fall minimum) is obviously determined by the variation of ozone concentration in the lower stratosphere.

Figure 5 suggest that the gradients described above do not continue into the Inner Arctic (represented by the ice floe station T3). Because the presently available data stem from a rather nonhomogeneous set of observations (with respect to sonde type, year of observation, and number of soundings) no speculations on the possible reasons for such an effect shall be made at this time.

In the equatorial zone [41] and in the subtropics, ozone concentrations are low up to the tropical tropopause (situated around 100 mb) from where a smooth increase leads to a maximum in the neighborhood of 26 and 24 km, respectively. In middle and high latitudes, however, the upward increase of ozone concentration starts at the much lower tropopause of these regions and a secondary maximum is indicated in spring, whose level is decreasing with increasing latitude; the secondary minimum at which the further rise to the

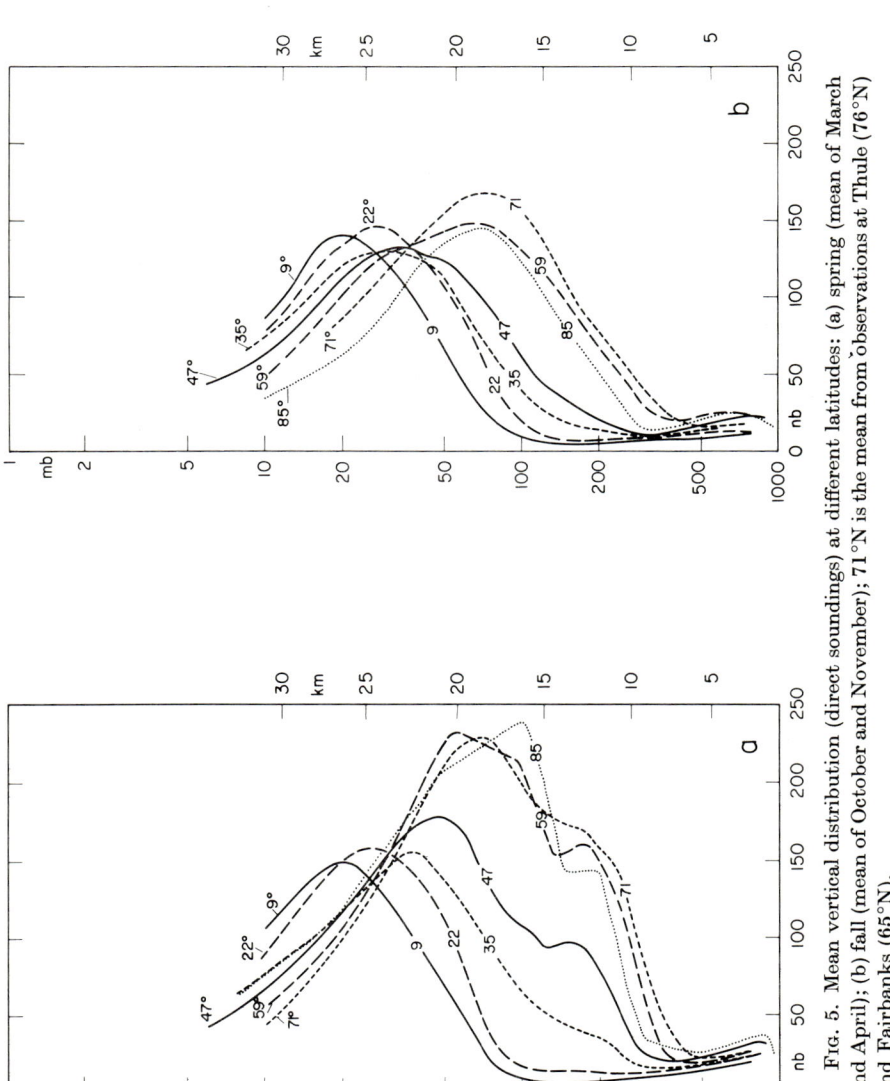

FIG. 5. Mean vertical distribution (direct soundings) at different latitudes: (a) spring (mean of March and April); (b) fall (mean of October and November); 71°N is the mean from observations at Thule (76°N) and Fairbanks (65°N).

main maximum begins seems to be a poleward tilted extension of the tropical tropopause, thus indicating the importance of large-scale horizontal mixing processes in determining the observed ozone distribution. This feature is largely missing in fall after a period of much reduced horizontal transport in summer.

The level of the main maximum is continuously falling from the equator to the pole from 26 to 16 km in spring and from 26 to 19 km in fall.

1.2.3. Seasonal Variations. Reasonably well-established time cross sections can at present only be drawn for a number of mid-latitude stations; even those are still not fully representative, because the time span of the available observations is yet too short for a full elimination of the influences of the rather pronounced year to year variations (see Section 1.2.5).

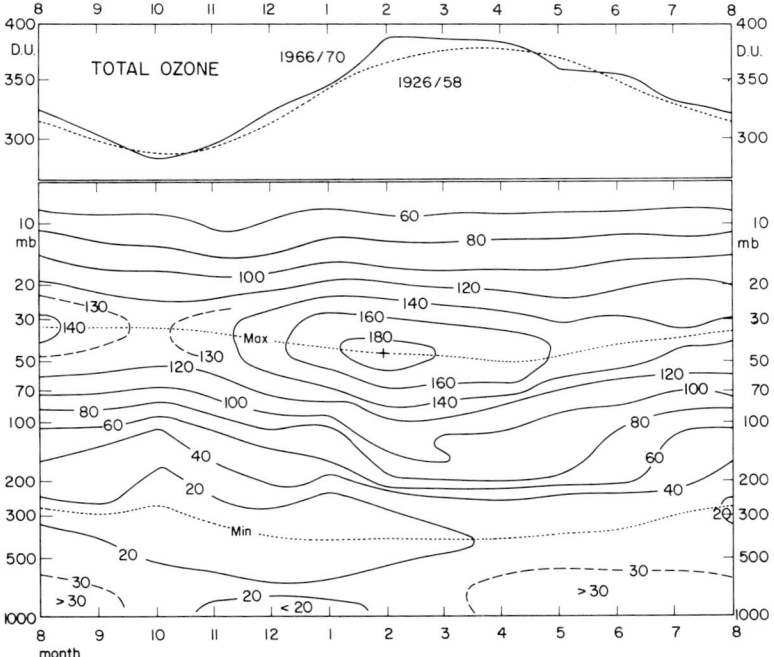

Fig. 6. Time cross section of vertical ozone distribution in mid-latitudes from four years of observations at Thalwil and Payerne, Switzerland, 47°N [42].

The mid-latitude time cross section shown in Fig. 6 is constructed from the sounding material of two Swiss stations [42] only 150 km apart comprising about 500 ascents rather evenly spread over a four year period. It shows that the seasonal increase during winter is not confined to the lower stratosphere, where the strong day-to-day fluctuations connected with weather (i.e., with large-scale mixing processes) occur; a very rapid rise is also observed at the level of the ozone maximum around 40 mb, amounting from November to January to one-third of the initial value. The level of the ozone maximum decreases steadily from October through April by 2 km and rises again through summer. The time of the seasonal maximum is delayed from February at the level of the highest ozone concentrations (or even January around 25 km) to April in the lower stratosphere. This delay is slightly smeared out in the data presented here, because the February mean of the total amount was considerably above normal in the observational period used to construct Fig. 6 (an event produced by excessive ozone content in the lower stratosphere). The tropospheric maximum is further delayed into early summer.

Although a considerable percentage of the ozone produced photochemically in the stratosphere is destroyed each year at the ground [43–45] after its downward transport through the photochemically protected regions (see Section 2.2.4), the minimum concentration is not found at the surface (except in stagnant air), but close to the tropopause, where the strong tropospheric mixing produces almost constant mixing ratio. The variation of the position of that minimum with season (lowest from December through April and highest from July through October) varies parallel to that of the monthly tropopause minima.

The seasonal variation shows in the middle stratosphere (around 10 mb) a double maximum (Fig. 7). The main peak is observed in late summer and an additional more transient maximum occurs in January. While the summer peak is obviously produced photochemically, it is thought that the January maximum is a result of ozone transport. The accumulated surplus ozone may be destroyed photochemically when toward the end of the winter these processes are accelerated with increasing solar height and length of day. Going downward in the atmosphere the January maximum becomes increasingly stronger compared to the summer one; at the 30 mb level a single winter peak is observed and summer brings only an interruption in the decrease of ozone concentration, prevailing through spring and into fall. While the rapid rise in winter is certainly an ozone transport effect (as at lower levels), it is not yet quantitatively established to what extent photochemical processes contribute to the erosion of the excess ozone accumulated during the cold season; the present uncertainties of the photochemical theory, as discussed in Section 3.2, inhibit the final solution of that problem (see also Section 4).

The extensive observational series at Boulder, Colorado [42,46], (7° latitude further south) does, however, not show the secondary maximum at the 10-mb level; it developes only below 20 mb. Whether this discrepancy is an effect of the difference in latitude (stronger photochemical action also in mid-winter), or whether it has to do with the fact that the Swiss series

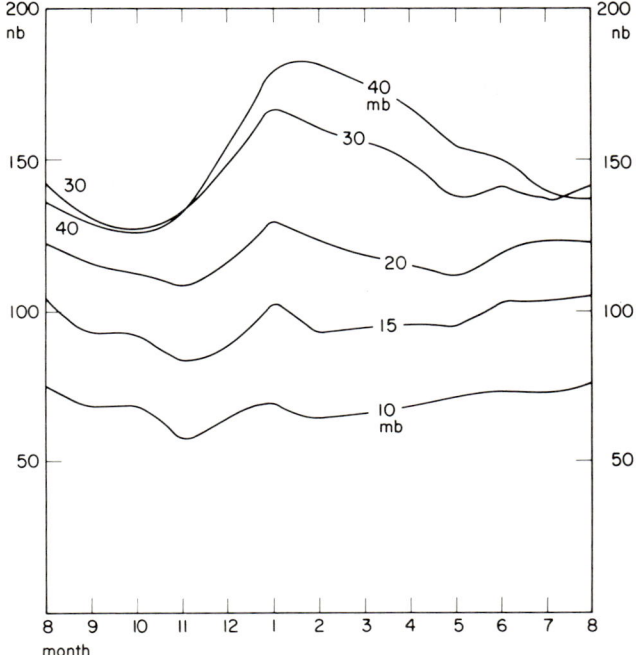

FIG. 7. Seasonal variation of ozone concentration at different levels in the middle stratosphere at 47°N (computed from the same material as Fig. 6).

(1966–1970) comprises several major mid- or early-winter stratospheric warmings, while there was none of importance during the period of observation in Boulder (1963–1966), has yet to be shown by further observations.

The seasonal variation of the vertical ozone distribution in the southern hemispheric mid-latitudes is in general rather similar to that described above [47]; there are, however, certain differences (Fig. 8). The seasonal variation at the level of the ozone peak is, in the Southern Hemisphere only about two-thirds of that in the north and the maximum is reached about one month later in the season. The phase lag between that level and the lower stratosphere seems to be missing in the south, where, however, around 15 km the seasonal variation is at least as large as in the north. Whether the rather big differences indicated for the middle stratosphere (lower values in the south) are real remains to be shown by further observations; it is rather difficult to explain them at that level where at least in summer values close to photochemical equilibrium should be expected; similar differences at that altitude are, however, also shown for high latitudes.

Rather big differences between the two hemispheres are shown by Figs. 9 and 10 for the seasonal variation of the vertical ozone distribution in polar

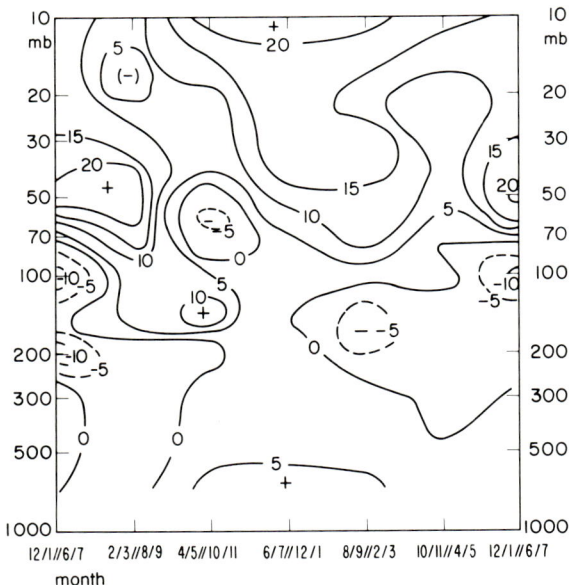

FIG. 8. Time cross section of the difference between mid-latitude vertical ozone distribution in the two hemispheres (North minus South). Bimonthly means for Boulder (40°N)/Albuquerque (35°N) and Aspendale (38°S) are compared for corresponding seasons.

regions [48–50]. The spring maximum is less pronounced in the Antarctic and especially it is delayed by almost two months compared to the north. The phase shift is even larger in the lowest part of the stratosphere, where the Antarctic maximum comes in summer instead of in the spring as in the north. In contradiction to what is found in southern mid-latitudes, the increase at the low levels comes very late in the polar regions.

Both polar cross sections show a rather rapid decrease of ozone concentration around 25 km after the spring maximum. This indicates that the maximum is not of photochemical nature also at that level; however, the decrease may at least be partly photochemical (destruction of surplus ozone imported by the general circulation); the more rapid decay over the Antarctic, where the maximum occurs later, i.e., at higher sun (shorter relaxation time), would support this assumption.

Caution is indicated in placing too much emphasis, at present, on the discussion of smaller-scale features of high-latitude ozone climatology, because the number of observations is still limited and the set rather inhomogeneous. In order to smooth out the noise introduced by these circumstances all observations poleward of 64° latitude have been combined in the mean values used for this discussion.

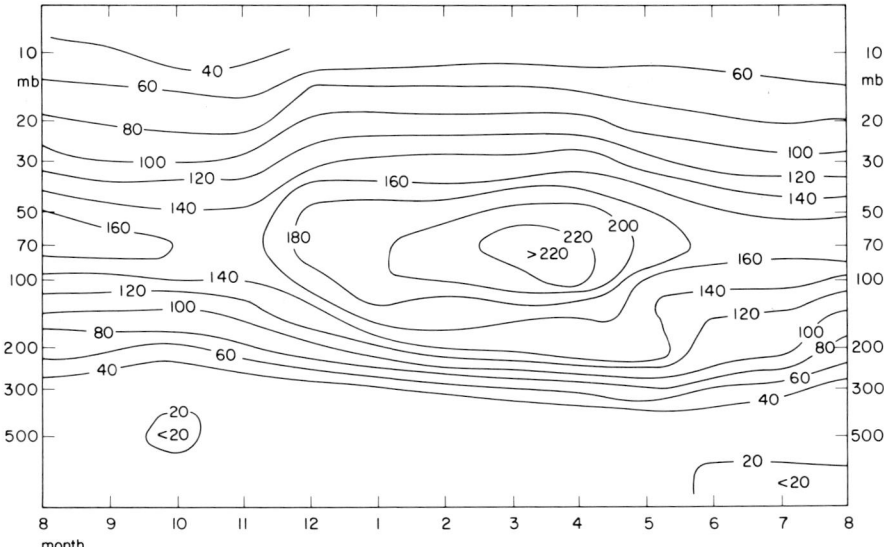

FIG. 9a. Time cross section of vertical ozone distribution in the Arctic [mean from T3 (86°N), Thule (76°N), and Fairbanks (65°N)].

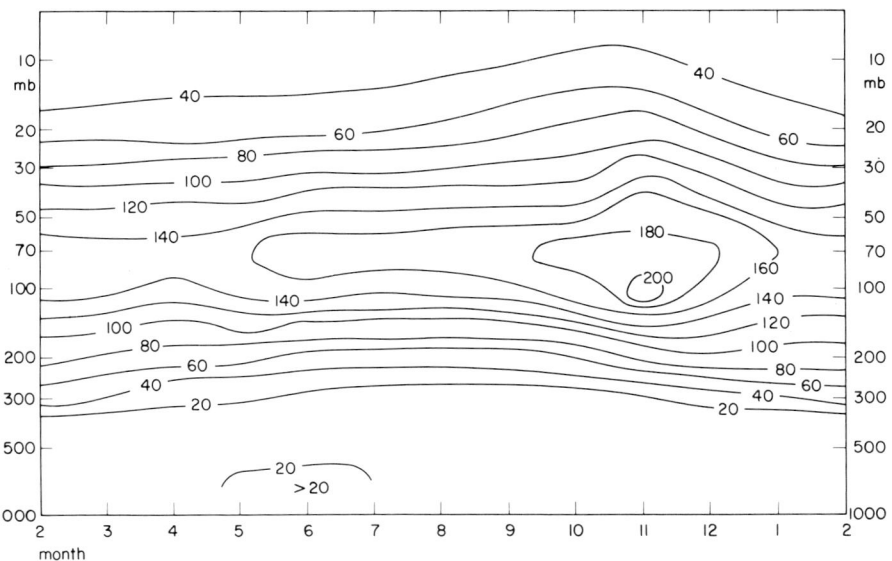

FIG. 9b. Time cross section of vertical ozone distribution in the Antarctic [mean from Amundsen-Scott (90°S), Byrd (80°S), Hallet (72°S), Wilkes (66°S), and Siowa (69°S)].

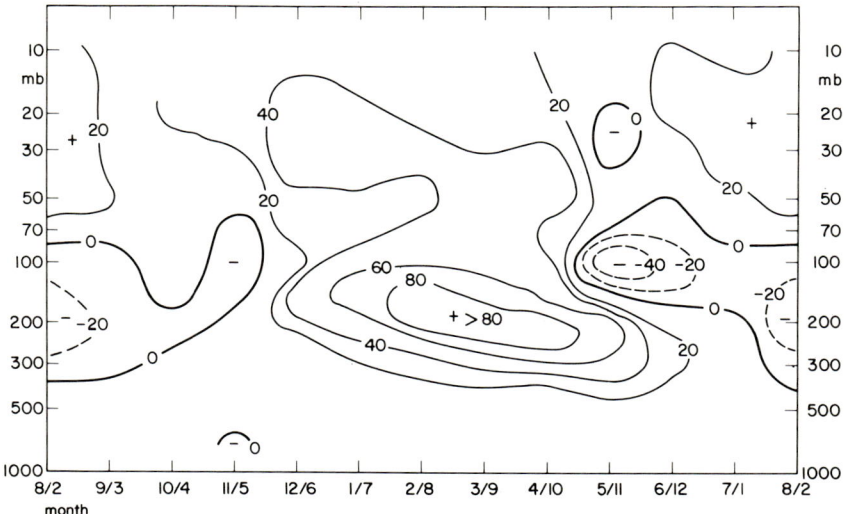

Fig. 10. Time cross section of the difference Arctic–Antarctic.

The pole-to-pole cross sections of vertical ozone distribution, shown in Fig. 11 were drawn for two-month periods in order to obtain a better data coverage and to smooth out some of the noise introduced by the considerable inhomogeneity of the set of observations available at present; considerable interpolation was nevertheless needed for completing the isolines over the Southern Hemisphere. The two cross sections are 7 and 5 months, respectively apart; like this maximum and minimum ozone concentrations can be shown for the Northern Hemisphere and, due to the delay of the winter–spring increase in the southern polar regions, largely also for that hemisphere. These cross sections demonstrate quite clearly that the very low ozone content of the high-reaching tropical tropopause is the reason for the minimum of total ozone observed at low latitudes. The band of maximum ozone concentration with lower position over the poles and maximum height near the equator is most strongly developed over the polar regions of the spring hemisphere and shows minimum strength over mid-latitudes in fall (<140 nb).

A considerable asymmetry between the hemispheres that is more pronounced in spring than in fall can be seen on these cross sections. Although the Antarctic maximum, which is somewhat less intense than that in the north (200 nb instead of 230), is positioned at a slightly lower level, high values here do not extend as far downward as over the Arctic. The quite pronounced secondary maximum observed in the Northern Hemisphere from mid-latitudes into the inner Arctic, which is apparently connected with the effects of large scale horizontal mixing, is clearly much less developed in the south.

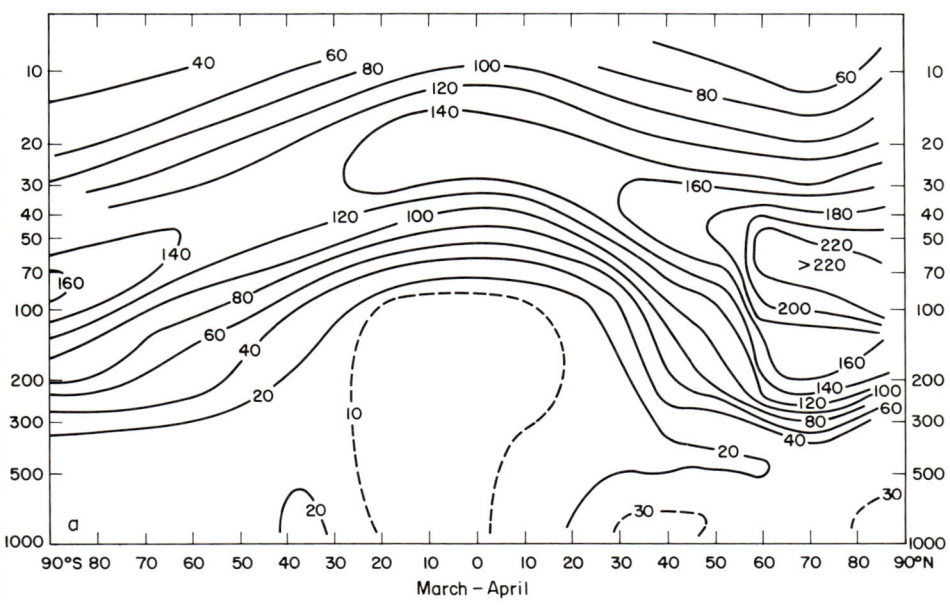
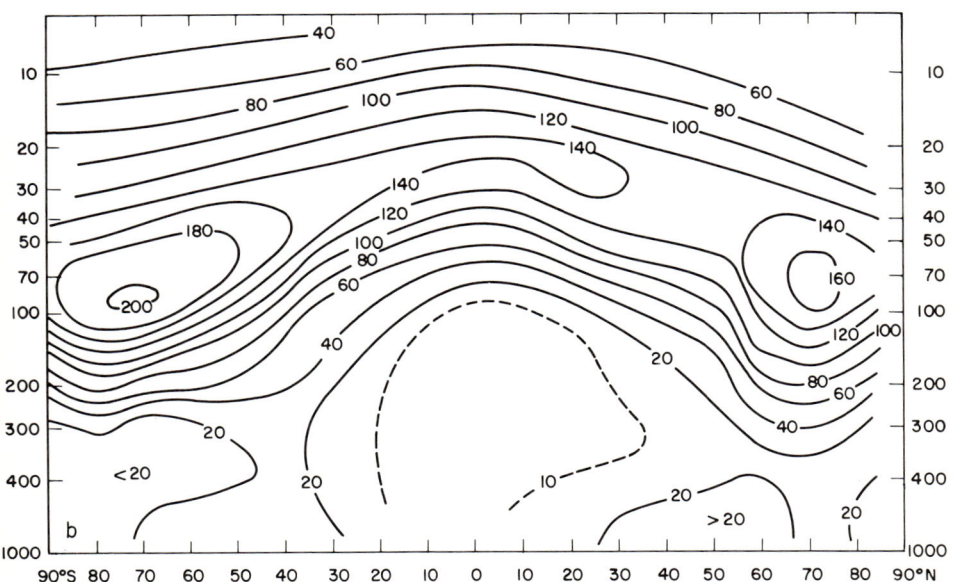

Fig. 11. Pole-to-pole cross section of vertical ozone distribution: (a) spring (March–April); (b) fall (October–November).

The considerably smoother transition between 10 and 15 km from the low tropical to the high polar values of ozone concentration shown in spring for the Southern Hemisphere fits into this picture, although it has yet to be shown whether it is not partly produced by the sparseness of data on that side of the equator.

Tropospheric values seem to be higher in the Northern Hemisphere, whereby maximum concentrations near the surface are found in lower mid-latitudes on both sides of the equator, in agreement with the observation that injections of stratospheric air into the troposphere tend to occur below the jetstream in the equatorward direction [51,52]

Lower values are indicated in the south also in the middle stratosphere; this observation, which is of considerable interest in connection with the

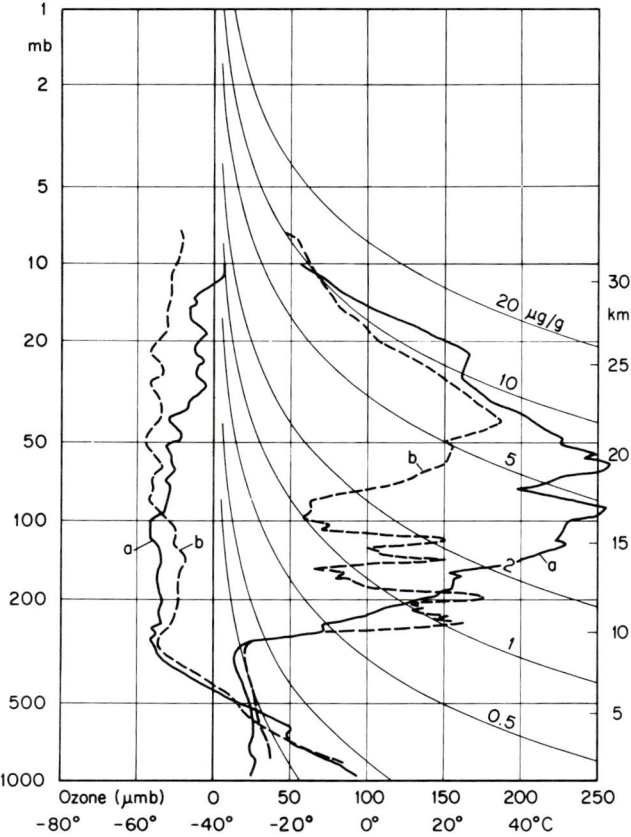

FIG. 12a. Winter–spring vertical distributions in Arctic (a) and Polar air (b). (a) 17 February 1970 at Payerne, Switzerland, 47°N; (b) 4 March 1966 at Boulder, Colorado, 40°N.

photochemical theory, especially with respect to the magnitude of the relaxation time (being, as demonstrated in Section 3.2, still rather uncertain) has yet to be substantiated by many more soundings.

1.2.4. Day-to-Day Variations of Vertical Distribution. As already indicated by the high short-time variability of total ozone mentioned in Section 1.1, the vertical ozone distribution shows astonishingly high changes from one day to the next (some time within a few hours) especially in midlatitudes during the period of high ozone concentration in winter and spring [53]. The pronounced double peak shown by curve b in Fig. 12a is typical for winter and spring vertical distributions in mid-latitudes (it is most prominent in polar air masses and under cyclonic conditions); in stable high-pressure areas

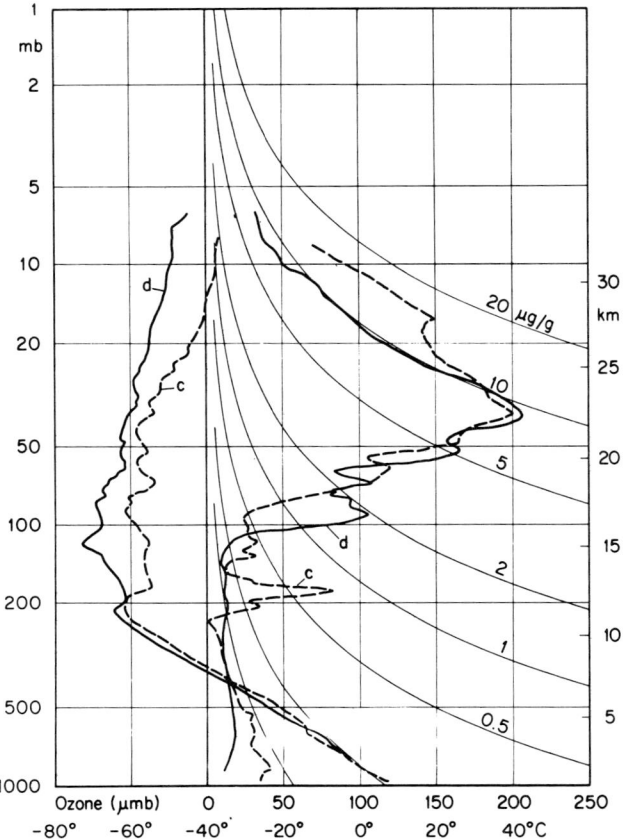

FIG. 12b. Winter–spring vertical distributions in subtropical air. (c) 20 March 1970 at Payerne, Switzerland, 47°N; (d) 10 February, 1964 at Boulder, Colorado, 40°N.

and in air of subtropical origin it may be much reduced (curve c in Fig. 12b). Its complete removal as shown by curve d for an ascent in Boulder at 40° N has never been observed at Zürich at 47° N in the first half of the year during four years of observations. Mainly in air of Arctic origin the two separated peaks may merge into one (Fig. 12a, curve a) situated around 100 mb, near the position of the saddle which normally separates the two peaks at a level which indicates a poleward extension of the tropical tropopause. While the main maximum around 40 to 50 mb is rather steady (relative standard deviation around 10 %), the secondary one in the lower stratosphere is, as demonstrated by Figs. 12 and 13, extremely variable in position as well as in intensity; it is often split into several peaks and very thin layers of high ozone content may be observed (Fig. 13).

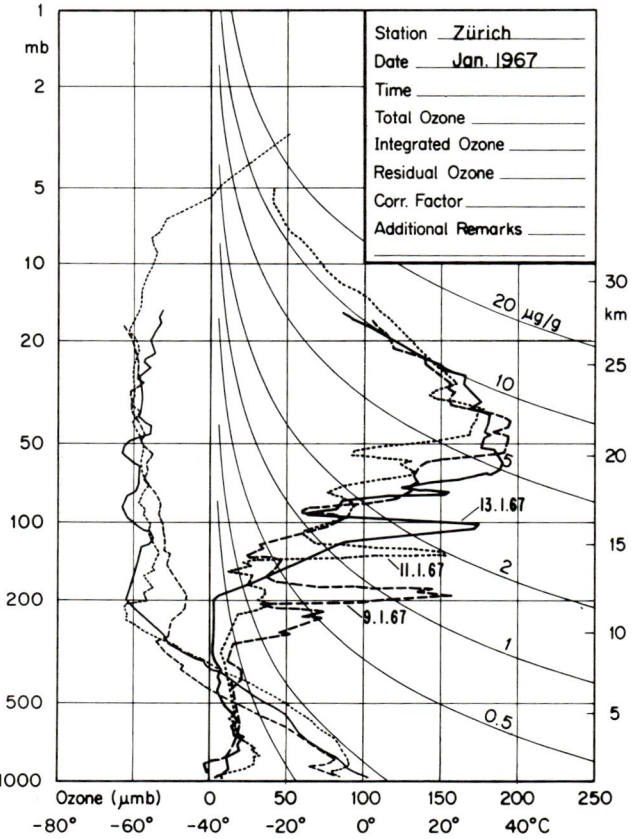

FIG. 13. Sandwich structure of the stratosphere: narrow maxima in the lower stratosphere in three successive soundings.

Such rapid variations cannot be of photochemical origin (as will be shown in Sections 2 and 3); they must be produced by air motions. The pronounced layer-wise structure strongly indicates the importance of large scale quasi-horizontal exchange processes for the ozone distribution, especially in the lower stratosphere (see also Section 5); the effects may be accentuated by the vertical motions mostly connected with the meridional displacement.

Especially during the time of high ozone concentration in the lower stratosphere (late winter and spring) the tropopause level is very well marked by the ozone distribution (see Figs. 12 and 13). While the well-mixed troposphere shows (above the ground layer) almost constant mixing ratio and thus a decrease in ozone partial pressure to a minimum in the upper troposphere, a sharp upward increase begins normally right at the tropopause. On some

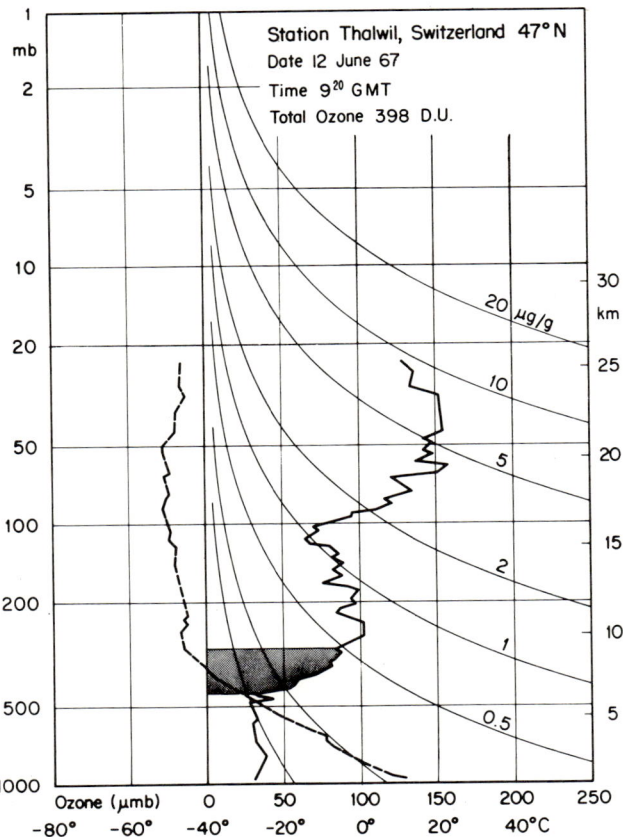

FIG. 14. Transfer of stratospheric air into the troposphere: strong increase of ozone concentration with height in the upper troposphere.

occasions, however, the rise in ozone concentration starts well below the (conventional) tropopause (Fig. 14). Beginning transfer of air from the stratosphere into the troposphere is indicated by such soundings. In a later stage of the development (Fig. 15) such intrusions [52] are observed as pronounced ozone peaks in the middle troposphere; due to the relatively strong vertical mixing at that level, they are rather short lived.

Rapid variations of the ozone concentration in the middle stratosphere are relatively rare and are tied to the large-scale circulation phenomena known as sudden stratospheric warmings (Fig. 16); the influence of vertical motions in producing both temperature and ozone changes is quite pronounced; but also at that level they are combined with large meridional displacements, as can easily be deduced from the combined study of ozone and temperature distribution.

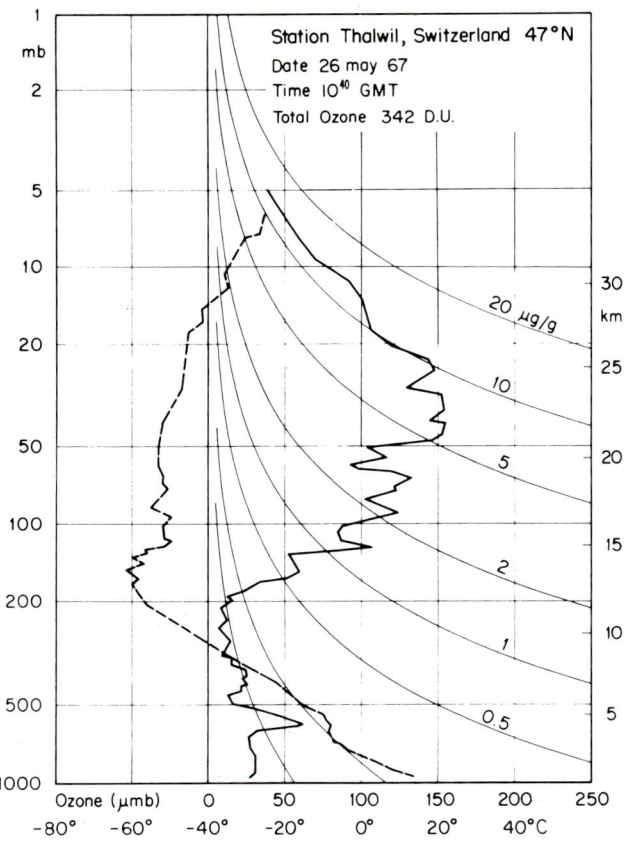

FIG. 15. Layer of initially stratospheric air in the lower troposphere: pronounced ozone peak below 500 mb.

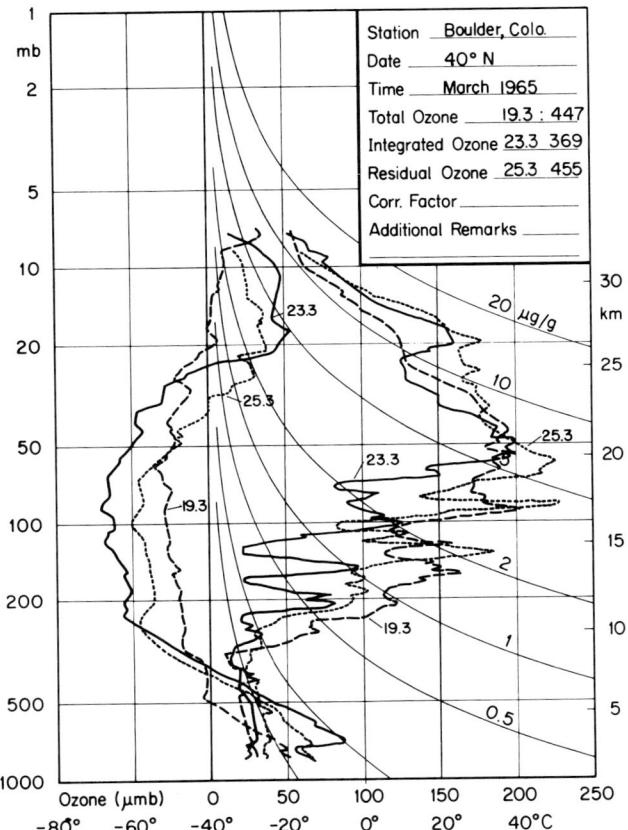

Fig. 16. Sudden stratospheric warming progressing downward over Boulder, Colorado. Comparison between the ozone variations in the middle and the lower stratosphere shows that the changes are much more regular (i.e., on a larger scale in space and time) at the higher level.

In late summer and fall the secondary ozone maximum in the lower stratosphere is mostly much reduced or even missing (Figs. 17 and 18) and the tropopause may on many occasions be only marginally marked by the ozone distribution. However, also at that season pronounced low stratospheric peaks are observed at the time of strong polar air advection or in the region of a cut-off low.

In tropical regions (with high-reaching troposphere) no pronounced secondary maxima are observed and the details in the vertical structure are (normally) minor (Fig. 19), indicating reduced importance of large-scale

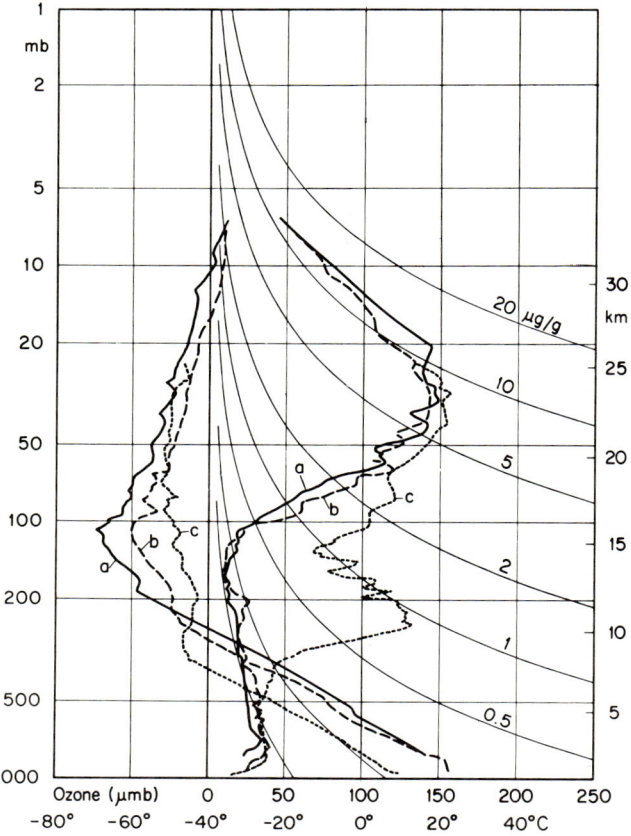

Fig. 17. Vertical ozone distribution in summer: (a) and (b) are typical summer distributions in subtropical air; (c) shows still a pronounced secondary maximum in polar air (in a strong upper air trough). (a) 18 August 1965 Boulder, Colorado, 40°N; (b) 16 August 1967 Thalwil, Switzerland, 47°N; (c) 8 July 1969 Zürich, Switzerland, 47°N.

horizontal mixing. At high latitudes strong single maxima around 100 mb are rather frequent, but the secondary peaks, typical for mid-latitude winter and spring distribution, are quite often found also in polar regions indicating the extension of large scale exchange processes into the inner Arctic.

1.2.5. Year-to-Year Variations. It has been shown in the preceeding section that the ozone distribution is intimately connected with weather; it is thus

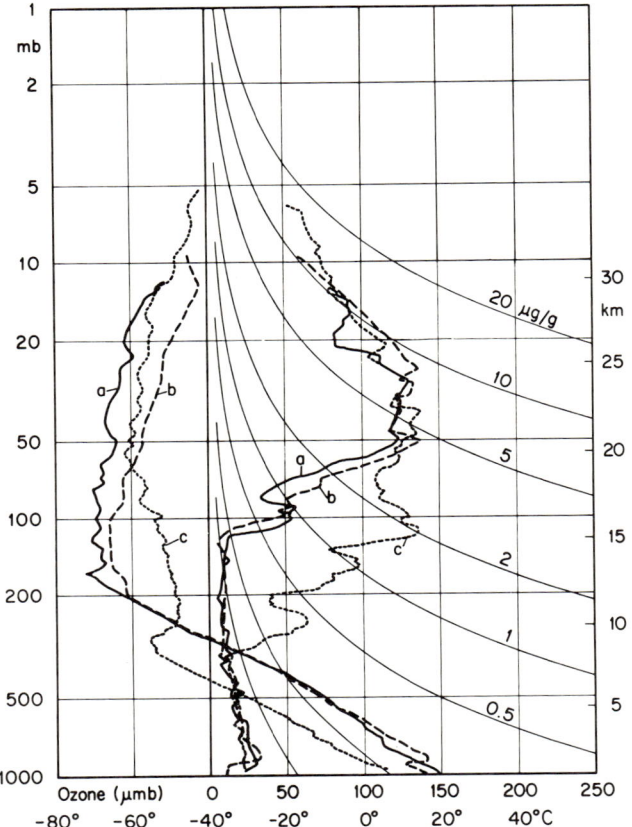

Fig. 18. Vertical ozone distribution in fall: in a strong flux of subtropical air (a) and in a stable anticyclone (b) low ozone concentrations are found almost up to the 100 mb level; high values in the lower stratosphere are, however, occasionally observed in a strong upper air trough. (a) 2 November 1970 Payerne, Switzerland, 47°N; (b) 9 October 1967 Thalwil, Switzerland, 47°N; (c) 21 October 1970 Payerne, Switzerland, 47°N.

not surprising that considerable variations of monthly mean values are observed from one year to the next (Figs. 20 and 21). They are strongest in the lower stratosphere at the level of highest interdiurnal variability and apparently connected to large-scale anomalies of the general circulation as displacements or intensity changes of the quasi-stationary upper air troughs and ridges, whose mean position apparently determines the longitudinal variation of ozone concentration (Section 1.1). There are some indications

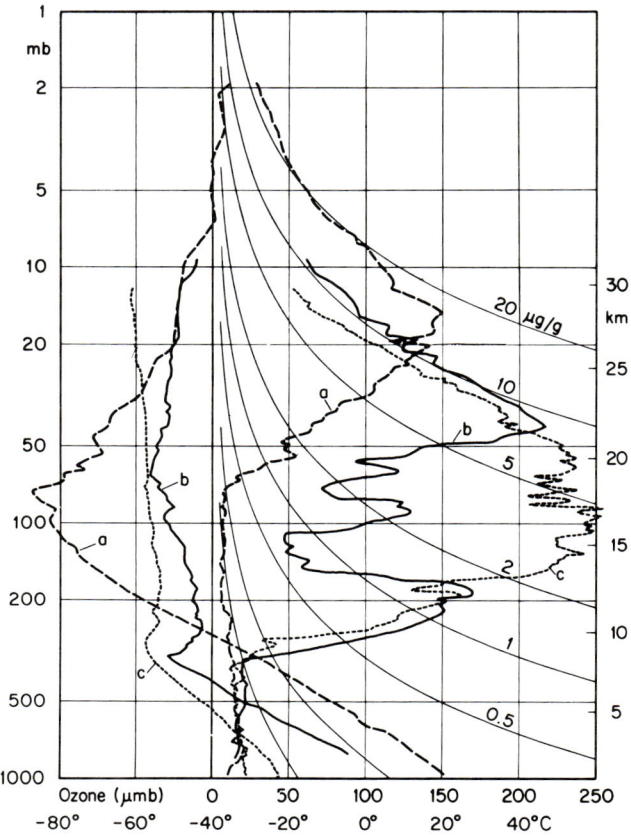

FIG. 19. Single ozone distributions at different latitudes. (a) Equatorial distribution: 20 March 1963, Balboa, Canal Zone, 10°N; (b) Mid-latitudes (polar air): 14 February 1964, Boulder, Colorado, 40°N; (c) Arctic distribution: 20 March 1963, Thule, Greenland, 76°N.

of certain systematic variations at higher levels, but none of them is really confirmed beyond doubt for middle- and high-latitude stations.

A biennial variation of total ozone in equatorial and subtropical regions (showing an 180° phaseshift between the two and being connected with the 26-month cycle of the tropical stratospheric circulation) has been established [54]. Extension of such relationship to mid-latitudes has also been claimed [46,47,55], but is not yet proved beyond doubt; further research on a

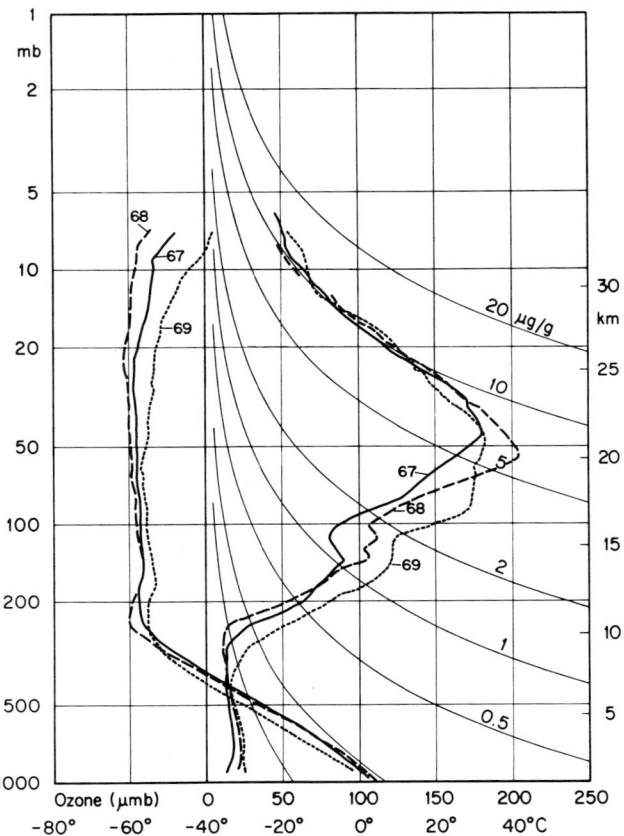

Fig. 20. Year-to-year variations of monthly mean ozone distribution over mid-latitude stations in late winter (February, Thalwil/Payerne, Switzerland, 47°N).

hemispheric basis rather than for single stations, which are at least in the Northern Hemisphere too much affected by "meteorological noise" is needed for this purpose.

Secular variations in the upper stratosphere, seemingly connected with the solar cycle and being of great interest in connection with the photochemistry of these layers have also been reported but are yet doubtfull, too (see Section 6).

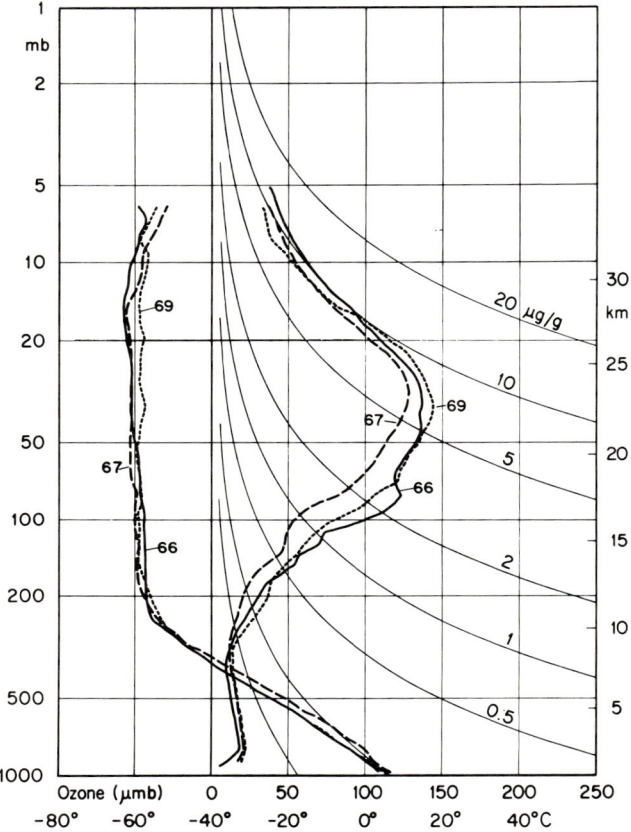

Fig. 21. Year-to-year variations of monthly mean ozone distribution over a mid-latitude station in fall (November, Thalwil/Payerne, Switzerland, 47°N).

2. The Classical Photochemical Theory: "Pure Oxygen Atmosphere"

2.1. Introduction

The observed vertical distribution of ozone in the atmosphere—maximum concentration in the stratosphere—is only possible if this trace constituent is continuously formed and destroyed; otherwise complete mixing had to be expected.

It was proposed by Chapman in 1930 [56–58] that the formation and largely also the destruction of ozone was of photochemical nature. Although other mechanisms have been suggested on several occasions (production in connection with auroras or by cosmic rays [55,59,60]) any quantitatively important contribution from such processes (at least for the stratosphere) has never been proved and it may still safely be assumed that photochemical action determines together with transport processes the three-dimensional distribution of ozone in the stratosphere.

The photochemical theory put forward by Chapman was developed for a pure oxygen atmosphere (other atmospheric constituents acting possibly as a third body in certain reactions). Most of the further studies in the following three decades followed this line [61–68]. For mesospheric conditions the validity of this assumption was already questioned in 1950s [69], but only during the last few years has it become increasingly evident that reactions including other trace constituents may also be important below the stratopause [70–72].

However, the quantitative importance of these additional reactions at least under stratospheric conditions, is still debated because of the uncertainty of many parameters entering this more complicated system [72,73]. Since most of the important features of ozone photochemistry and also of its interactions with transport processes can be discussed taking into account the classical system alone—many things can even be demonstrated more clearly because of the simplicity of that theory—this chapter shall be confined to the discussion of ozone photochemistry in a pure oxygen atmosphere.

2.2. Equilibrium Theory

2.2.1. Basic Reactions.[3]

Ozone formation in the atmosphere is initiated by photodissociation of molecular oxygen:

[2*] $O_2 + h\nu \longrightarrow O + O \quad f_2$

To provide the necessary dissociative energy the absorbed light quantum must be of wavelengths shorter than 2420 Å. Actual ozone formation follows then by the three-body reaction

[2] $O_2 + O + M \longrightarrow O_3 + M \quad k_2$

---

[3] In the following discussion reactions are numbered in square brackets. The numbers are the same as the indices of the reaction rates k; the primary photochemical reactions are indicated by a star, their number is the same as the index of the dissociation rate f. This nonsequential numbering is used in order to avoid a change of indexing from one article to another which is very awkward, especially in the more complicated theory developed in Section 3.

The third body which will predominantly be O_2 or N_2 is needed for simultaneous conservation of energy and momentum. The ozone molecule is not only formed but also destroyed by photodissociation:

[3*] $\qquad\qquad O_3 + h\nu \longrightarrow O_2 + O \qquad f_3 \qquad \lambda \leq 11{,}800 \text{ Å}$

Because of the relative instability of the three-atomic modification visible and even near infrared radiation provides the necessary energy for this process.

The odd oxygen particles (O and O_3) produced by the initiating photodissociation or by the subsequent reactions, are destroyed, i.e., reconverted into normal O_2 by two processes:

[3] $\qquad\qquad O + O_3 \longrightarrow 2 O_2 \qquad k_3$

[1] $\qquad\qquad O + O + M \longrightarrow O_2 + M \qquad k_1$

For an easy quantitative treatment of the system, these five principal reactions may be subdivided in two categories (see also Fig. 22):
  (a) reactions producing or destroying odd oxygen particles [2*], [1], and [3].
  (b) reactions converting odd oxygen particles into each other [2] and [3*].

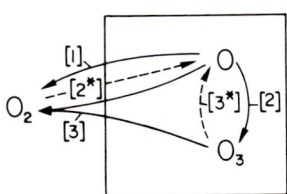

FIG. 22. Reaction scheme of the classical photochemical theory; dashed arrows, primary photochemical reactions; full arrows, secondary photochemical reactions.

Numerical estimates show that the reactions of type (b) are orders of magnitude faster than those of type (a). Thus an equilibrium between O and O_3 is very rapidly established, practically independently of formation and destruction of odd oxygen particles:

(2.1) $\qquad\qquad f_3 n_3 = k_2 n_1 n_2^2$

with

(2.2) $\qquad\qquad k_2 \equiv k_2' + k_2'' n_{N_2}/n_2$

k_2' reaction rate with O_2 as a third body, k_2'' reaction rate with N_2 as a third body. This representation is useful as long as n_{N_2}/n_2 is practically constant. i.e., up to about 90 km.

The dissociation rate f_3 and its height dependence are discussed in Section 2.2.3. From Eq. (2.1) the ratio of odd oxygen particle concentrations is obtained:

(2.3) $$n_1/n_3 = f_3/k_2 n_2^2$$

This ratio is rapidly decreasing downward through the atmosphere.

Because the equilibrium between O and O_3 is quickly assumed, the balance between production and destruction can be taken for the sum of the odd oxygen particles

(2.4) $$2f_2 n_2 = 2k_3 n_1 n_3 + 2k_1 n_1^2 n_2$$

whereby k_1 is defined similarly as k_2. Numerical estimates show, that below the stratopause the second term on the right-hand side of Eq. (2.4) can be neglected and a simple equation for ozone concentration can thus be derived for stratospheric conditions.

2.2.2. *Ozone Equilibrium in the Stratosphere.* By inserting Eq. (2.3) into (2.4) and neglecting the second destruction term we obtain

(2.5) $$n_3 = n_2^{3/2}(kf_2/f_3)^{1/2}$$

with

(2.6) $$k \equiv k_2/k_3$$

Figure 23 shows how the different height dependence of the quantities on the right-hand side of Eq. (2.5) leads to an ozone distribution with a maximum in the stratosphere (around 25 km). In the upper stratosphere, between 50 and 35 km the two dissociation rates dwindle almost equally rapidly with decreasing altitude (for reasons discussed in Section 2.2.3); k increases downward, i.e., toward lower temperatures (because k_3 is much more strongly temperature dependent than k_2). Thus the ozone concentration must rapidly increase with increasing density (n_2) and also the (number)-mixing ratio n_3/n_2 is rising in the downward direction. Below 35 km the altitude dependence of f_3 becomes rather small while f_2 is falling off more and more rapidly. For this reason the theoretically computed mixing ratio

(2.7) $$n_3/n_2 = (n_2 kf_2/f_3)^{1/2}$$

yields a maximum between 30 and 35 km and the ozone concentration itself is highest in the neighborhood of 25 km. The accurate position of these maxima depends on season and latitude (i.e., sun height) and on the values used for a number of parameters, which are not yet accurately known (k and its temperature dependence, spectral energy distribution in the solar uv, absorption coefficients, etc., see Section 2.6).

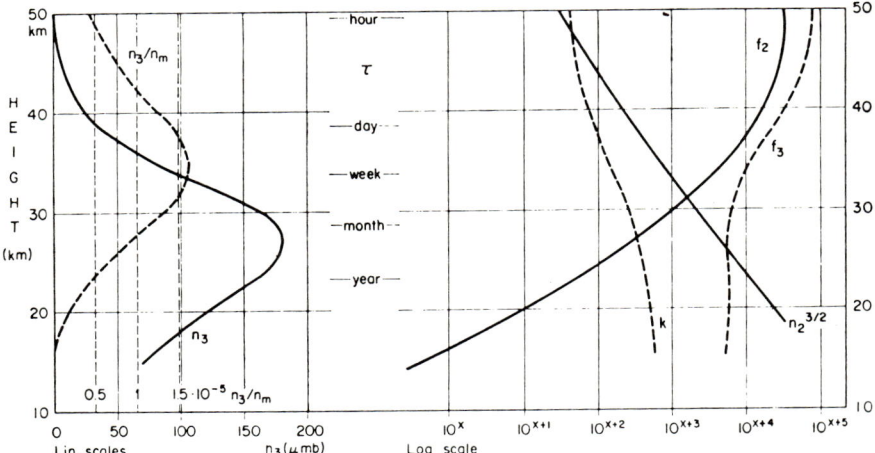

FIG. 23. Vertical distribution of ozone and ozone-to-air mixing ratio derived from the "classical" photochemical theory (left side) and vertical distribution of the parameters entering into the equilibrium equation (2.5) on the right. Mean values of the relaxation times are indicated in the middle of the graph. The numerical values for x are $f_2 = -14$, $f_3 = -7$, $k = -21$, $n_2^{3/2} = 26$.

2.2.3. The Dissociation Rates and Their Height Dependence. The dissociation rates f_2 and f_3, i.e., the number of dissociating quanta absorbed per oxygen and ozone molecule respectively and per second are given by

$$(2.8) \quad f_2 = (1/L) \int_0^{\lambda_{max}} \left(\alpha_{2,\lambda} + \alpha'_{2,\lambda} \frac{d_2}{H}\right) I_{0,\lambda}$$

$$\exp -\{(\alpha_{2,\lambda} + \alpha'_{2\lambda} d_2/2H)d_2' + \alpha_3 d_3'\} d\lambda$$

$$(2.9) \quad f_3 = (1/L) \int_0^{\lambda_{max}} \alpha_{3,\lambda} I_{0,\lambda} \exp -(\alpha_{3,\lambda} d_3') d\lambda$$

$$(2.10) \quad d_2' = \mu d_2 \quad \text{and} \quad d_3' = \mu d_3$$

are the slant path lengths of the sunlight above the point under consideration (μ = Bemporad function). H is the height of the homogeneous atmosphere (8×10^5 cm). The complication in the formula for f_2 is due to the deviation from Beer's law in the Herzberg continuum between about 2000 and 2420 Å discovered by Heilpern [74] and confirmed by Ditchburn and Young [75]. α_2' gives the contribution by the term which is proportional to $(p/p_0)^2$, a contribution which is relatively minor above 20 km, i.e., at the levels where photochemical processes are really important. The importance of the deviation from Beer's law in the oxygen absorption is further discussed in Section 2.6.2.

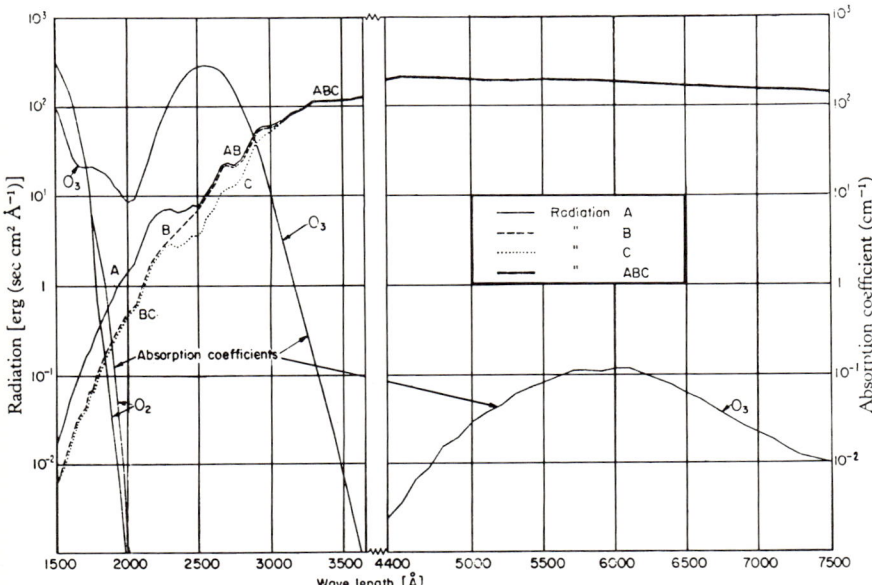

Fig. 24. Absorption coefficients of ozone and oxygen and extra-terrestrial spectral intensity distribution of solar radiation. The two curves for oxygen absorption between 1750 and 2000 Å give upper and lower limits for the rapidly varying coefficients in the Schuman–Runge bands. Intensity distribution A according to Detwiler et al. [85] (called radiation 1); B according to Brewer and Wilson [84]; in curve C Brewer and Wilson's data (measured at 2100 Å) are brought only gradually to join Detwiler's distribution near 3000 Å (called radiation 2).

Figure 24 shows the absorption coefficients of oxygen and ozone as a function of wavelength and Fig. 25 demonstrates the extinction of uv solar radiation on its way through the atmosphere (computed for a mean ozone distribution and a fixed solar height). All wavelengths shorter than about 1800 Å (except Ly-α) are completely absorbed above the mesopause and need not to be further discussed in this context.

While below about 2000 Å the cut-off of solar radiation is completely due to the absorption by molecular oxygen, ozone is above about 2200 Å the only important absorber and eliminates most of the radiation around 2500 Å already in the upper stratosphere, i.e., well above the ozone maximum.

Between these two regions of strong absorption, we find a window, where photochemically active radiation is allowed to penetrate relatively deeply into the atmosphere (although not to the ground). It is this part of the spectrum which dissociates oxygen below the stratopause and which is thus responsible for the formation of the main ozone layer.

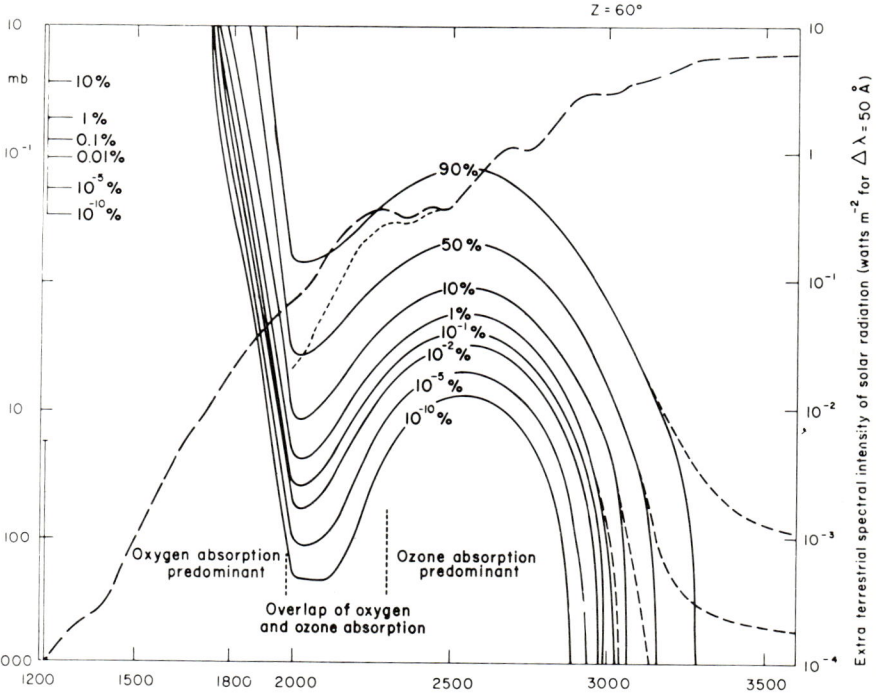

FIG. 25. Attenuation of solar radiation on its way through the atmosphere (solid lines) and extraterrestrial intensity distribution after Detwiler et al. [85] (long-dashed lines) as functions of wavelength. For comparison, the dotted line indicates the changes to Detwiler's intensity curve suggested by the recent measurements of Brewer and Wilson [84]. Short dashed lines: taking into account attenuation by scattering.

The overlapping of ozone and oxygen absorption in this window region is an important feature of ozone photochemistry. About 80–90% of the (potentially ozone formative) radiation in that part of the spectrum is in reality absorbed by the ozone itself and thus destroys it. This property together with the specific features of the height distribution of f_3 (ozone dissociation rate) acts as a strong damping mechanism on photochemically produced ozone variations.

Whereas—in the stratosphere—the ozone producing radiation stems from a relatively narrow radiation band with rather uniform properties, the destroying radiation is composed of two separate and rather differently behaving parts: one in the ultraviolet and one in the visible. It may be added that while f_2 is a function of both d_2 and d_3, f_3 is influenced by ozone alone (the contribution to it from the overlapping region of the two absorptions being negligible).

The UV part, which gives the main contribution to f_3 in the upper stratosphere (90 % and more) is mainly extinguished above 35 km. Below that level ozone dissociation is predominantly produced by visible light (Chappuis bands) with a small contribution from the near UV above 3000 Å, i.e., by radiation which is only weakly absorbed, but has a high intensity in the solar spectrum. Thus the height dependence of f_3 (Fig. 23) and the influence of sun height and ozone concentration on this quantity is rather slight around the level of the main ozone maximum. This together with the already mentioned ozone influence of f_2 produces some of the main features of the ozone layer. First, the existence of a pronounced maximum is, as demonstrated by Fig. 23, a product of this particular behavior of f_2 and f_3. Secondly, the fact that f_2 is, especially below 35 km, strongly depressed by an increase in d_3 (overlaying ozone amount), whereas f_3 is relatively insensitive to it, provides a strong damping effect on the influence of the variation of external parameters on the total equilibrium amount of ozone [76]. Without this mechanism a uniform 10° temperature decrease in the whole ozone layer would raise the ozone amount by about 30 % (because of the resulting increase in k); the actual decrease is only 13 % in summer at 45° lat. Similarly, any effect of a possible variation of solar intensity or spectral distribution will be decreased.

Further a strong influence of the solar height on the equilibrium amount of ozone below 35 km and by this on total ozone, is introduced by these properties of the dissociation rates. While in the upper stratosphere the important ratio f_2/f_3 is only slightly dependent on the sun's zenith angle, it falls rapidly away with decreasing sun height below that level.

The (classical) photochemical equilibrium theory thus predicts an ozone maximum near the equator and minimum values in polar regions, as well as a seasonal variation with a summer maximum [both in full contradiction to the observed distribution of total ozone (Fig. 26)]. At the same time the mean total amount over the globe and the mean vertical distribution given by this theory are in reasonably good agreement with the measurements.

Because the ozone concentration at a given level is, owing to the dependence of f_2 and f_3 on d_3, a function of the ozone distribution above that altitude the equilibrium value of d_3 has to be computed by an iterative process. For an expedient solution of the numerical problem f_3 is tabulated as a function of d_3' and f_2 as a function of d_2' and d_3' beforehand, and the appropriate values of the dissociation rates are obtained by interpolation. This procedure avoids the enormous computational task of determining f_2 and f_3 from Eq. (2.8) and (2.9), respectively, for each single step; it is, however, only possible if the dissociation rates are unique functions of d_2' and d_3'. This condition is not exactly fulfilled for f_2; because of the deviation from Beer's law the result depends also on μ. However, above 20 km the error produced by using a constant value for μ (preferably 2) for the tabulations is small, because the differences against the exact evaluation from Eq. (2.8) are partly canceling each other.

Below 30–35 km the actual value of f_2 can differ considerably from that given by Eq. (2.8) because of Rayleigh scattering [65] (slight decrease at high and considerable increase at low sun). Corrections can also be tabulated. The scattering influence on f_3 is, however, insignificant, because below 30–35 km ozone dissociation is predominantly by visible light. Only the fraction yielding excited oxygen atoms (see Section 2.6.3) is again influenced by scattering [72] because only short wavelengths (less than about 3100 Å) contribute.

2.3. Nonequilibrium Theory

Although in a first approximation the theory as discussed in the previous section gives a reasonable explanation for the existence of the ozone layer, one must deduce from its failure to explain correctly the variations of ozone with latitude and season (Fig. 26) and, of course, the considerable day-to-day fluctuations observed in middle and high latitudes, that the bulk of atmospheric ozone cannot be in photochemical equilibrium. This is easily understood if the time constant of such equilibrium is investigated. From Eqs. (2.3) and (2.4) the change with time of the deviation (n_3') from the equilibrium ($n_{3,0}$) is obtained

(2.11) $$\frac{d(n_{3,0} + n_3')}{dt} = \frac{dn_3'}{dt} = 2f_2 n_2 - \frac{2(n_{3,0} + n_3')^2 f_3}{k_2 n_2^2}$$

Substituting for $n_{3,0}$ the equilibrium values we obtain for small values of n_3'

(2.12) $$dn_3'/dt = -n_3' 4(f_3 f_2/kn_2)^{1/2} = -n_3'/\tau$$

where

(2.13) $$\tau = \tfrac{1}{4}(kn_2/f_2 f_3)^{1/2} = \tfrac{1}{2}(n_{3,0}/2f_2 n_2)$$

is the relaxation time of a deviation from equilibrium (often called time of half-restoration).

For greater deviations, the accurate solution of Eq. (2.11) has to be used [65]:

(2.14) $$n_3(t) = \frac{n_3(t_0)(4f_2 f_3/kn_2)^{1/2} + 2f_2 n_2 \tanh[(4f_2 f_3/kn_2)^{1/2}(t-t_0)]}{(4f_2 f_3/kn_2)^{1/2} + (2f_3 n_3(t_0)/kn_2^2)\tanh[(4f_2 f_3/kn_2)^{1/2}(t-t_0)]}$$

A more general expression for the time of half-restoration which is also

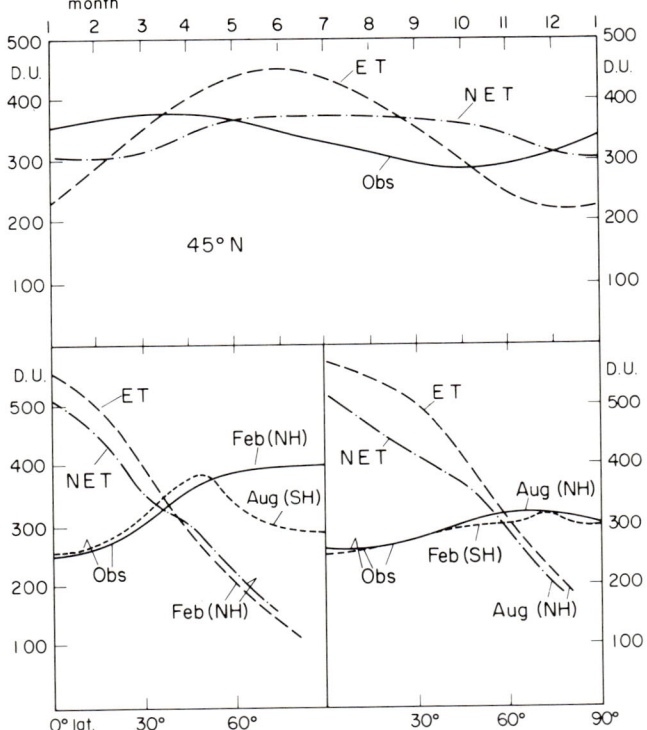

FIG. 26. Discrepancy between observed and theoretically computed total ozone distribution. Top: seasonal variation at 45°N; bottom: latitude dependence in late winter and late summer. ET, Equilibrium theory; NET, Nonequilibrium theory; Obs, observed; NH, Northern Hemisphere; SH, Southern Hemisphere.

applicable for large deviations from the equilibrium can be derived from the formula

$$\tau = \frac{1}{4}\left(\frac{k_2 n_2}{f_2 f_3}\right)^{1/2} \ln\left(\frac{n_3' + 2en_{3,0}}{n_3' + 2n_{3,0}}\right) \qquad (2.15)$$

where ($e =$ base of the natural logarithms). It follows from this that for large positive deviation τ is smaller, for high negative deviation larger than for small disturbances (surplus ozone is more rapidly destroyed than a deficiency is restored).

Obviously the relaxation time is strongly increasing with decreasing altitude from about one hour at the stratopause to several months or a year and more at the level of the ozone maximum and below, with values which are at the same level larger in winter than in summer (Figs. 23 and 37). This means that equilibrium will never exist (except accidentially) in the lower stratosphere. Below the level of the ozone maximum the distribution will largely be determined by air motions and it is obvious to assume that the descrepancy between observed and theoretically computed distribution of the total amount is a result of redistribution of ozone by the general circulation.

This assumption is strongly supported by the observational results on vertical distribution obtained during the past decade (see Section 1) which show above about 25 km the theoretically expected increase from high to low latitudes and from winter to summer (Figs. 5 and 7) however, the behavior of total ozone is determined by the considerably stronger reversed gradients in the lower stratosphere, i.e., in the region where redistribution by circulation dominates because of the very long relaxation times. At these levels the mixing ratio ozone/air (or n_3/n_2) becomes a conservative property of an air parcel (considerably more conservative than potential temperature). Ozone can thus be used as a tracer.

The application of such methods in synoptical studies has only become fully possible by the use of chemical ozone sondes giving full details and not being confined to fair weather conditions. They are especially suited for research on the transfer of air from the stratosphere into the troposphere and vice versa [52]. They also give a good idea of the effects of large scale quasi-horizontal mixing on the ozone distribution (production of a fine structure by many successive shear processes).

For such research (at least as long as studies are confined to the lower stratosphere) the ozone-air mixing ratio can be assumed to be fully conservative; photochemical processes are not to be taken into account. If, however, one tries to reveal from ozone observation some aspects of the general circulation in the stratosphere—existence of a meridional circulation or intensity and behavior of large scale mixing—higher levels have to be included and photochemistry has to be taken into account. It is in such studies, where differences become important, that the remaining uncertainties of the theory are strongly felt. They will be discussed in Section 2.6.

2.3.1. Nonequilibrium Photochemical Ozone Distribution in an Atmosphere at Rest. For obtaining an idea of the interplay of photochemical action and transport processes the nonequilibrium equation (2.14) may be used to compute first the world-wide ozone distribution in an atmosphere at rest. At each level—working downward through the atmosphere—the ozone variation throughout the year has to be evaluated in appropriate time steps. Each single result is not only a function of solar height, overlaying ozone, etc., as in the equilibrium theory, but is also influenced by the initial conditions, the more the lower in the atmosphere the computation is made (increasing relaxation time). If after 12 months the initial value is not met, the computation

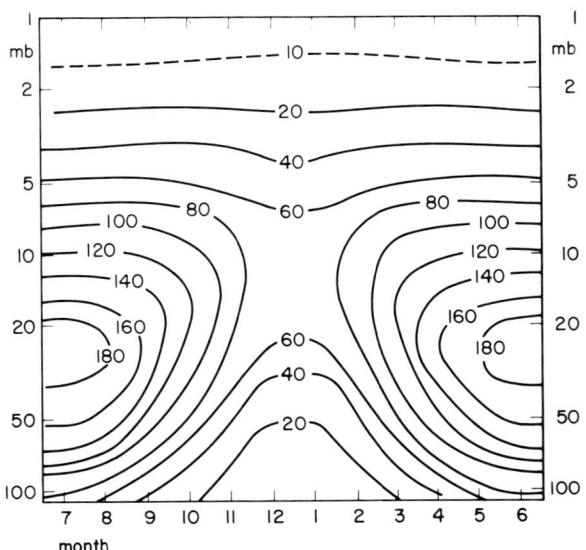

FIG. 27a. Time cross section of vertical ozone distribution at 45°N computed with the classical equilibrium theory.

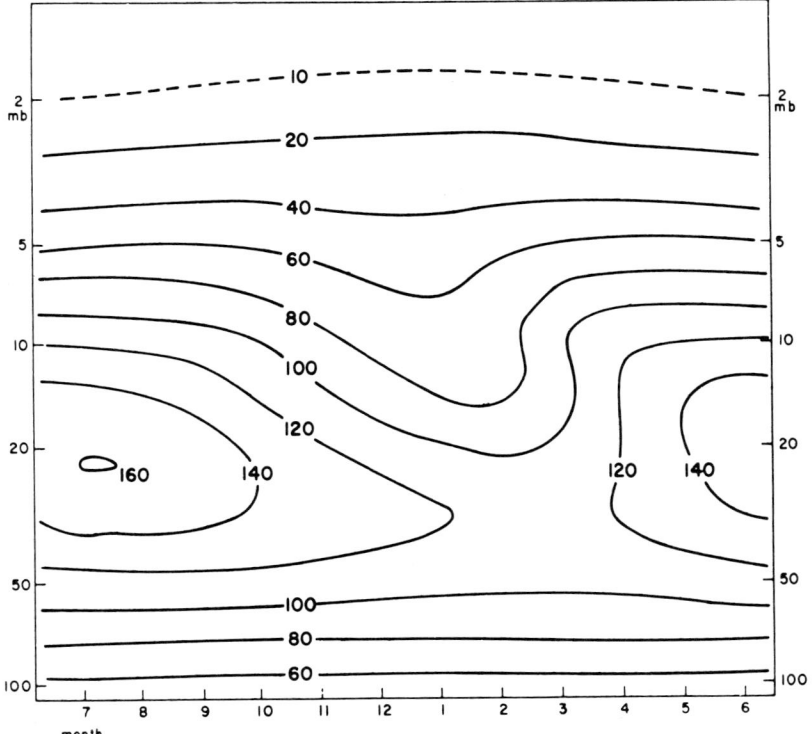

FIG. 27b. Time cross section of vertical ozone distribution at 45°N computed with the classical nonequilibrium theory.

has to be repeated with an adjusted starting concentration and this iteration must be continued until a reasonably good fit is accomplished.

Figures 27a and 27b show the results of such computations compared with those obtained from the equilibrium theory for 45°N. The considerable asymmetry shown in the seasonal variation in the middle stratosphere and just above the level of the ozone maximum is a result of a relaxation time of the order of a few months; the still higher values around 20 km and below lead to ozone concentrations which are almost independent of season.

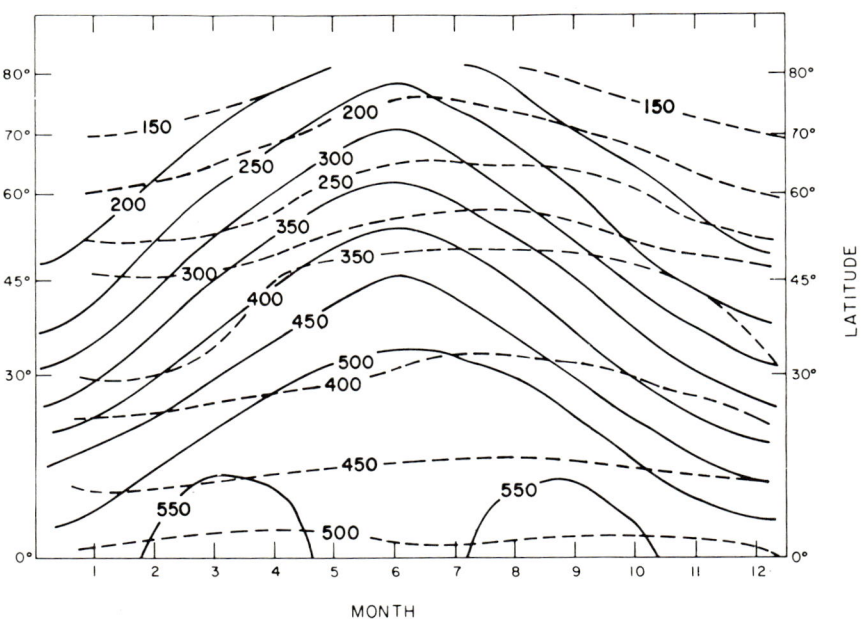

FIG. 28. Variation of total ozone with season and latitude as calculated from the photochemical theory (full lines equilibrium, dashed lines nonequilibrium theory).

As is demonstrated by Fig. 28 the discrepancy between theory and observation with respect to the world-wide distribution of the total amount remains although its magnitude is slightly reduced. Thus transport phenomenon must be included for a full explanation.

2.4. *Influence of Motion on Photochemistry*

There are two ways to combine the influence of photochemistry and air motions depending on their relative importance.

2.4.1. Predominance of Photochemistry.

In case that the transport gives a relatively small correction of a primarily photochemically determined ozone concentration an additional term (transport-term) is introduced in Eq. (2.11):

$$\text{(2.16)} \qquad \frac{\partial n_3}{\partial t} = 2f_2 n_2 - \frac{n_3^2 f_3}{n_2^2 k} - \text{div } \mathbf{F}_3$$

where $\mathbf{F}_3$ is the ozone flux vector

$$\text{(2.17)} \qquad \mathbf{F}_3 = \mathbf{v} \cdot n_3 - n_2 \mathscr{K} \text{ grad } n_3/n_2$$

$\mathscr{K}$ = mixing tensor. With $B \equiv -\text{div } \mathbf{F}_3$ the following modified equilibrium equation results:

$$\text{(2.18)} \qquad n_3 = n_2^{3/2} [k(f_2 + B/2n_2)/f_3]^{1/2}$$

Thus it depends on the ratio of f_2 to $B/2n_2$ whether the photochemical or the transport effects are predominant in determining the equilibrium. A corresponding correction is introduced in the nonequilibrium theory. If $(f_2 + B/2n_2) < 0$, equilibrium is not possible, however, the nonequilibrium equation can be derived also for this case which is at lower levels physically well possible; however, under such conditions the formulation given in the following section is preferable and the modified nonequilibrium equation is therefore not derived here.

2.4.2. Predominance of Transport.

In this case it may be useful to introduce the photochemical terms as a correction. Without such processes the mixing ratio of an individual air parcel is constant

$$\text{(2.19)} \qquad \frac{d}{dt}\left(\frac{n_3}{n_2}\right) = 0$$

and thus the local changes of the mixing ratio is given by

$$\text{(2.20)} \qquad \frac{\partial}{\partial t}\left(\frac{n_3}{n_2}\right) = -V \text{ grad } \left(\frac{n_3}{n_2}\right)$$

In the presence of photochemical processes this is converted to

$$\text{(2.21)} \qquad \frac{d}{dt}\left(\frac{n_3}{n_2}\right) = 2f_2 - \frac{2n_3^2 f_3}{kn_2^3}$$

and

$$\text{(2.22)} \qquad \frac{\partial}{\partial t}\left(\frac{n_3}{n_2}\right) = -V \text{ grad } \left(\frac{n_3}{n_2}\right) + 2f_2 - \frac{2n_3^2 f_3}{kn_2^3}$$

2.5. Mesospheric Ozone

2.5.1. Pure Photochemistry.
In the upper part of the mesosphere the recombination of atomic oxygen cannot be neglected any more; if the assumptions used in the derivation of Eq. (2.4) were still valid, this would just result in an additional term in the equilibrium equation (2.7)

(2.23) $$n_3 = n\,^{3/2}_2[f_2 k/f_3(1+A)]^{1/2}$$

with

(2.24) $$A \equiv f_3 k_1/n_2 k_2 k_3$$

Table I shows A and also n_1/n_3 as a function of height. We find that at about 55–60 km (depending on the value of some not exactly known parameters)

TABLE I. Height dependence of the ratio of atomic oxygen to ozone and of the additional term A in Eq. (2.24)

p (mb)	A	n_1/n_3
0.64	0.001	0.33
0.32	0.003	1.11
0.16	0.009	3.92
0.08	0.037	10.5
0.04	0.145	29.8
0.02	0.61	83.0
0.01	3.2	212
0.004	8.0	1340
0.002	6.9	8160

the atomic oxygen concentration which is much smaller than that of ozone in the stratosphere, becomes equal to the latter and above that level O becomes increasingly more abundant than O_3 (this means also, that at these levels the internal equilibrium between O particles must adjust to the instantaneous concentration of atomic oxygen and not to that of ozone as in the stratosphere; thus transport of O instead of O_3 may become of importance.)

However, it can easily be shown that the simplification used in the derivation of Eqs. (2.4) and (2.23) becomes increasingly useless with growing height at mesospheric levels. Undoubtedly, it is still applicable in daytime, where ozone goes to equilibrium with the existing concentration of atomic oxygen within a minute; at night, however, where the primary photochemical reactions [2*] and [3*] are absent, this is not the case. In principle, this restriction applies also to the stratosphere. Only at those levels all existing atomic oxygen is very rapidly—within seconds—converted into ozone after sunset. After this also all secondary photochemical processes ([2] and [3]) vanish, because they all include atomic oxygen; this means that the ozone present at sunset—below 40 km only negligibly increased by the conversion of atomic oxygen—is "frozen in" for nighttime; thus the theory can in the stratosphere be confined to daytime.

This does not hold for the mesosphere, especially above about 60 km. Because the speed of the conversion (by a three-body collision) of O into O_3 decreases rapidly with dwindling density its relaxation time given approximatively by

(2.25) $$\tau_c = 1/k_1 n_2^2$$

becomes relatively large in the mesosphere. This means that the destruction of odd oxygen particles by reactions [3] and [1] continues during part or, near the mesopause, during the full night, i.e., during a period where the equilibrium condition given by Eq. (2.3) is not fulfilled. The derivation presented in Section 2.2.1 and also used to obtain Eq. (2.23) is thus not applicable any more above 55–60 km.

The daily variation of n_1 and n_3 in the mesosphere and the mean daytime level of odd oxygen particles can thus only be derived by numerical integration of two coupled differential equations

(2.26) $$dn_1/dt = 2f_2 n_2 + f_3 n_3 - k_2 n_1 n_2^2 - k_3 n_1 n_3 - 2k_1 n_1^2 n_2$$

(2.27) $$dn_3/dt = -f_3 n_3 - k_3 n_1 n_3 + k_2 n_1 n_2^2$$

although some special results may again be reached by simplified calculations as shown later in this section. The result of such computations is shown in Fig. 29 and in Table II. First of all, a pendular motion between n_1 and n_3 is a characteristic feature of odd oxygen in the mesosphere; it leads to a secondary maximum in the height distribution of n_3 around 70 km at night (compare, however, Section 3.5). Such a maximum is shown by some of the few existing nighttime measurements near that level [27,77,78].

Further, we see from Fig. 30 and Table II that also the daytime concentrations of O and O_3 above about 70 km become considerably lower than indicated by the (daytime) equilibrium equation whereby this effect is more pronounced in winter than in summer. It stems from the fact, that at these levels (especially around the mesopause) the equilibrium is with fair approximation, given by a balance between daytime production and nighttime destruction of odd oxygen particles.

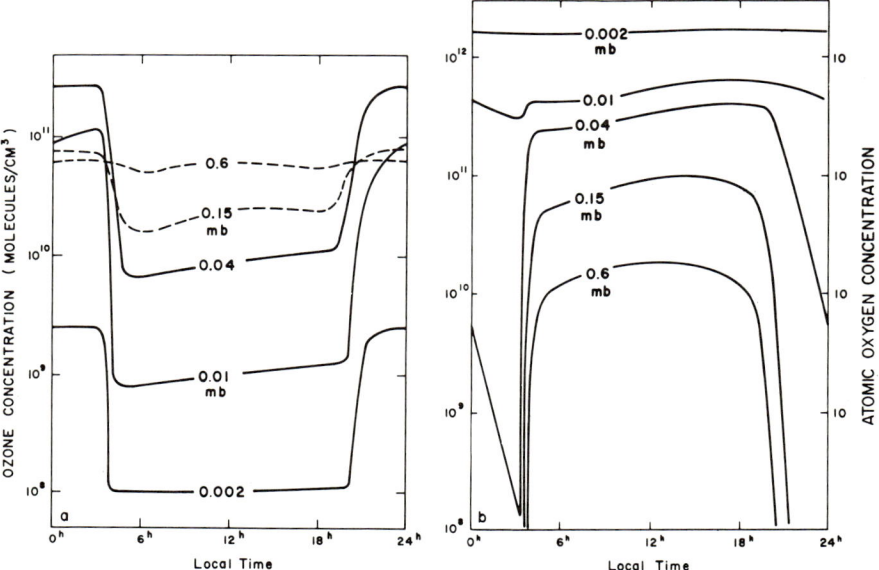

Fig. 29. (a) Day–night variation of atomic oxygen concentration at different levels in the mesosphere (summer, 45° lat). (b) Day–night variation of ozone concentration at different levels in the mesosphere (summer at 45° lat); both calculated from the classical photochemical theory.

The destruction of such particles by reaction [3]

$$\text{(2.28)} \qquad \frac{d\tilde{n}}{dt} = -2n_1 n_3 k_3$$

is rather considerably more efficient at night than in daytime, because with the conversion of the more abundant O to O_3 progressing after sunset, the product $n_1 n_3$ is increased. Even around 80 km, where the main destruction in daytime is by recombination of atomic oxygen, reaction [3] is at night much more efficient than reaction [1] in destroying odd oxygen particles.

The depression of the actual level of n_1 and n_3 against the values computed from the daytime equilibrium equation is at a maximum around the mesopause level during winter at high latitudes. At the levels where atomic oxygen is not further completely removed by conversion to O_3 by reaction [2] a steady state of ozone concentration is reached after some time. As is seen from Eq. (2.27) n_3 may in the absence of photodissociation ($f_3 = 0$) not become larger than $n_1 n_2^2 k_2 / n_1 k_3$, thus

$$\text{(2.29)} \qquad n_3 \leq k n_2^2$$

PHOTOCHEMISTRY OF ATMOSPHERIC OZONE

TABLE II. Mesospheric concentration, production, and destruction of odd oxygen particles

Pressure (mb)	Height (km)	Concentration $\bar{n} = n_1 + n_3$ actual	Concentration day time equilib.	Production of odd oxygen particles per cm³ day	By $O + O_3$ day	By $O + O_3$ dusk and dawn	By $O + O_3$ night	By $O + O + M$ day	By $O + O + M$ dusk and dawn	By $O + O + M$ night	Destruction total production	Destruction total concentration	τ (days)
\multicolumn{14}{c}{summer 45° lat.}
0.6	54	7.85×10^{10}	7.85×10^{10}	5.37×10^{11}	98.2	1.2	—	0.6	—	—	1.2	8.2	0.13
0.3	59.6	7.76	7.76	4.04	97.6	1.6	—	0.8	—	—	1.6	8.3	0.3
0.15	64.8	1.17×10^{11}	1.27×10^{11}	3.27	91	6	0.04	3	—	—	6	16.6	0.6
0.08	69.1	1.92	2.46	2.87	72.3	16.6	2.8	8	0.3	0.3	19.7	29.6	0.8
0.04	73.5	3.49	6.01	2.76	31.1	18.5	29.8	18.8	1.5	1.9	50.1	40.1	1.1
0.02	77.5	4.3	1.08×10^{12}	2.68	6.5	7.9	65.0	16.8	1.9	4.4	76.8	48.8	2.6
0.01	81.4	5.31	1.55	2.37	2.5	7.5	69.5	14.2	1.9	10.1	83.3	39.1	6
0.004	86.6	9.09	2.24	1.86	2.4	5.9	52.1	25.7	3.8	19.5	71.9	13	
0.002	90.7	1.67×10^{12}	2.86	1.42	3.6	4.0	16.6	48.7	7.6		47.7	3.8	
\multicolumn{14}{c}{winter 45° lat.}
0.3	56.0	5.69×10^{10}	5.97×10^{10}	1.52×10^{11}	95.7	3.7	—	0.6	—	—	4.3	10	0.09
0.15	61.2	6.97	8.06	1.11	84.9	13	0.2	1.9	—	—	13.2	21	0.24
0.08	65.7	1.03×10^{11}	1.44×10^{11}	8.87×10^{10}	61.4	29.9	3.9	4.4	0.4	—	32.4	28.6	0.54
0.04	70.4	1.34	3.05	7.31	19.5	20.5	53.5	5.4	0.9	0.2	75.1	44.9	0.87
0.02	74.8	1.17	6.04	6.68	2.6	4.7	89.0	2.5	0.5	0.7	94.9	55.5	0.84
0.01	79.1	1.11	9.69	6.03	0.6	3.3	93.2	1.3	0.3	1.3	98.1	54.3	0.9
0.004	84.7	2.49	1.51×10^{12}	4.97	0.7	3.4	87.7	2.9	0.7	4.6	96.4	20	2.4
0.002	89.1	5.37	1.99	4.08	1.7	2.9	69.7	9.1	2.2	14.4	89.4	6.3	6.7

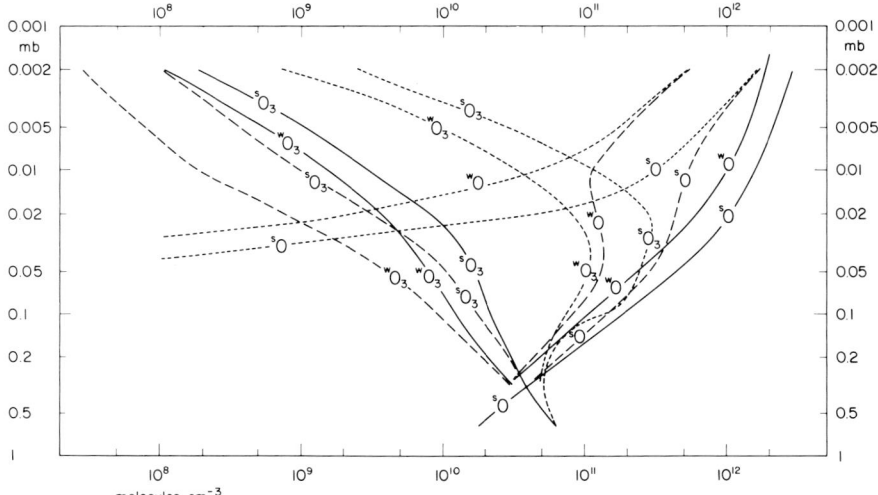

FIG. 30. Vertical distribution of ozone and atomic oxygen in the mesosphere computed from the classical photochemical theory for 45° lat. Index s on the upper left—summer; w—winter. Full lines, noon time equilibrium values; dashed lines, noon time concentration for nonequilibrium (considering daily variations of solar height and night time processes); dotted lines, concentrations at the end of the night.

If the atomic oxygen concentration is not considerably altered until this equilibrium is reached and if it is assumed that the daytime ozone content is negligible compared to this nighttime value (both assumptions being valid above 80 km), the conversion after sunset is given by

$$n_3(t) = k n_2^2 (1 - \exp-[(t-t_0)/\tau_3]) \tag{2.30}$$

with

$$\tau_3 = 1/k_3 n_1 \tag{2.31}$$

This means that the equilibrium is just barely reached within a midlatitude night around 80 km and it is obtained within a few hours at 90 km.

We may thus summarize that up to about 65 km the nighttime value of n_3 reaches 80% or more of the daytime value of the total odd oxygen particle concentration (which starts increasing upward above 55–60 km), whereas above 75–80 km it is at the end of the radiation-free time closely given by Eq. (2.29) and thus rapidly decreasing with increasing height. A secondary nighttime maximum of n_3 (Fig. 30) is situated, as already pointed out, in the gap not covered by either of these approximations. This secondary maximum is, however, hardly shown by the expanded theory including H̃ particles (Section 3); further measurements are thus important because they could give information on the relative importance of that contribution (depending on some, not exactly known, parameters) and possibly on some influences of motions.

2.5.2. *Nonequilibrium and Influence of Motion in the Mesosphere.* Relaxation times for a number of specific processes in the odd oxygen particles photochemistry have been introduced in the preceding section. The time to obtain equilibrium for the total number of oxygen particles can still be adequately described by

(2.32) $$\tau_1 = \tfrac{1}{2}(n_1 + n_3)/2f_2 n_2$$

$\tilde{n} \equiv n_1 + n_3$ has a minimum close to the level where $n_1 = n_3$, i.e., where the removal of odd oxygen particles is (relatively) most efficient. This is around 55–60 km as seen from Table II. Because $\partial f_2/\partial z$ is relatively small in this region, the position of $\tau_{\min}$ is found at an only slightly higher position than that of $\tilde{n}_{\min}$. $\tau_{\min}$ is of the order of one hour. Owing to the upward increase of τ, the daytime equilibrium becomes depressed by the nighttime destruction of odd oxygen particles, as already discussed.

Around the mesopause the relaxation time is again of the order of a day and at 90 km several days. Thus some transport influence must be expected again (as in the lower part of the stratosphere).

The steep gradient of the mixing ratio n_1/n_2 will cause vertical mixing to produce a considerable downward flux of atomic oxygen. Its accurate magnitude can only be estimated because the mixing coefficient Kzz and its dependence on altitude is not well known. Assuming $Kzz = 10^6$ cm^2 sec^{-1} at the mesopause an atomic oxygen flux of about 2×10^{12} atoms cm^{-2} sec^{-1} is obtained. If the mixing power should be not much less than this, a certain increase of odd oxygen particles will result from turbulence in the upper part of the mesosphere (of the order of 30–40% under the assumed conditions at 80 km). Below 70 km such influence is presumably missing. Any turbulent diffusion of ozone in daytime can be neglected because the fluxes are (due to the smaller gradients) much smaller and ozone is also almost instantaneously adjusted to the existing concentration of the more abundant atomic oxygen.

The downward flux of atomic oxygen—if as large as indicated above, i.e., if the turbulent mixing around the mesopause is as high as assumed—would produce a heating of the order of 3° per day (which is about the same as the energy absorbed by O_2, but only one-half to one-third of the heat absorbed by ozone) in the top part of the mesosphere. The downward flux of O corresponds to a latent heat transport, the source being the absorption of solar energy in the thermosphere [79–81].

Vertical motion has similar effects. A downward motion of 2 cm/sec yields around the mesopause about the same atomic oxygen flux as the assumed distribution of large scale mixing. From Eq. (2.20) the local change of any trace substance produced by vertical motions becomes approximately (if local density changes and horizontal transport are neglected):

$$(2.33) \qquad \frac{\partial [X]}{\partial t} = -w n_2 \frac{\partial}{\partial z} \frac{[X]}{n_2} = -w \left(\frac{[X]}{H} + \frac{\partial [X]}{\partial t} \right)$$

where H is the scale height at that level. With a distribution of atomic oxygen as computed from this theory and given by Table II, it is found that also the influence of vertical motions is decreasing downward through the mesosphere but less fast than that of transport by turbulent mixing. Whereas a vertical motion of -2 cm/sec. would increase the atomic oxygen content at 80 km by about 40%, the rise produced by the same vertical speed at 60 km is only about 5%.

2.5.3. Polar Night Conditions.

In mid-latitude nights atomic oxygen practically vanishes below about 70–75 km and it is strongly reduced in another 5 km layer; even around 85 km the reduction is of the order of 20 %. During the polar night, however, almost complete destruction reaches to above 90 km in an atmosphere at rest. In reality the losses are at first replaced by turbulent mixing from above down to the mesopause (at 75 km the concentration would, however, already be drastically depressed) and n_1 would only gradually decrease with the proceeding exhaustion of the reservoir above. This process shall not be discussed here in detail because it is of somewhat academic nature as there is also influx by large scale quasi-horizontal mixing (and possibly meridional circulation) and also because it is considerably complicated in the presence of $\tilde{H}$ particles (see Section 3). It is, however, evident, that at this level atomic oxygen and, to a certain extent, ozone would be valuable tracers for following transport processes if measurement of odd oxygen particles concentration would become possible. In this connection the possible production of atomic oxygen by auroral electrons [59,60], which might considerably alter polar night conditions, especially above the mesopause, and which could also complicate the day/night-variations discussed here and in Section 3, should be further investigated.

2.6. Uncertainties in the Theory

Until about ten years ago the theoretical knowledge (from photochemical calculations) of the vertical ozone distribution was considered to be more, or at least as reliable as the observations. Starting in the mid 1950s, and especially with the IGY, the number and the quality of direct and indirect measurements of the vertical ozone distribution was rapidly increasing and the data began to approach a level where quantitative combination of theory and observations in tracer studies of the general circulation could be considered. Because that meant mainly taking differences between the two, considerably higher demands for accuracy had to be made and the theory, especially many parameters entering the computations, had to be reconsidered under this aspect.

In this section only the remaining uncertainties of a theory in a pure oxygen atmosphere (as defined in Section 2.1) will be discussed, while the still more difficult problem of the possible influence of additional reactions with other trace substances is left to Section 3.

All variables on the right side of Eq. (2.5) are only known as a function of location, altitude and time within certain accuracy limits. With respect to n_2, which is below 90 km proportional to density and thus a unique function of vertical temperature distribution and surface pressure, the limits have become quite narrow. However, k and f_2 as well as f_3 are by far not as well established as one might wish.

2.6.1. Reaction Rates. Widely different values of k_2 and k_3 and by this of $k = k_2/k_3 = k_0 e^{+b/T}$ are found in the literature with respect to their absolute values as well as with respect to their temperature dependence, which is mainly determined by the activation energy of reaction [3]. A comparative discussion is given by Campbell and Nudelmann [82].

A change in temperature dependence (exponential factor b) influences the vertical gradient of ozone in the upper stratosphere (increase of $|\partial n_3/\partial z|$ with increasing value of b), because of the strong vertical temperature gradient in this region (Fig. 31). Below 35 km the influence is less pronounced (small temperature gradient).

A change of the preexponential factor k_0 alters in the upper stratosphere the overall level of the ozone concentration ($n_3 \sim k_0^{1/2}$) has, however, not much influence on the vertical gradient, because an increase in d_3 influences f_2 and f_3 almost equally in this region; it will, however, change that gradient below 35 km because below this level f_2 is more strongly depressed than f_3 by an increase of d_3. If all other parameters were well known, it would probably be the best solution to determine k (i.e., k_0 and b) by trying to obtain the best fit between theory and observations in the upper stratosphere, where ozone concentration can be expected to be close to equilibrium (at least for monthly mean values) [72, 83]. However, as discussed below, the situation is complicated by the uncertainty of other parameters (and especially by the questionable validity of the assumption of a pure oxygen atmosphere); further our observational knowledge of the ozone distribution in the upper stratosphere is not too accurate; apart from a few scattered rocket measurements it is still mainly based on the indirect Umkehr method.

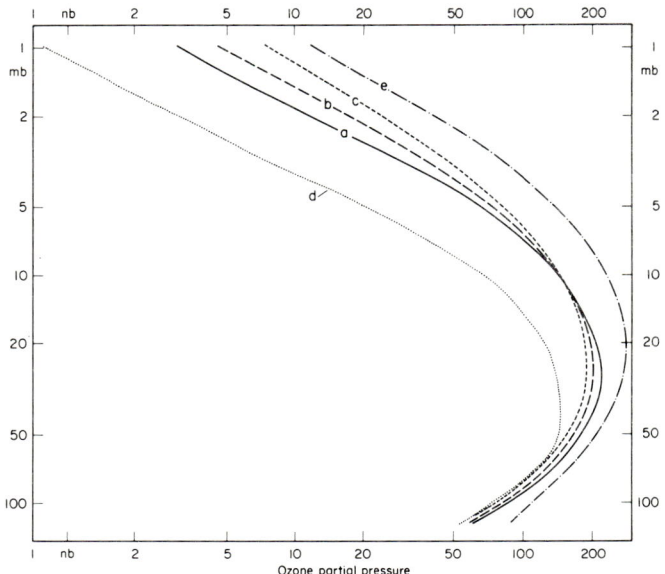

Fig. 31. Influence of different assumptions on the equilibrium constant $k = k_2/k_3 = k_0 e^{b/T}$ on the calculation of vertical ozone distribution. Curve a,b $= 2700°$; curve b,b $= 1700°$; curve c,b $= 700°$ with k_0 chosen to yield equal k values at $T = 230°$K. k_0 was taken 10 times smaller in curve d and 10 times larger in curve e than in a, while b was also $2700°$.

2.6.2. Dissociation Rates.

As is seen from the formulation given in Section 2.2.3, the dissociation rates are, apart from their described dependence on d_2 and d_3, influenced by the solar spectral energy distribution, by the absorption coefficients of oxygen and ozone and possibly by quantum efficiencies differing from unity.

Down to about 3000 Å the sun's spectrum can be observed from the earth's surface; measurements at shorter wavelengths are, however, only possible from rockets and satellites. The specific temperature distribution at the top of the sun's photosphere causes the intensity decrease toward shorter wavelengths, to be in this region much more rapid than would be expected from the visible energy distribution, which resembles closely that of a black body of 5800°K (such a distribution in the uv would produce a thicker ozone layer with a center of gravity at higher altitudes than observed) (Fig. 32) [65].

While above 3000 Å the spectral distribution is well established and no variations (at least of any importance to photochemistry) with time (e.g. within the solar cycle) have been reported, considerable differences remain between the different observations with respect to the intensity of the

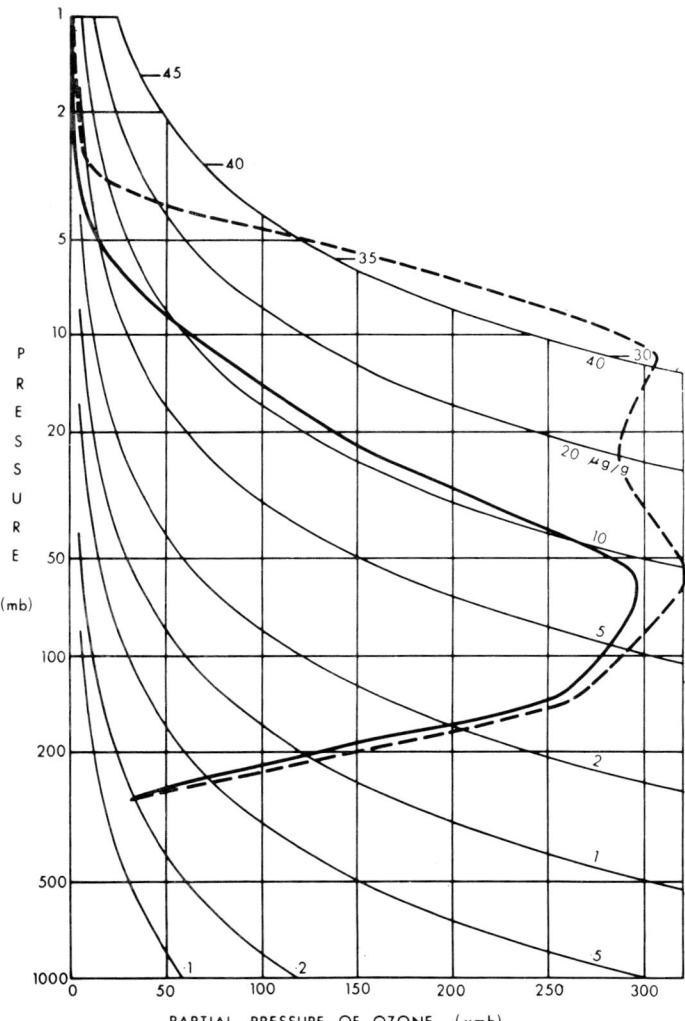

Fig. 32. Influence of UV intensity distribution on the ozone layer. (a) Assumption of 6000°K black-body radiator also in the UV; (b) reduced UV intensity (as actually observed). Results are for vertical solar incidence and are computed with earlier assumptions on the deviation from Beer's law and on stratopause temperatures [65]; also curve (b) is therefore not strictly comparable with more recent results.

ozone-producing radiation around 2100 Å. Values measured by Brewer and Wilson [84] by balloon ascents in this so-called window region (including considerable extrapolation) are about three times lower than those given by Detwiler et al. [85] obtained from rockets, the data which were most widely used in the past in photochemical work.

The use of the Brewer–Wilson values reduces the total amount of ozone computed theoretically by about 20 % and lowers the center of gravity slightly. The relaxation times are increased by about 20 %, too, in the middle stratosphere and still more at higher levels; only below the level of the maximum is that change minor. The potential productivity of the ozone source region (mainly the tropical stratosphere above 20–25 km up to the level of maximum mixing ratio) which has to balance the losses at other places especially near the ground, is altered considerably more than theoretical total amount and relaxation times are changed [86].

Our knowledge of the ozone absorption coefficients seems at present well established by the measurements of Vigroux [4] and Inn and Tanaka [87]. Changes which might yet turn up will thus hardly be large enough to have a significant influence on photochemical results, except possibly if they should be in the region of the overlapping between ozone and oxygen absorption (2000–2300 Å).

The most critical absorption coefficients with respect to stratospheric ozone distribution are those of oxygen in the window region. This is clearly shown by the difference between some earlier photochemical computations, which assumed the validity of Beer's law—which is correct for all other regions of ozone and oxygen absorption in the visible and the uv—also for this particular region and more recent ones taking into account the considerable deviation from Beer's law indicated by Heilpern [74]. Later measurements by Ditchburn and Young [75] substantiated Heilpern's findings although the numerical values were slightly changed.

Taking into account the deviation from Beer's law lowers the center of gravity of the ozone layer and increases the total amount, because in this case more dissociating quanta are absorbed at levels of higher density where they produce according to Eq. (2.5) more ozone. The difference, which is not very big, between the absorption coefficients given by Heilpern and those published by Ditchburn and Young, now generally used, produces a notable difference in results and their reevaluation would thus be of importance.

2.6.3. Excited Oxygen Atoms. The difference between theory and observation with respect to vertical gradient in the upper stratosphere, which is rather difficult to eliminate if one wants to stay within reasonable limits of the not exactly known parameters [83] led to the search for additional reactions which might be of importance.

Hunt [88] looked into the possibility that secondary reactions of the excited (^{1}D)-oxygen atom (O*) produced by the dissociation of oxygen and mainly ozone by energetic quanta might be of importance. The reaction system given in Section 2.2.1 is enlarged as follows:

[2*a]	$O_2 + h\nu$	⟶	$O + O^*$	$a_2 f_2$
[2*b]		⟶	$O + O$	$(1 - a_2) f_2$
[3*a]	$O_3 + h\nu$	⟶	$O_2 + O^*$	$a_3 f_3$
[3*b]		⟶	$O_2 + O$	$(1 - a_3) f_3$
[2]	$O_2 + O + M$	⟶	$O_3 + M$	k_2
[5]	$O_2 + O^* + M$	⟶	$O_3 + M$	k_5
[3]	$O_3 + O$	⟶	$2O_2$	k_3
[6]	$O_3^* + O$	⟶	$2O_1$	k_6
[4]	$O^* + M$	⟶	$O + M$	k_4

a_2 and a_3 give the fraction of dissociating quanta yielding an excited oxygen atom.

While the two body reaction $O_3 + O^*$ is expected to be much faster when excited atoms are involved ($k_6 \gg k_3$) (because of the necessary activation energy), not much difference with respect to the ozone producing three body collision [2] can be expected. Reaction [5] may thus be neglected. The simple reactions scheme shown by Fig. 22 must be replaced by the more complicated one given by Fig. 33. Under stratospheric conditions the production of excited oxygen atoms by dissociation of O_2 is zero ($a_2 \approx 0$); only in the upper mesosphere there is a considerable contribution by $Ly-\alpha$ which shall not be taken into account in the following deduction. Again, the internal reactions in the box in Fig. 33 (conversion of odd oxygen particles into each other) are much faster than those producing and destroying them; thus an internal

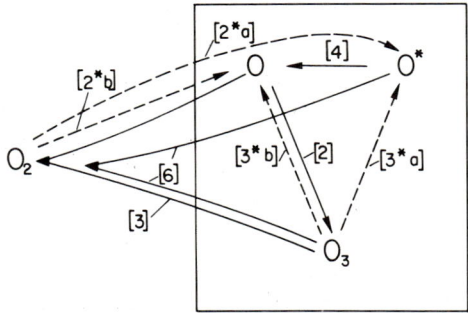

FIG. 33. Reaction scheme in a "dry atmosphere" including excited oxygen atoms, dashed arrows, primary photochemical reactions; full arrows, secondary photochemical reactions.

equilibrium (during sunlit hours) between n_1, n_1^*, and n_3 is attained irrespective of the level of production or destruction.

(2.34) $$k_4 n_1^* n_2 = a_3 f_3 n_3$$

and

(2.35) $$k_2 n_1 n_2^2 = f_3 n_3$$

where k_4 is similarly defined as k_2 by Eq. (2.1). This leads to the following ratios:

(2.36) $$n_1^*/n_3 = a_3 f_3 / k_4 n_2$$
(2.3) $$n_1/n_3 = f_3 / k_2 n_2^2$$
(2.37) $$n_1^*/n_1 = a_3 k_2 n_2 / k_4$$

The balance equation for formation and destruction of odd oxygen particles is expanded to

(2.38) $$2 f_2 n_2 = 2 k_3 n_3 n_1 + 2 k_6 n_3 n_1^*$$

which yields considering (2.36) and (2.37)

(2.39) $$n_3 = n_2^{3/2} \left[f_2 k / f_3 \left(1 + a_3 \frac{k_6}{k_3} \frac{k_2}{k_4} n_2 \right) \right]^{1/2}$$

If

(2.40) $$a_3 \frac{k_6}{k_3} \frac{k_2}{k_4} n_2 \ll 1$$

this reduces to Eq. (2.5), i.e., the additional reactions are of no practical importance. The deactivation rate k_4 is the least-well-established parameter in this equilibrium condition. At the time when Hunt discussed this possibility [88] the values given in the literature varied between 10^{-10} and 10^{-15} cm³ sec⁻¹. In that latter case a strong influence (increasing downward with density) would result. However, such low deactivation rates are quite improbable, because they would lead to a much higher red air glow than observed (the red air glow ($\lambda = 6300$ and 6364 Å) is produced by the radiative deactivation of O(^{1}D) in the upper atmosphere [89] and is thus competing with the deactivation by impact (reaction [4]). With the values of k_4 presently considered as most probable ($10^{-11} < k_4 < 10^{-10}$ cm³ sec⁻¹ [90,91] condition (2.40) is fulfilled and the influence of the reactions with excited oxygen atoms becomes small in a pure oxygen atmosphere. It will, however, be shown in Section 3, that they become of dominating importance in the presence of water vapor.

Primary as well as secondary photochemical reactions produce also excited oxygen molecules. Studies by Hunt [92] and already earlier investigations

[93] have, however, proved that they are of no quantitative importance in the photochemical theory of ozone and atomic oxygen in a pure oxygen atmosphere, although they might instigate many additional reactions not discussed here.

The question of a contribution to ozone photochemistry by excited ^{1}S-oxygen atoms which might be produced through the absorption by ozone of wavelengths shorter than 2340 Å has recently been raised by Crutzen [94]. Although the fraction of quanta yielding such particles is certainly much smaller than a_3 because this transition is spin-forbidden, the concentration of $O(^1S)$ could at least at certain levels become higher than that of $O(^1D)$ because of the much faster deactivation of the latter. The problem, which could be of importance in the theory of a "moist" stratosphere (next Section), shall not be further discussed there because the necessary quantitative information is not available at present.

3. Ozone Photochemistry in a "Moist" Atmosphere

3.1. Introduction

Bates and Nicolet [69] demonstrated in 1950 that active radicals produced from water vapor have a considerable influence on atmospheric distribution of odd oxygen particles ($\tilde{O}$). It was then assumed that such particles (H, OH, HO$_2$) were only produced by direct photodissociation of water vapor [reaction 1*]. Since the dissociation energy is 5.2 eV, only wavelength shorter than 2390 Å will contribute; further the absorption cross sections of water vapor are very small above 1950 Å; thus f_1 becomes almost negligible below the stratopause; and it was therefore thought that the additional reactions becoming possible on this basis would only be of importance in the mesosphere.

It was not before 1964, that Hampson [70] suggested—on the basis of laboratory measurements of McGrath and Norrish [95]—that the water molecule might also be split by reaction [21] with excited (^{1}D)-oxygen atoms (O*). Such excited atoms are produced by photodissociation of ozone by wavelengths shorter than about 3100 Å and also by absorption of oxygen in the Schumann–Runge continuum ($\lambda \leq 1749$ Å) and at still shorter wavelengths. While the latter reactions give only a contribution through $Ly-\alpha$ in the upper part of the mesosphere, the predominant part of ozone photodissociation in the upper stratosphere above 35 km produces excited O atoms and, although this fraction (a_3) decreases rapidly below 35–40 km, there is still an important contribution down to the lower stratosphere and even to the earth's surface. H radicals ($\tilde{H}$) as the ensemble of H, OH, and HO$_2$ will

TABLE III. Tabulation of reactions pertaining the photochemistry of atmospheric ozone and atomic oxygen.

a	b		c		d	e	f
S M	[3*a]	$O_3 + h\nu$	$\rightarrow$	$O_2 + O$	$a_3 f_3$		[96]
S M	[3*b]		$\rightarrow$	$O_2 + O$	$(1 - a_3) f_3$		
S M	[2]	$O_2 + O + M$	$\rightarrow$	$O_3 + M$	k_2	$8.2 \times 10^{-35} \exp(890/RT)$	[90, 91]
S M	[4]	$O^* + M$	$\rightarrow$	$O + M$	k_4	3.8×10^{-11}	
—	[5]	$O_2 + O^* + M$	$\rightarrow$	$O_3 + O^* + M$	k_5	$\approx k_2$	
— M	[2*a]	$O_2 + h\nu$	$\rightarrow$	$O + O^*$	$a_2 f_2$		[97]
S M	[2*b]		$\rightarrow$	$O + O$	$(1 - a_2) f_2$		
S M	[1]	$O + O + M$	$\rightarrow$	$O_2 + M$	k_1	2.7×10^{-33}	[96, 98]
S	[3]	$O_3 + O$	$\rightarrow$	$2O_2$	k_3	$3.4 \times 10^{-11} \exp(-4500/RT)$	[99]
S	[6]	$O_3 + O^*$	$\rightarrow$	$2O_2$	k_6	3×10^{-10}	
—	[7]	$O + O^* + M$	$\rightarrow$	$O_2 + M$	k_7	$\approx k_1$	
—	[5*]	$HO_2 + h\nu$	$\rightarrow$	$OH + O$	f_5		[69]
S M	[10]	$O_3 + H$	$\rightarrow$	$HO_2 + O$	k_{10}	$2.0 \times 10^{-10} \exp(-4000/RT)$	[100]
S M	[11]	$O_3 + H$	$\rightarrow$	$OH + O_2$	k_{11}	2.6×10^{-11}	[100]
S M	[12]	$O_3 + OH$	$\rightarrow$	$HO_2 + O_2$	k_{12}	5×10^{-13}	[71]
S M	[13]	$O_3 + HO_2$	$\rightarrow$	$OH + 2O_2$	k_{13}	1×10^{-14}	[101]
—	[14]	$O + OH + M$	$\rightarrow$	$HO_2 + M$	k_{14}	1.4×10^{-31}	[100]
S M	[15]	$O + OH$	$\rightarrow$	$H + O_2$	k_{15}	5×10^{-11}	[101]
S M	[16]	$O + HO_2$	$\rightarrow$	$OH + O_2$	k_{16}	1×10^{-11}	[100]
—	[17]	$H + O_2$	$\rightarrow$	$OH + O$	k_{17}	$1 \times 10^{-9} \exp(-16800/RT)$	[102]
S M	[18]	$H + O_2 + M$	$\rightarrow$	$HO_2 + M$	k_{18}	$3 \times 10^{-32} (273/T)^{1/3}$	[69]
—	[19]	$H + HO_2$	$\rightarrow$	$OH + M$	k_{19}	8×10^{-32}	[69]
—	[20]	$H + HO_2$	$\rightarrow$	$2OH$	k_{20}	3×10^{-12}	
(S) M	[1*]	$H_2O + h\nu$	$\rightarrow$	$H + OH$	f_1		
S M	[21]	$O^* + H_2O$	$\rightarrow$	$2OH$	k_{21}	1×10^{-11}	[71]
(S) M	[22]	$O^* + H_2$	$\rightarrow$	$OH + H$	k_{22}	1×10^{-11}	[71]
(S) M	[23]	$OH + OH$	$\rightarrow$	$H_2O + O$	k_{23}	2×10^{-12}	[100]

PHOTOCHEMISTRY OF ATMOSPHERIC OZONE

		Reaction		Rate	Ref.
S M	[24]	$OH + HO_2 \longrightarrow$	$H_2O + O_2$	k_{24}	1×10^{-11}	[100]
S(M)	[25]	$OH + H_2O_2 \longrightarrow$	$H_2O + HO_2$	k_{25}	4×10^{-13}	[103]
S M	[26]	$HO_2 + HO_2 \longrightarrow$	$H_2O_2 + O_2$	k_{26}	1.5×10^{-12}	[100]
S M	[4*]	$H_2O_2 + h\nu \longrightarrow$	2OH	f_4		[103]
(S)	[27]	$O + H_2O_2 \longrightarrow$	$OH + HO_2$	k_{27}	1×10^{-15}	[103]
M	[28]	$H + HO_2 \longrightarrow$	$H_2 + O_2$	k_{28}	3×10^{-12}	[104]
	[29]	$O + H_2 \longrightarrow$	$OH + H$	k_{29}	$7 \times 10^{-11} \exp(-10200/RT)$	
[S][M]	[30]	$H_2 + OH \longrightarrow$	$H_2O + H$	k_{30}	1×10^{-13}	[100]
[S][M]	[31]	$H + H_2O_2 \longrightarrow$	$H_2 + HO_2$	k_{31}	$3 \times 10^{-11} \exp(-8300/RT)$	[103]
—	[32]	$H + OH \longrightarrow$	$H_2 + O$	k_{32}	2.5×10^{-31}	[100]
—	[33]	$H + OH + M \longrightarrow$	$H_2O + M$	k_{33}	$2 \times 10^{-10} \exp(-4000/RT)$	[100]
—	[34]	$H + HO_2 \longrightarrow$	$H_2O + O$	k_{34}	$1.2 \times 10^{-32}(273/T)^{0.7}$	[69]
—	[35]	$H + H + M \longrightarrow$	$H_2 + M$	k_{35}	3×10^{-11}	[102]
S	[36]	$CH_4 + O* \longrightarrow$	$CH_3 + OH$	k_{36}		[105]
S	[41]	$NO + O_3 \longrightarrow$	$NO_2 + O_2$	k_{41}	$1.7 \times 10^{-12} \exp(-2600/RT)$	[106]
S	[42]	$NO_2 + O \longrightarrow$	$NO + O_2$	k_{42}	$3.2 \times 10^{-11} \exp(-1050/RT)$	[106]
S	[6*]	$NO_2 + h\nu \longrightarrow$	$NO + O$	f_6		

[a] S means that this reaction is important for stratospheric, M for mesospheric photochemistry; S or M in round brackets: only marginal importance; square brackets: only of importance for the photochemistry of H_2 or H_2O.
[b] Reaction number (* = primary photochemical reaction).
[c] Reaction.
[d] Designation of reaction or dissociation rate.
[e] Value of dissociation rate according to Hesstvedt [90].
[f] Original references on reaction rates.

In the formulation used in this paper all rates concerning three body reactions must be multiplied with 4.75 (or another number depending on the relative efficiency of O_2 and N_2 as third bodies), because n_2 has been inserted as concentration of the third body.

be called, are also to be expected in the stratosphere and their possible influence must be considered in the photochemical theory of ozone.[4]

As it is seen from Table III the number of reactions has become manifold compared to the "classical" theory and if the full set of equations is used we have to deal with nine variable trace-substances (O, O*, O_3, H, OH, HO_2, H_2, H_2O, and H_2O_2) instead of the initial two (O_3 and O). Thus there are nine interrelated equilibrium equations and this full set can only be solved numerically.

It can, however, be shown that at certain levels the problem can nevertheless be treated analytically with a fair approximation if the relative importance of different reactions is considered [72,107,108]. This not completely accurate procedure gives a much better insight in the photochemical processes than the simple numerical results and allows thus more easily to check on the errors introduced by the uncertainty of a number of reaction rates and other parameters. Such an approximative theory will thus be developed for stratospheric conditions in the next section.

3.2. Semiquantitative Theory of Stratospheric Ozone

Table IV is a condensed form of the complete tabulation of reactions given in Table III, showing only those which are of importance to the ozone photochemistry below the stratopause. Included are the reaction rates and the number of reactions per cubic centimeter and second at three different levels. These 17 remaining reactions are distributed into five groups:

  I: converting odd oxygen particles ($\tilde{O}$) into each other
  II: producing and destroying such particles
  III: converting H radicals ($\tilde{H}$) into each other
  IV: producing and destroying such particles (including H_2O_2)
  V: conversion between active hydrogen particles ($\tilde{H}$) and H_2O_2

There is an overlap between groups II and III; i.e., $\tilde{H}$ particles are important in determining the $\tilde{O}$-particles budget (i.e., in the stratosphere the ozone concentration). This means, as is also shown by Fig. 34, that two equilibria (between O_2 and the O particles on one side and H_2O and $\tilde{H}$ particles on the other side) are interrelated with each other; the second of those equilibria is further complicated by the interference of H_2O_2. Fortunately, however, there is a pronounced difference in speed between the reactions belonging to the different groups (I, II/III, and IV), which allows, as in the derivation of

---

[4] $\tilde{H}$ particles are also produced by reaction of methane with excited O atoms [reaction 36]. As the effect is the same as that of a correspondingly higher value of the not very well-known water vapor concentration, this sequence is not further considered here.

TABLE IV. Reactions relating to ozone photochemistry below stratopause in a "moist" atmosphere

a		b			c			d			e
					1.38 mb	7.81 mb	44.2 mb	1.38 mb	7.81 mb	44.2 mb	
[2]	$O_2 + O + M \rightarrow O_3 + M$	k_2		$cm^6 sec^{-1}$	4.2(−34)	5.3(−34)	6.3(−34)	2.9(+8)	1.4(+9)	7.1(+8)	I
[3_a*]	$O_3 + h\nu \rightarrow O^* + O_2$	$a_3 f_3$		sec^{-1}	2.8(−3)	3.4(−4)	8.5(−5)	2.5(+8)	6.2(+8)	1.1(+8)	
[3_b*]	$O_3 + h\nu \rightarrow O + O_2$	$(1 − a_3)f_3$		sec^{-1}	4.7(−4)	4.6(−4)	4.4(−4)	4.2(+7)	8.3(+8)	5.9(+8)	
[4]	$O^* + M \rightarrow O + M$	k_4		$cm^3 sec^{-1}$	3.8(−11)	same	same	2.5(+8)	6.2(+8)	1.1(+8)	
[2*]	$O_2 + h\nu \rightarrow O + O$	f_2		sec^{-1}	3.0(−10)	5.0(−11)	1.2(−12)	4.7(+6)	4.9(+6)	7.4(+5)	II
[3]	$O_3 + O \rightarrow 2O_2$	k_3		$cm^3 sec^{-1}$	9.1(−15)	3.0(−15)	1.3(−15)	3.9(+6)	2.5(+6)	9.1(+3)	
[6]	$O_3 + O^* \rightarrow 2O_2$	k_6		$cm^3 sec^{-1}$	3.0(−10)	same	same	9.4(+3)	7.4(+4)	1.7(+3)	
[1]	$(O + O + M) \rightarrow O_2 + M$	k_1		$cm^6 sec^{-1}$	2.7(−33)	same	same	5.0(+3)	3.2(+1)	2.5(−2)	
[11]	$O_3 + H \rightarrow OH + O_2$	k_{11}		$cm^3 sec^{-1}$	2.6(−11)	same	same	6.6(+4)	1.6(+3)	1.4(−1)	III
[12]	$O_3 + OH \rightarrow HO_2 + O_2$	k_{12}		$cm^3 sec^{-1}$	5.0(−13)	same	same	1.2(+5)	1.2(+6)	3.7(+5)	
[13]	$O_3 + HO_2 \rightarrow OH + 2O_2$	k_{13}		$cm^3 sec^{-1}$	1.0(−14)	same	same	1.3(+4)	1.0(+6)	3.6(+5)	
[15]	$O + OH \rightarrow H + O_2$	k_{15}		$cm^3 sec^{-1}$	5.0(−11)	same	same	3.1(+5)	1.5(+4)	7.0(+1)	
[16]	$O + HO_2 \rightarrow OH + O_2$	k_{16}		$cm^3 sec^{-1}$	1.0(−11)	same	same	3.5(+5)	1.4(+5)	6.8(+2)	
[18]	$H + O_2 + M \rightarrow HO_2 + M$	k_{18}		$cm^6 sec^{-1}$	1.2(−32)	1.3(−32)	1.4−(32)	2.4(+5)	1.4(+4)	7.0(+1)	
[1*]	$H_2O + h\nu \rightarrow H + OH$	f_1		sec^{-1}	2.6(−10)	7.5(−12)	2.5(−16)	4.8(+1)	8.7(+0)	1.8(−3)	IV
[21]	$O^* + H_2O \rightarrow 2OH$	k_{21}		$cm^3 sec^{-1}$	1.0(−11)	same	same	3.2(+2)	8.1(+2)	1.5(+2)	
[22]	$(O^* + H_2) \rightarrow OH + H$	k_{22}		$cm^3 sec^{-1}$	1.0(−11)	same	same	2.0(−1)	3.2(−3)	6.4(−8)	
[23]	$OH + OH \rightarrow H_2O + O$	k_{23}		$cm^3 sec^{-1}$	2.0(−12)	same	same	1.3(+1)	3.2(+0)	6.4(−1)	
[24]	$OH + HO_2 \rightarrow H_2O + O_2$	k_{24}		$cm^3 sec^{-1}$	1.0(−11)	same	same	3.7(+2)	7.8(+2)	1.4(+2)	
[25]	$OH + H_2O_2 \rightarrow H_2O + HO_2$	k_{25}		$cm^3 sec^{-1}$	4.0(−13)	same	same	2.2(+0)	4.3(+1)	6.7(+0)	
[26]	$HO_2 + HO_2 \rightarrow H_2O_2 + O_2$	k_{26}		$cm^3 sec^{-1}$	1.5(−12)	same	same	3.1(+2)	5.1(+3)	1.1(+5)	V
[4*]	$H_2O_2 + h\nu \rightarrow 2OH$	f_4		sec^{-1}	1.4(−4)	6.3(−5)	3.6(−5)	3.0(+2)	5.1(+3)	1.1(+3)	

a: reactions; b: designation and dimension of rate constants; c: rate constants at three different levels (same means that this rate is independent of the level); d: speed of reaction in molecules $cm^{-3} sec^{-1}$ (under c and d the second number (integer with sign) gives the power of ten); e: reaction group.

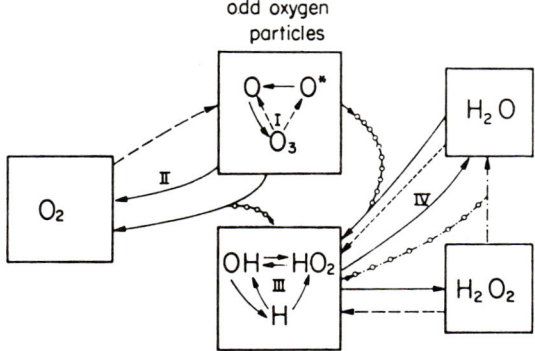

Fig. 34. Reaction scheme of the interplay of the Õ- and H̃-systems in the stratosphere.

the classical theory, a stepwise procedure yielding considerable simplifications.

The conversion between Õ particles is still orders of magnitude faster (in daytime) than any other process (when the number of reactions per cubic centimeter and second are compared). Thus the initial equilibrium between such particles remains unchanged as already given by Eq. (2.3), (2.36), and (2.37). Secondly, the conversion of H̃ particles into each other (mostly combined with destruction of Õ particles) is about two orders of magnitude faster than their formation and destruction; these equilibria are influenced by the concentration of Õ particles, which is, because of the overlaps of groups II and III, on the other hand a function of H̃ particle content. However, except at the top of the stratosphere, the H̃-particle equilibria depend only on the ratios of the Õ-particle concentrations which are fixed by the internal reactions in group I.

The concentrations of H, OH, and HO_2 radicals will be designated by x, y, and z, respectively, and those of H_2O, H_2, and H_2O_2 by w_1, w_2, and w_3. Then the equilibrium between x and y is determined by reactions [11], [15], and [18]:

(3.1) $$k_{11} n_3 x + k_{18} n_2^2 x = k_{15} n_1 y$$

thus

(3.2) $$\frac{x}{y} = \frac{k_{15} n_1}{k_{11} n_3 + k_{18} n_2^2} = \frac{k_{15} f_3 n_3}{k_2 n_2^2 (k_{11} n_3 + k_{18} n_2^2)}$$

This means that

(3.3) $$x/y < n_1/n_3$$

x is thus decreasing very rapidly with decreasing altitude (see Fig. 39) and reactions with hydrogen atoms become of only marginal importance for stratospheric ozone content.

The equilibrium between y and z is given by reactions[5] [11], [12], [13], [15] and [16]:

(3.4) $$k_{11} n_3 x + k_{16} n_1 z + k_{13} n_3 z = k_{12} n_3 y + k_{15} n_1 y$$

Considering (2.3) and (3.2) this yields

(3.5) $$\frac{z}{y} = \frac{k_{12} + k_{15} \dfrac{n_1}{n_3}\left(1 - \dfrac{k_{11} n_3}{k_{11} n_3 + k_{18} n_1^2}\right)}{k_{13} + k_{16}(n_1/n_3)}$$

$$= \frac{k_{12} + k_{15} \dfrac{f_3}{k_2 n_2^2}\left(1 - \dfrac{k_{11} n_3}{k_{11} n_3 + k_{18} n_2}\right)}{k_{13} + k_{16}(f_3/k_2 n_2^2)}$$

For most levels this rather complicated relation becomes much simpler, namely

(3.6) $$R \equiv z/y \approx k_{12}/k_{13} \approx 35$$

below about 35 km, where $n_1 \ll n_3$, i.e., where the destruction of O_3 predominates; and

(3.7) $$R \equiv z/y \approx k_{15}/k_{16} \approx 5$$

above about 45 km, where mostly atomic oxygen is destroyed by the OH–HO_2 cycle. Between these levels there is a continuous transition. The actual numbers may yet differ from those shown here, because there is still a considerable uncertainty about some of the reaction rates involved, especially about k_{13}. Some authors (Crutzen [94] and Kaufman [109]) even question whether reaction [13] occurs (there are no measurements on it). The absence of an OH–HO_2 cycle via ozone destruction would considerably change the deduction used in this section and the changes thereby introduced are briefly discussed in Section 3.2.2.

If now reaction [25] which is only about one-tenth as effective in destroying $\tilde{H}$ particles (via H_2O_2) than reaction [24] is disregarded, y can be computed easily from the equilibrium assumption for the formation and destruction of $\tilde{H}$ particles:

(3.8) $$2k_{21} w_1 n_1^* = 2k_{24} yz = 2k_{24} Ry^2$$

---

[5] The contribution of the photodissociation of H_2O_2 (i.e., the conversion HO_2 to OH via H_2O_2 is neglected.

Considering Eq. (2.36), this yields

(3.9) $$y = \left(n_3 \frac{w_1}{n_2} \frac{k_{21} a_3 f_3}{k_{24} k_4 R} \right)^{1/2}$$

whereby w_1/n_2 is 4.75 times the number mixing ratio of water vapor.

The omission of reaction [25] is presumably the most aggravating simplification made in this semiquantitative theory; however, its effect on the numerical results is certainly smaller than the remaining uncertainties about many of the reaction rates and other parameters, except under the conditions discussed in the next section.

The H_2O_2 concentration can be derived from Eqs. (3.6) and (3.9) by neglecting with only a small error reaction [25] against the photodissociation of hydrogen peroxide [4*]

(3.10) $$f_4 w_3 = k_{26} z^2$$

and

(3.11) $$w_3 = R \frac{a_3 f_3}{f_4} \frac{k_{21} k_{26}}{k_4 k_{24}} \frac{w_1}{n_2} n_3$$

The hydrogen peroxide content of the stratosphere is thus proportional to the ozone concentration and the water vapor mixing ratio.

The equilibrium concentration of ozone may now be obtained by balancing the odd oxygen particle production through photodissociation of oxygen [2*] by the destruction through reactions [3], [12], [13], [15], and [16]

(3.12) $$2 f_2 n_2 = n_3 \left(\frac{2 f_3}{k n_2^2} n_3 + \frac{f_3}{k_2 n_2^2} (k_{15} y + k_{16} z) + k_{12} y + k_{13} z \right)$$

Omitting here reaction [11] is much more justified that not taking into account reaction [25] in Eq. (3.8). Because $k_{15} y \approx k_{16} z$ above 45 km and $k_{12} y \approx k_{13} z$ below 35 km this can be simplified (again with a certain error in the intermediate layer) to

(3.13) $$f_2 n_2 = n_3 \left(\frac{f_3}{k n_2^2} n_3 + \frac{f_3}{k_2 n_2^2} k_{15} y + k_{12} y \right)$$

Since $y \approx n_3^{1/2}$ [Eq. (3.9)] this leads to a fourth-order equation for the ozone concentrations:

(3.14) $$f_2 n_2 = n_3 [A n_3 + (B+C) n_3^{1/2}]$$

with

(3.15) $$A = f_3 / k n_2^2$$

(3.16) $$B = \frac{f_3}{k_2 n_2^2} \left(\frac{a_3 f_3 k_{21}}{R k_4 k_{24}} \frac{w_1}{n_2} \right)^{1/2}$$

(3.17) $$C = \left(\frac{a_3 f_3 k_{21}}{R k_4 k_{24}} \frac{w_1}{n_2} \right)^{1/2}$$

Considerable insight into the relative importance of the different processes for the ozone equilibrium can be obtained from this equation without solving it formally.

3.2.1. *Discussion of Results.* The first term in Eq. (3.14) (with coefficient A) represents the $\tilde{O}$ destruction in a pure oxygen atmosphere; the other two give the contribution from water vapor derivatives, namely the term with B through reactions with atomic oxygen and that with C of ozone with $\tilde{H}$ particles. If the reaction rates assembled by Hunt [71] in his thorough treatise of the ozone photochemistry in a "moist" atmosphere are used, the first term becomes almost negligible (Fig. 35d); the destruction of $\tilde{O}$ particles is fully dominated by the reactions with H radicals, whereby the term with B predominates in the upper stratosphere, the term with C below 30–35 km. The ozone concentration below 25–30 km becomes very much depressed by these high additional contributions to $\tilde{O}$ particle destruction and is far below the observed values (Fig. 36a–d).

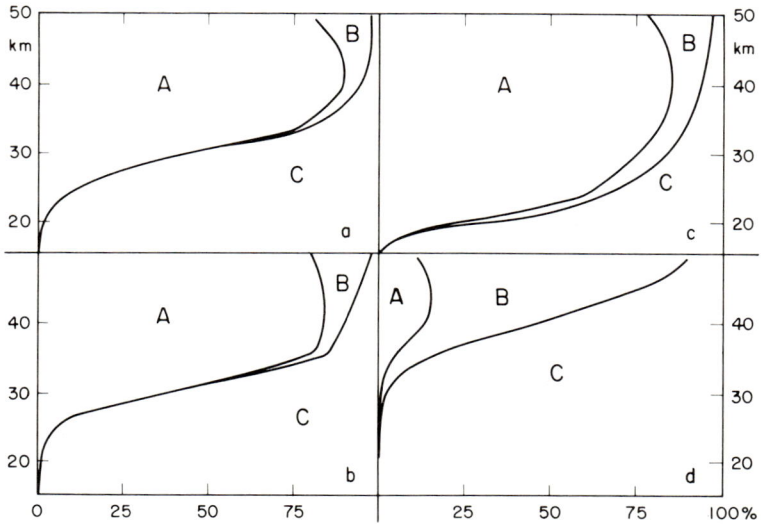

FIG. 35. Contribution (in percent) to odd oxygen particles destruction by different mechanisms (A, $O + O_3$, B, $O + \tilde{H}$, C, $O_3 + \tilde{H}$). (a) Reaction rates after Hesstvedt [90], water vapor mixing ratio 3×10^{-6}; (b) Reaction rates after Hesstvedt [90]; water vapor mixing ratio 5×10^{-6}; (c) Reaction rates after Hesstvedt [90]; but $k_{13} = 0$; mixing ratio 5×10^{-6}; (d) Reaction rates after Hunt [71]; mixing ratio 3×10^{-6}.

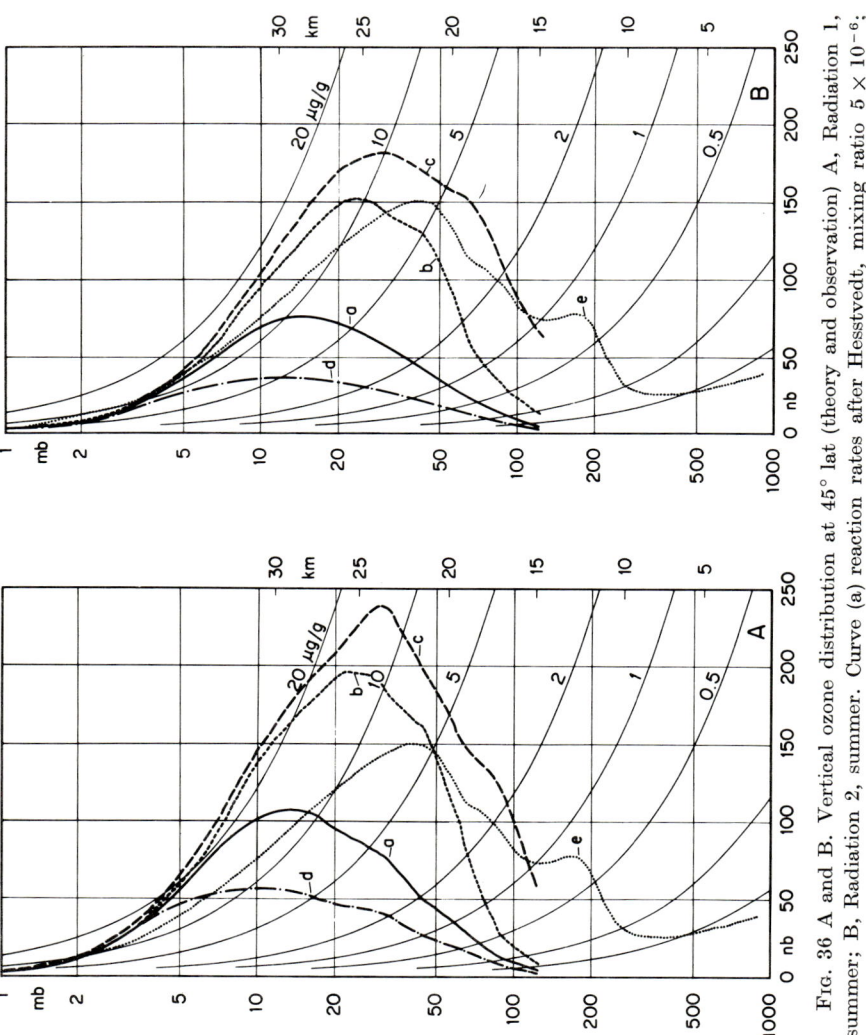

FIG. 36 A and B. Vertical ozone distribution at 45° lat (theory and observation) A, Radiation 1, summer; B, Radiation 2, summer. Curve (a) reaction rates after Hesstvedt, mixing ratio 5×10^{-6}; (b) reaction rates after Hesstvedt, but $k_{13} = 0$; mixing ratio 5×10^{-6}; (c) classical theory; (d) reaction rates after Hunt; (e) observed.

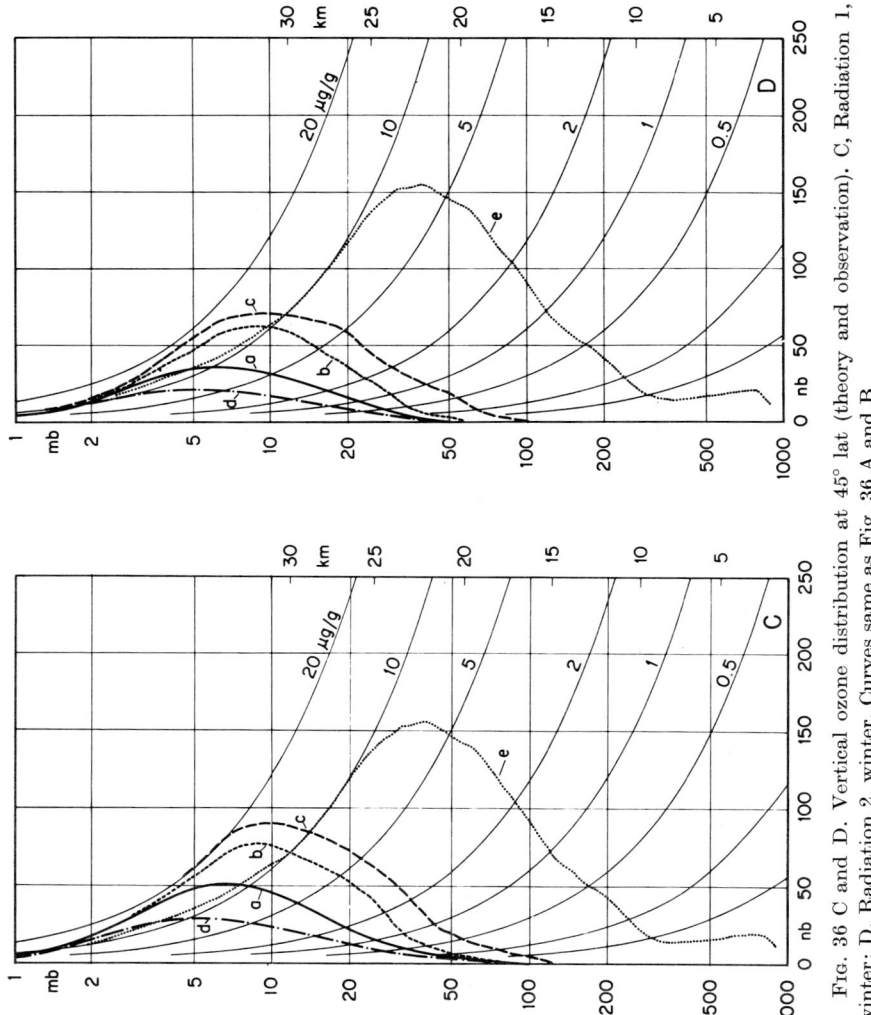

Fig. 36 C and D. Vertical ozone distribution at 45° lat (theory and observation). C, Radiation 1, winter; D, Radiation 2, winter. Curves same as Fig. 36 A and B.

Any discrepancy between theory and observations below about 25 km can be easily explained in the classical theory by ozone redistribution by air motions, because of the very long relaxation times at these levels. This does, however, not apply in the photochemistry of a "moist" stratosphere (at least if the parameter values given by Hunt [71] are used).

By a derivation similar to that in Eq. (2.11)–(2.13) it can be shown that the relaxation time is still given by

$$(3.18) \qquad \tau = \frac{1}{m} \frac{n_{3,0}}{2 f_2 n_2}$$

whereby m is equal to the power under which n_3 appears in the term which is predominant in ozone destruction in Eq. (3.14). The τ values obtained from this formula with Hunt's parameter set turn out to be much shorter than in the classical theory (Fig. 37). Not only is $n_{3,0}$ much smaller than in a dry atmosphere, but at the same time f_2 is considerably increased because of the decrease of absorption by ozone of the oxygen dissociative radiation around 2100 Å (dependence of f_2 on d_3). With relaxation times as low as some days or a few weeks, even below 20 km, differences between theory and observation as shown in Fig. 36 cannot develop, and it would even be very hard to visualize how the well-known discrepancy between observed and theoretically computed total ozone distribution with respect to latitude and season could exist. This does not prove, however, that $\tilde{H}$-particle reactions are of no importance for the ozone photochemistry, but it indicates that some of the reaction rates quoted by Hunt must be in error [71]. When the set of such rates more recently compiled by Hesstvedt [90] is applied, the results are considerably different from those quoted above. Mainly a higher value for k_4 (3.8×10^{-11} cm^3 sec^{-1}) has been used leading to a decrease in the O* concentration and thus of all values of the whole $\tilde{H}$ system, while also lower numbers for $k = k_2/k_3$ were assumed. Under such conditions $\tilde{O}$-particle removal by reaction [3] (representing destruction in the classical theory) and the corresponding terms for a moist atmosphere (mainly [12], [13], [15], and [16] become of about equal importance) (Fig. 35a–b). The relaxation times are at the same time considerably increased compared to those obtained with Hunt's parameter set (Fig. 37) although they are still notably smaller than those computed for the classical theory in the middle stratosphere and below; "protected regions," i.e., regions where n_3 can stay high above the equilibrium value for an extended length of time, thus exist in the lower stratosphere. This is of big importance for the interplay between transport processes and photochemistry as it will be discussed in Section 4.

At the levels for which the appropriate curve in Fig. 37 crosses the hatched

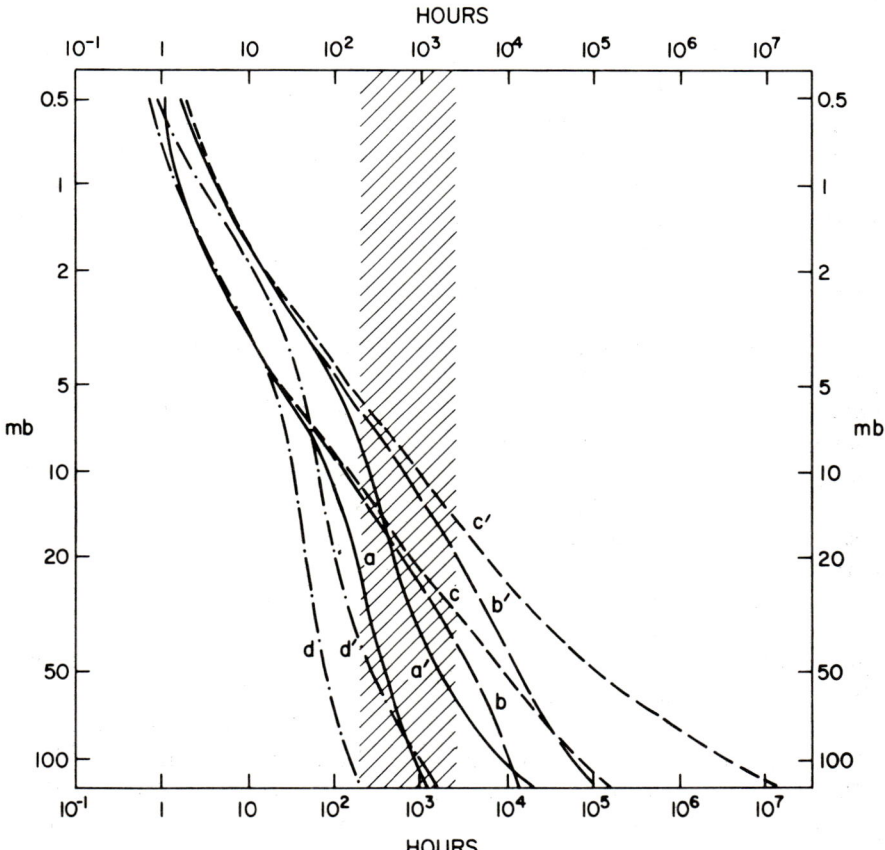

Fig. 37. Relaxation times as a function of pressure (height); summer (a–d) and winter (a'–d'). (a) Reaction rates after Hesstvedt, mixing ratio 3×10^{-6}; (b) Reaction rates after Hesstvedt, but $k_{13} = 0$; mixing ratio 3×10^{-6}; (c) classical theory; (d) reaction rates after Hunt, mixing ratio 3×10^{-6}. At all levels where the applicable curve is to the left of the hatched zone (mean) ozone values are close to photochemical equilibrium; if it is to the right of this zone the influence of air motions is predominant; in the hatched zone the two are of comparable importance.

zone photochemistry and transport processes compete in determining the ozone concentration. At higher altitudes monthly mean values must be close to photochemical equilibrium; below the ozone distribution is mainly determined by the general circulation. While in the classical case the motion influence dominates in winter also in the middle stratosphere (and in summer at least up to its lower border), this layer becomes the transition region for winter conditions in the "moist" theory (when Hesstvedt's rates are used) and in summer there is some photochemical influence down to the 100-mb level. (With Hunt's rates photochemical equilibrium would in summer prevail until below 20 km and even in winter deviations from it would stay moderate down to that level in obvious contradiction to the observations.) If in the "moist" theory k_{13} is assumed to be very small (as discussed below), the borders between the three regions of practically no, moderate, or dominant transport influence are only slightly lowered in comparison to the classical case.

3.2.2. The Problem of Reaction [13]. With the use of the reaction rates proposed by Hesstvedt [90] the ozone photochemistry of a moist stratosphere obviously yields results which lie within reasonable limits and seem to be compatible with the observed behavior of the trace gas (Fig. 36), although certain discrepancies still remain (they are discussed in the next section). Nevertheless the assumed set of reaction rates (as well as the values of the dissociation rates) are certainly not final. The most questionable assumption is that on reaction [13] ($O_3 + HO_2$). As already mentioned this process has not been really observed and reputated reaction chemists warn against its inclusion [109]. While this is not of much consequence in the upper stratosphere, where mainly atomic oxygen reacts with $\tilde{H}$ particles, the whole photochemistry would be considerably changed below about 35 km.

If reaction [13] should be non-existent (or very slow) the simple chain by which $\tilde{O}$-particles are destroyed is broken and the reconversion of HO_2 into OH would, according to Eq. (3.4), also at low levels, be produced by reaction [16] with atomic oxygen and thus become very slow (because n_1 is small). However, under such conditions the conversion via H_2O_2 by the subsequent reactions [26] and [4*] which was neglected in Eq. (3.4) (being at least one order of magnitude slower than the faster processes on its right-hand side) must also be considered and becomes even dominant below about 25 km. At the same time reaction [25] can also not be disregarded any more (its contributions to $\tilde{H}$-particle destruction amounts to almost one-third around 25 km). Such complications prevent the deduction of a relatively simple analytical solution of the form of Eq. (3.14).

While in the upper stratosphere the results obtained by numerical solution correspond quite closely to those computed before for the "moist" case [Eq. (3.13)] they are closer to those of the classical theory [Eq. (2.5)] between 30 and 20 km with respect to ozone concentration as well as relaxation times, but deviate again from them at still lower levels (Figs. 36 and 37).

3.2.3. Other Remaining Uncertainties. Figure 36 indicates, as already mentioned, reasonably good agreement between theory and observation (when Hesstvedt's set of rates is used) at the levels where according to Fig. 37 the photochemical influence predominates. The best fit is obtained with the Brewer–Wilson [84] radiation data (radiation 2) and the agreement might be improved by a k_{13} value somewhat lower than assumed by Hesstvedt and Hunt (but not zero). If, however, the results are plotted on a logarithmic scale (Fig. 38) some remaining discrepancies emerge with respect to the upper stratosphere, which were not easily visible in the linear representation. The theoretical results considered in Fig. 38 show a stronger vertical gradient than the observational data. The effect is least with Hunt's set of rates having a predominance of the $\tilde{H}$ system in odd oxygen particle destruction also in the upper stratosphere which means only a rather minor temperature influence on the n_3 concentration; further, the discrepancy is larger in summer than in winter. The difference could be reduced if a smaller temperature dependence of k_3 would be introduced in Hesstvedt's rate set or if the vertical temperature gradient would be less than assumed.

Also it must be realized that the observed vertical ozone gradient (based mainly on Umkehr data) is not too well known. The remaining observational uncertainty amounts probably at least to half of the discrepancy shown in Fig. 38. It seems thus not impossible to reconcile theory and observation on the basis of the system described in this chapter.

Crutzen [94] has shown that the difficulties with respect to the upper stratosphere are removed when additional odd oxygen particle destruction by reaction with nitrogen oxides are considered (see Section 4). However, as long as the $\widetilde{NO}$ content in the stratosphere is not known directly from observations but rather inferred on not-too-well established theoretical grounds [110], which means that its height distribution can be readily adjusted to yield the correct result for the photochemistry of ozone, this solution is not completely convincing.

3.2.4. Vertical Distribution of the Different Trace Constituents in the Stratosphere. The concentrations of the different trace constituents of the interrelated $\tilde{O}$ and $\tilde{H}$ systems and the neutral hydrogen compounds are plotted as a function of pressure ($\approx$ height) in Fig. 39 for equilibrium conditions at 45°N in June and December in an atmosphere at rest. A constant number mixing ratio of 5×10^{-6} was assumed for water vapor.

The reason for higher winter values of ozone in the upper stratosphere is mainly due to the temperature dependence of k_3. Near the stratopause also f_2/f_3 is higher in winter because there f_3 depends more strongly on μ

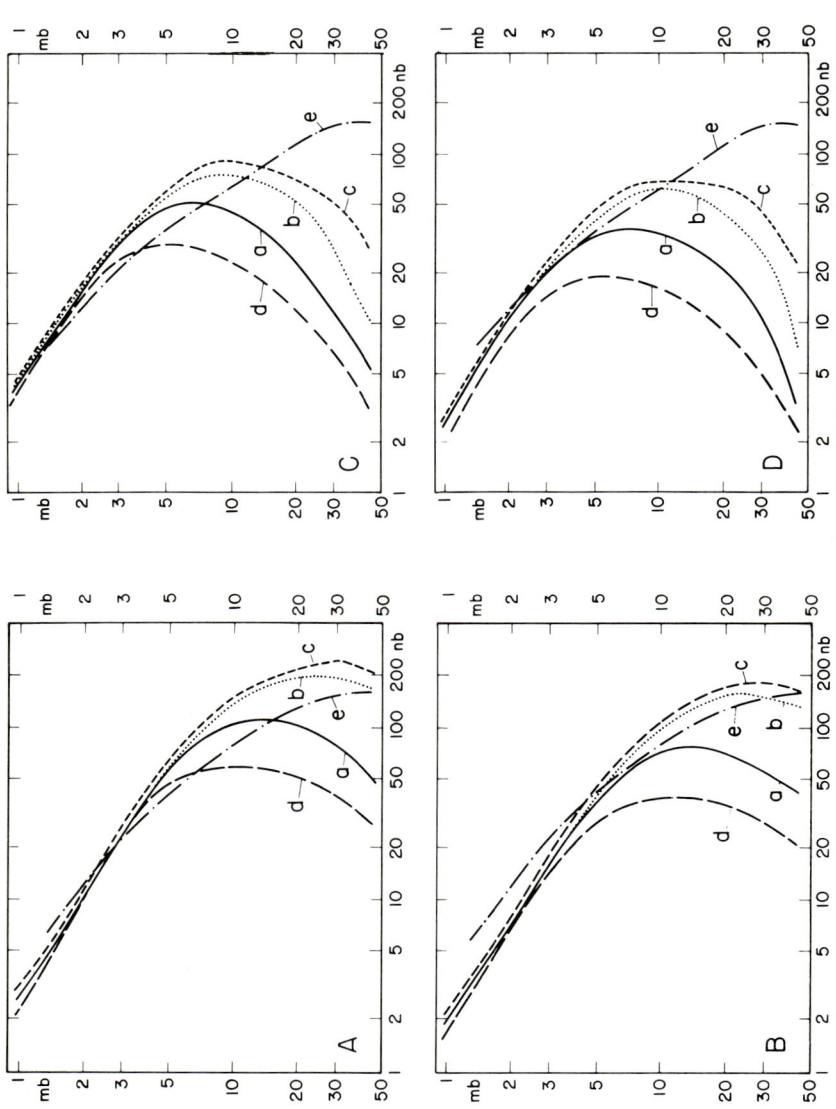

Fig. 38. Vertical ozone distribution at 45° lat (theory and observation) in logarithmic representation. Notation as in Fig. 36.

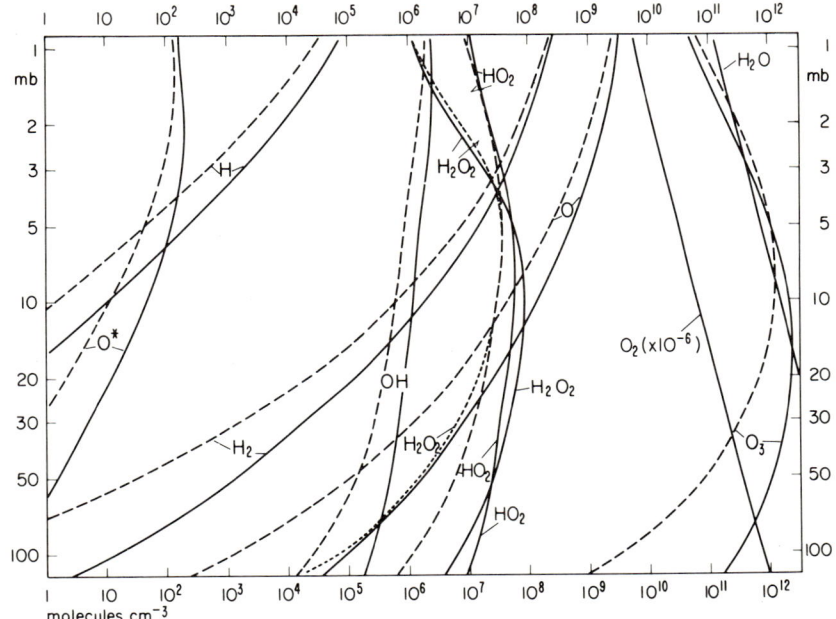

Fig. 39. Vertical distribution of $\tilde{O}$, $\tilde{H}$ and neutral hydrogen compounds in the stratosphere in summer (full lines) and winter (dashed lines); the dash-dotted line gives for comparison the molecular oxygen distribution (shifted by six powers of ten).

(Bemporad function) than f_2, but at lower elevations the dependence of this ratio is more and more reversed, leading to the winter depression of n_3 increasing rapidly in the downward direction.

The atomic oxygen concentration is lower in winter at all pressure levels, because near the stratopause Eq. (2.8) of the classical theory is a good approximation and the decrease of the product $f_2 f_3$ (instead of the ratio in the case of n_3) overrides the temperature influence. In the lower stratosphere the ratio between the winter and summer values is almost the same for n_1 and n_3 because f_3 [in Eq. (2.3)] is practically the same throughout the year at these levels.

The concentration of excited (^{1}D)-oxygen atoms is first slightly increasing below the stratopause with the increasing mixing ratio ozone/air but decreases more and more rapidly below about 40 km (in winter more than in summer) as a consequence of the strong height dependence of the factor $a_3 f_3$ in Eq. (2.36).

As suggested by Eq. (3.3) the concentration of atomic hydrogen is decreasing even more rapidly with increasing pressure than that of atomic oxygen. The OH content shows an upward increase which is, however, rather slow, while HO_2 has a maximum in the middle stratosphere. This behavior is explained by the proportionality to $n_3^{1/2}$ and by the height dependence of a_3 and R. The hydrogen peroxide concentration has, as has to be expected from Eq. (3.10) and (3.11), a still more pronounced maximum in the middle stratosphere slightly above the level of the highest ozone concentration. The molecular hydrogen content (which was not included in the simplified theory in Section 3.2) dwindles very rapidly with decreasing height. This is a result of the fact that the speed of the hydrogen destroying reactions [22] and [30] decreases only slowly downward while reproduction occurs practically only through reaction [28] and is thus proportional to the very rapidly diminishing concentration of atomic hydrogen.

It must, however, be emphasized that some of these equilibrium values computed for an atmosphere at rest may be considerably different from the actual concentration of these trace constituents. It has already been discussed (see Section 3.2.1) that below 25–35 km—depending on season and latitude —the ozone concentration will be altered considerably by transport process. Most of the other trace constituents are tied to the ozone concentration rather closely (by processes having relaxation times of a day or less). Their deviation from equilibrium will thus follow that of ozone at least during daytime with only minor aberration.

This is, however, not true for molecular hydrogen and water vapor which have very long relaxation times. It has therefore been assumed in this theory that the H_2O content is determined by transport processes, yielding constant mixing ratio under these conditions (more or less in agreement with the observations). The H_2 equilibrium distribution derived on this basis is certainly rather meaningless; considering the relaxation times of more than a year, air motions will determine these concentrations. Hesstvedt [111] has calculated the H_2 distribution in the stratosphere produced by combined action of photochemistry and vertical mixing, in assuming the mixing ratio at the tropopause to be equal to that observed in the troposphere. Further comments on this problem are set forth in Section 3.5.

3.3. Nighttime Processes in a "Moist" Stratosphere

As already discussed in Section 2.5.1 (classical theory) all photochemical processes (also secondary) are under stratospheric conditions terminated right after sunset, when all atomic oxygen is converted to ozone within less than a minute. Also in a "moist" stratosphere atomic oxygen is removed rapidly; the reproduction by reaction [23] is too small to be of any importance (the nighttime equilibrium value of n_1 basing on this reproduction is more than eight orders of magnitude less than the daytime concentration). Nevertheless a destruction of ozone continues by reactions with $\tilde{H}$ particles, albeit not indefinitely because these particles also gradually disappear after sunset; partly they recombine to water vapor reaction [24] but most of them are converted to H_2O_2 by reaction [26] from which reservoir they are very rapidly restored after sunrise by photodissociation [4*]. It is for this reason that nighttime destruction of $\tilde{H}$ particles must not be included in an approximative theory of "moist" ozone photochemistry as discussed in Section 3.2.

According to Eq. (3.6) which applies at night to the whole stratosphere and (3.12) the ozone destruction during the radiation free time becomes

(3.19) $$dn_3/dt = -2k_{13} zn_3$$

HO_2 is removed according to

(3.20) $$dz/dt \approx -2k_{26} z^2$$

(neglecting reaction [24]; hence

(3.21) $$z(t) = \frac{z_0}{1 + 2k_{26} z_0(t - t_0)}$$

whereby $z_0 = z(t_0)$ is the HO_2 concentration at sunset.

Combining (3.19) and (3.21) the relative loss of ozone during the radiation free time is obtained as

(3.22) $$\frac{\Delta n_3}{n_3} = -\frac{k_{13}}{k_{26}} \ln[1 + 2k_{26} z_0(t - t_0)]$$

With the parameter values given in Table III this predicts about a 2% loss of stratospheric ozone content during a 12-hr night, i.e., a value which should be observable with the Dobson spectrophotometer if a long enough series of observations (preferably at low latitudes) is taken in order to eliminate weather type fluctuations. The result could better be used to check on the ill-known value of k_{13} than on the $\tilde{H}$ content of the stratosphere [which appears under the logarithm in Eq. (3.22)].

3.4. Mesospheric Photochemistry in the Presence of Water Vapor

While the ozone photochemistry of the stratosphere is especially studied—apart from the inherent interest—to provide the full use of ozone as a tracer, corresponding research on mesospheric layers becomes of importance in trying to explain such features as the OH airglow (Meinel bands) [90], or the intensity distribution of the green line in the night sky light [81], noctilucent clouds [112], or the possible warming of the mesopause level by the latent heat contained in a flux of atomic oxygen [79–81]. Tracer studies could also be made again around 80 km, if it were possible to measure ozone or atomic oxygen concentration at these levels routinely.

The importance of additional reactions in the odd oxygen photochemistry under mesospheric conditions was already pointed out in 1950 in the classical paper of Bates and Nicolet [69]; an increasing amount of work on this subject has been done in recent years by Hampson [113], Hunt [71], and especially by Hesstvedt [81,90,112].

While the influence of the reactions between the $\tilde{O}$ and the $\tilde{H}$ systems on the former are relatively minor around the stratopause—if the pertinent reaction rates are of the order of those compiled by Hesstvedt [90]—(see Fig. 35) their influence increases again upward and becomes dominant in the upper mesosphere whereby an H–OH cycle is replacing the stratospheric $OH–HO_2$ cycle in destroying odd oxygen particles.

While as shown in the previous section the stratospheric $\tilde{O}$ photochemistry is rather strictly a daytime mechanism being—except for a rather unimportant contribution—(Section 3.3) "frozen in" at night also in the presence of $\tilde{H}$ particles, nighttime processes become increasingly important in the mesosphere. This was already pointed out in Section 2.5.1 (classical theory of mesospheric ozone), but the additional problems become much more complex in the presence of hydrogen compounds. The behavior is rather different at different mesospheric levels, which therefore have to be treated separately in the following description.

While up to the mesopause $n_1 \ll n_2$ is always valid and therefore molecular oxygen concentration can for a given pressure and temperature be treated as a constant in all computations, this valuable assumption is not useful anymore in the lower thermosphere. The present discussion is therefore confined to the region below about 90 km (at which level n_1 is about 2% of the total oxygen content).

3.4.1. Lower Mesosphere. In this region, between the stratopause and 65 to 70 km, the conversion of atomic oxygen to ozone after sunset is rapid (less than an hour, below 60 km within a few minutes). Nighttime destruction of $\tilde{O}$ particles through reaction with $\tilde{H}$ radicals, which mainly occurs—as shown in the next subsection—during the transition period, stays thus relatively small and therefore the nighttime concentration of ozone becomes approximately equal to the total of daytime odd oxygen concentration.

(3.23) $$n_{3,n} \approx n_{3,d} + n_{1,d}$$

Because the nighttime destruction of odd oxygen particles is relatively small and the relaxation time of the $\tilde{O}$ system short (its smallest values for the whole atmosphere are found in that layer, where also $\tilde{n}$ is at its minimum) the daytime photochemistry is not affected by the nighttime processes; except for the fact that the nighttime increase of n_3 becomes substantial, conditions are similar to those described for the stratosphere.

The contribution to $\tilde{O}$ particles destruction from reaction [3] (classical theory) and from $\tilde{H}$ particle reactions (mainly [15], [16], and [11]) are of the same order (the latter rising from about 20 to 80% from the stratopause to 65 km). The equilibrium between the $\tilde{H}$ particles is attained very rapidly at these levels (within a minute) and must thus in daytime always be well fulfilled. The relation (3.5) between y and z is simplified to

$$(3.24) \qquad \frac{z}{y} = \frac{k_{15}}{k_{16}} \frac{k_{18} n_2^2}{k_{11} n_3 + k_{18} n_2^2}$$

which yields, considering (3.2), which relation is unchanged

$$(3.25) \qquad x/z = n(k_{16}/k_{18} n_2^2)$$

increasing upward very rapidly.

While in the stratosphere the production of $\tilde{H}$ particles was exclusively through reaction of excited O atoms with water vapor, direct photodissociation of H_2O becomes increasingly important upward as does the destruction by the reaction between H and HO_2. Thus

$$(3.26) \qquad w_1(f_1 + k_{21} n_1^*) = z(k_{24} y + k_{28} x)$$

An analytical solution for y by considering Eq. (3.2) and (3.24) becomes relatively involved and shall not be further elaborated.

The $\tilde{O}$-particles destruction by $H + O_3$, which is included in (3.27) is only becoming of some importance at the top of the layer under consideration

$$(3.27) \qquad 2 f_2 n_2 = n_3 \left(\frac{2 f_3}{k n_2^2} n_3 + \frac{2 f_3}{k_2 n_2^2} k_{15} y + k_{11} x \right)$$

Considering Eq. (3.25), (3.27) may be written as

$$(3.28) \qquad f_2 n_2 = \frac{n_3 f_3}{k_2 n_2^2} \left[k_3 n_3 + k_{15} y \left(1 + \frac{1}{2} \frac{n_3 k_{11}}{k_{11} n_3 + k_{18} n_2^2} \right) \right]$$

This is above 60 km where the $\tilde{H}$ production is mainly by photodissociation of H_2O approximately of the form

$$(3.29) \qquad f_2 n_2 = A n_3^2 + E n_3$$

and considering that at this level also $E n_3 > A n_3^2$, that relation can be reduced rather crudely to

$$(3.30) \qquad n_1 \approx f_2 n_2 (k_{24}/w_1 f_1 k_{16} k_{15})^{1/2}$$

(Because all errors in this approximation have the same sign this result becomes, however, too high by at least a factor 1.6 and gives only a rough idea of the role of the contributing factors.)

The cross-over point between n_1 and n_3 is within the lower mesosphere around 60 km or slightly below, depending on the not-too-accurately known parameter values k_2 and f_3; at that point the oxygen concentration is given by

$$(3.31) \qquad (n_2)_{n_1=n_3} = (f_3/k_2)^{1/2}$$

and this relation can be used to derive the corresponding pressure level, because both f_3 and k_2 are only slightly height dependent at this altitude.

3.4.2. Upper Mesosphere. The daytime photochemistry in this region is considerably different from that in the lower mesosphere. First, the $\tilde{H}$ particles are in equilibrium with H_2 rather than with H_2O. This results from the fact, that with increasing height a rapidly rising fraction of the total $\tilde{H}$ content is atomic hydrogen and therefore the destruction of active hydrogen radicals is more and more by

[25] $\qquad\qquad H + HO_2 \longrightarrow H_2 + O_2$

instead of

[24] $\qquad\qquad OH + HO_2 \longrightarrow H_2O + O_2$

The conversion of H_2 to H_2O by

[19] $\qquad\qquad H_2 + OH \longrightarrow H + H_2O$

is slow and therefore—in an atmosphere at rest—molecular hydrogen will increasingly replace water vapor (if a fixed total hydrogen mixing ratio is assumed) (see Fig. 41B).

For this region $\tilde{H}$ production by

[22] $\qquad\qquad H_2 + O^* \longrightarrow OH + H$

exceeds that by reaction [21] and even by the direct photodissociation of water vapor [reaction 1*]. H_2O is almost completely replaced by H_2 as a source of $\tilde{H}$ radicals. In a first approximation (becoming more accurate with increasing altitude) the conversion of H to other $\tilde{H}$ particles is by

[11] $\qquad\qquad H + O_3 \longrightarrow OH + O_2$

instead of

[18] $\qquad\qquad H + O_2 + M \longrightarrow HO_2 + M$

z/y is therefore decreasing upward and the equations for the ratios between x, y, and z become simpler than at lower levels:

$$(3.32) \qquad x/y = (k_{15}/k_{11})(n_1/n_3)$$

(3.33) $$x/z = (k_{16}/k_{18})(n_1/n_2^2)$$

and

(3.34) $$y/z = (k_{16}/k_{15})(k_{11}n_3/k_{18}n_2^2)$$

At the same time the odd oxygen particle destruction results almost completely (over 95%) from reactions with $\tilde{H}$ particles. With this approximation a relatively simple analytical formula may be derived for daytime photochemical equilibrium considering two approximative balance equations for $\tilde{H}$ and $\tilde{O}$ production and destruction:

(3.35) $$k_{22} w_2 n_1^* = k_{25} xz$$

and

(3.36) $$2f_2 n_2 = k_{11} n_3 x + k_{15} n_1 y \approx 2k_{11} n_3 x$$

($k_{11} n_3 x$ and $k_{15} n_1 y$ are nearly equal, because the reactions concerning HO_2 become increasingly unimportant.)

Considering Eqs. (3.33) and (3.35) this yields

(3.37) $$x^2 = \frac{k_{22} k_{16} w_2 n_1}{k_{25} k_{18} n_2^2} n_1^*$$

thus

(3.38) $$x = \frac{n_3 f_3}{k_2 n_2^2} \left(\frac{k_2 k_{22} k_{16}}{k_{18} k_{25} k_4} a_3 \frac{w_2}{n_2} \right)^{1/2} = n_1 \left(\frac{k_2 k_{22} k_{16}}{k_{18} k_{24} k_{24}} a_3 \frac{w_2}{n_2} \right)$$

(the production of excited O atoms by O_2 photodissociation has been disregarded against that from ozone in this approximative derivation because it yields only a few percent of the total; in a more complete discussion considering nighttime destruction of $\tilde{O}$ particles (see below) it should, however, be included).

Introducing this result into Eq. (3.36) the equilibrium concentrations for ozone and atomic oxygen are obtained as

(3.39) $$n_3 = n_2^{3/2} \left(\frac{f_2 k_2}{f_3 k_{11}} \right)^{1/2} \left(\frac{k_{18} k_{25} k_4 n_2}{k_2 k_{22} k_{16} a_3 w_2} \right)^{1/4}$$

and

(3.40) $$n_1 = \left(\frac{f_2 f_3}{k_2 n_2 k_{11}} \right)^{1/2} \left(\frac{k_{18} k_{25} k_4 n_2}{k_2 k_{22} k_{16} a_3 w_2} \right)^{1/4}$$

These equations are formally rather similar to those obtained for the stratosphere using the classical theory, whereby k_3 is replaced by k_{11}; the almost height-independent additional factor (the fourth root) contains the parameters governing the $\tilde{H}$ particle level and the percentagewise distribution of these radicals; the numerical value of this factor is about 60. Around 80 km the results obtained from the approximative Eq. (3.39) and (3.40) differ only by a few percent from those computed with the full set of reactions by numerical solution.

However, as already stated in discussing the classical theory for the mesosphere (Section 2.5.1), the daytime equilibrium is not reached in the upper mesosphere (except under "polar day" conditions), because of the strong nighttime destruction of odd oxygen particles. This destruction occurs mainly during the period of the conversion of atomic oxygen into ozone, which process proceeds for several hours in the upper mesosphere or is not even finished at dusk near the mesopause. This nighttime destruction is by reactions [11] and [15], which are also dominating in the daytime process of odd oxygen particle removal; however, they become both more effective at night, because after sunset the relative increase of n_3 is much faster than the relative decrease of x and the same is true with y compared to n_1. At the 0.02-mb level (about 77 km) around 90% of the odd oxygen particles existing at sunset are removed during night. An approximate representation of this nighttime variation of the pertinent trace substances is given in Fig. 40.

After the complete destruction of atomic oxygen the further removal of ozone (whose concentration goes through a maximum sometimes in early night) is slowed down strongly. Because x has also become very small according to Eq. (3.32) the O_3 destruction is now by the much slower $OH-HO_2$ cycle. The ratio z/y is by this changed to

$$(3.41) \qquad z/y = k_{12}/k_{13} \approx 50$$

The nighttime destruction of a relatively high percentage of the odd oxygen particles leads at these levels, where the production of one day cannot restore the complete daytime equilibrium amount to a considerable reduction of the $\tilde{n}$ concentration also during the sunlit hours (quite similarly as described in Section 2.5.1 for the classical theory).

The conversion of H into first OH and then HO_2 has some further important consequences. The destruction of $\tilde{H}$ particles, which is slow in daytime—the relaxation time for the $\tilde{H}$ system is of the order of weeks at that level—is increased. The destruction by

[25] \qquad $H + HO_2 \longrightarrow H_2 + O_2$

is replaced by

[23] \qquad $OH + OH \longrightarrow H_2O + O$

and

[24] \qquad $OH + HO_2 \longrightarrow H_2O + O_2$

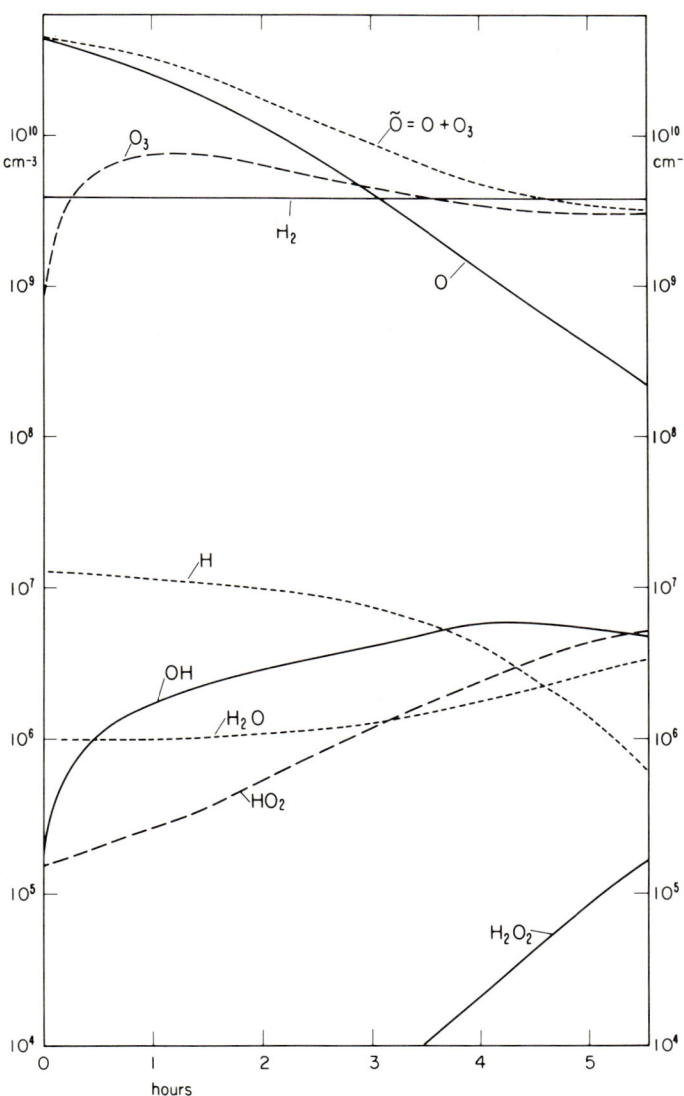

FIG. 40. Approximate variation of trace constituents during the first half of the night at the 0.02 mb level. Odd oxygen particle destruction is practically finished $5\frac{1}{2}$ hrs after sunset while $\tilde{H}$-particle destruction continues for most of the night (conversion into H_2O and H_2O_2).

which, due to the change in the $\tilde{H}$-particle ratios, work considerably faster. This leads also to a sizable nighttime loss in $\tilde{H}$ particles. This process has some time lag with respect to $\tilde{O}$ particle destruction (because the conversion of H to OH and HO_2 has to be rather advanced until it becomes effective and this is only after a considerable percentage of O is replaced by O_3; thus only the later stage of $\tilde{O}$-particle destruction is somewhat slowed down by the decrease of $\phi = x + y + z$). After the complete removal of O also the $\tilde{H}$ particle destruction is stopped, because with the ratio of z/y given by Eq. (3.41) which now applies, these particles are converted into H_2O_2 by reaction [33] (and thus reformed in a short time after sunrise by photodissociation reaction [4*]).

It is important to note that the $\tilde{H}$ particles lost at night are predominantly converted into H_2O and not into H_2 as in daytime. This leads to a considerable increase of water vapor content compared to the value computed for daytime equilibrium. On the other hand H_2O dissociation reaction [1*] is much faster in reproducing $\tilde{H}$ particles than

[28] $\qquad\qquad H_2 + O^* \longrightarrow OH + H$

Therefore the nighttime loss of $\tilde{H}$ particles should only produce a relatively small decrease of ϕ compared to the pure daytime equilibrium (except for the first hours of the day).

Near the end of the conversion of O into O_3 some O is reproduced by reaction [23] and n_1 is kept at the level

(3.42) $\qquad\qquad n_1 = k_{23} y^2 / k_2 n_2^2$

However, because as stated above, most $\tilde{H}$ particles are now converted into H_2O_2 via HO_2, this value does not remain constant but is also dropping off further.

3.4.3. Region Above the Mesopause. The secondary nighttime processes, which are in their very involved behavior (interrelated $\tilde{O}$ and $\tilde{H}$ particle losses) typical for the upper mesosphere, become somewhat simpler again in the region between 85 and 90 km, the highest which can yet be treated in assuming n_2 to be practically unaffected by the photochemical processes and thus the highest which shall be described here. Within a quarter of an hour or less ozone has reached its maximum level given by the balance of production through reaction [2] and destruction by [11]. This level is only by a factor five or less higher than the daytime value and because also n_1 is not lowered by more than a factor two during the night, y stays according to Eq. (3.32) too small for substantial contribution of reaction [23] to $\tilde{H}$ particle destruction. There is thus practically no day–night variation of ϕ as it is predicted by the theory for the upper mesosphere.

The rate of nighttime destruction of odd oxygen particles is, however, still several times higher than that in daytime and a considerable depression of the actual $\tilde{n}$ concentration during the sunlit hours against daytime equilibrium values is produced by this process. Because $\phi \sim n_1{}^*$ and thus to $\tilde{n}$, the effect is not quite as big as expected from the ratio of nighttime to the daytime destruction.

Further one of the basic assumptions of the photochemical theory, namely the much higher speed of the reactions of group I in Table IV converting $\tilde{n}$ particles into each other compared to all other processes is not valid anymore at this level (although they are still the fastest). The ratios between n_1, $n_1{}^*$ and n_3 are therefore not strictly independent of all other processes and Eqs. (2.3) and (2.37) are not as good an approximation of daytime equilibrium as in the upper mesosphere.

3.5. Influence of Vertical Mixing on Mesospheric Photochemistry

In Section 2.5.2 it was pointed out that vertical transport processes by mixing or possible vertical motions might have an influence on atomic oxygen distribution down to about 75 km. Hesstvedt [81,90] has intensively studied this problem for an oxygen–hydrogen mesosphere and has shown that the changes are considerably stronger in this more complex case.

The main reason for this is that not only the transport of atomic oxygen but especially that of water vapor and molecular hydrogen, and to some extent also atomic hydrogen, leads to a considerable redistribution of these substances with the consequence, that the relative total content of "neutral" hydrogen $[(w_1 + w_2)/n_2]$ is varying considerably with height under the combined effects of transport and photochemistry (Fig. 41B). This means that the height dependence of the effectiveness of the $\tilde{H}$ system in removing odd oxygen particles is also changed; i.e., the $\tilde{O}$ distribution is not only altered by the downward transport of atomic oxygen but also indirectly by the redistribution of hydrogen compounds. Only substances with high enough photochemical relaxation times can be effectively transported by air motions. If, however, photochemical processes are too slow over an extended vertical layer, the mixing ratio will become almost height independent and the vertical fluxes will be small. In the region between 60 and 100 km (in the upper mesosphere and above the mesopause) the relaxation times of a number of constituents change rather rapidly with height (from values of less than a day to several months and more), while the equilibrium values of the mixing ratios are also strongly height dependent; turbulent mixing which is thought to be quite strong in this region—although its magnitude is not well known—thus produces strong vertical fluxes of these substances and thereby considerable deviations from photochemical equilibrium concentrations which are in turn photochemically interrelated with each other.

298   H. U. DÜTSCH

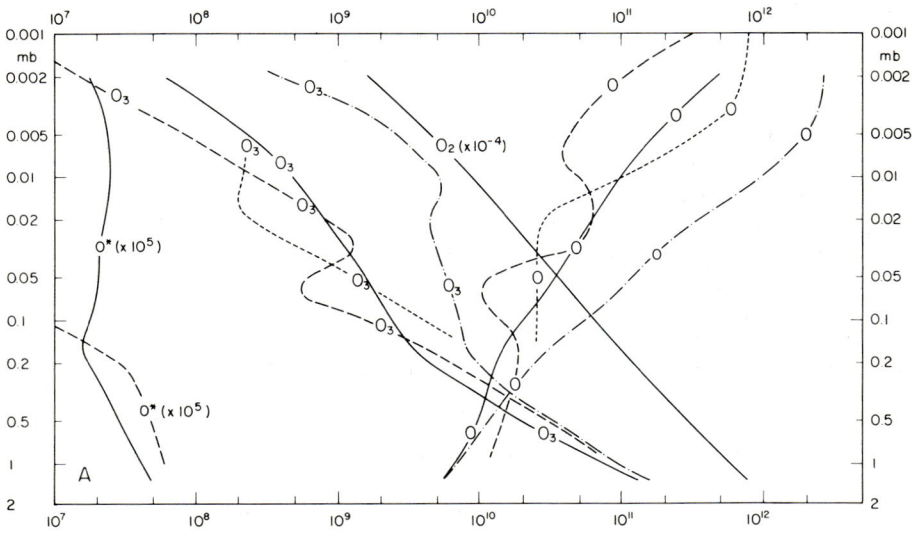

Fig. 41A.

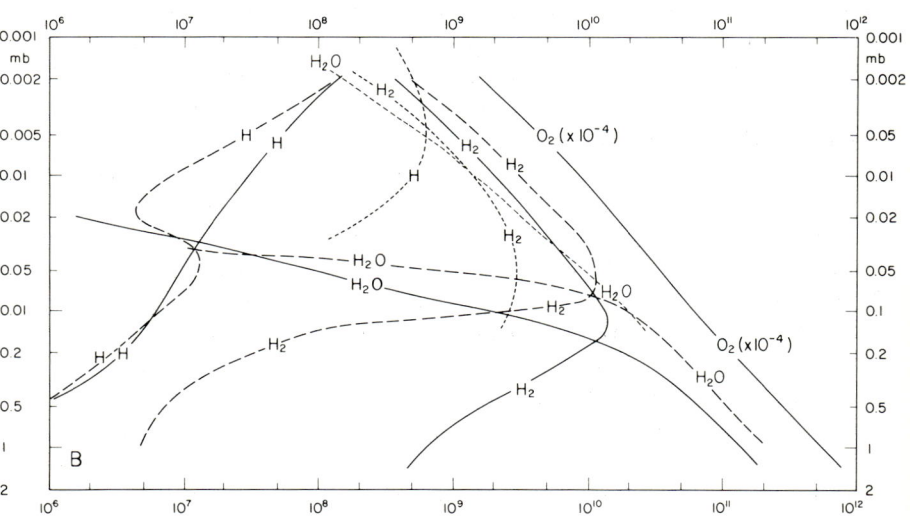

Fig. 41B.

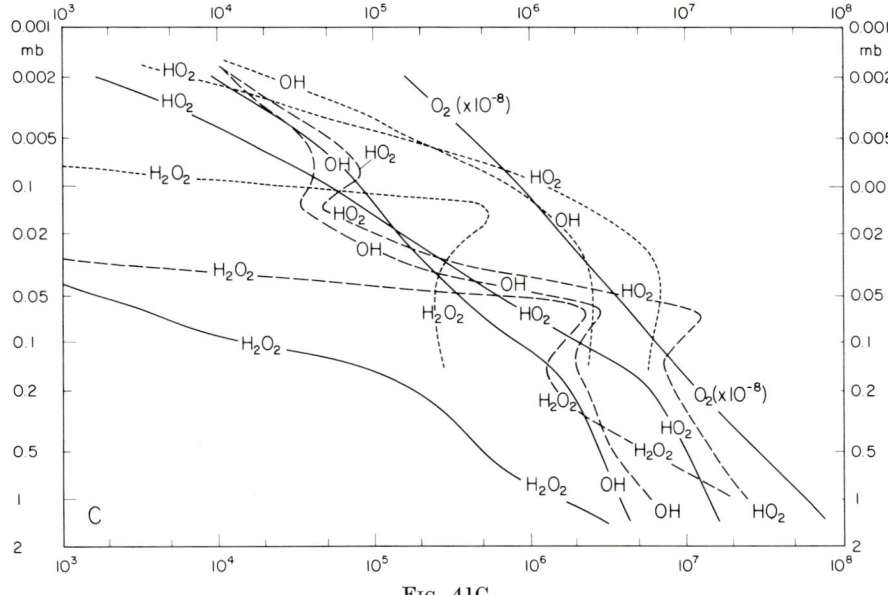

Fig. 41C.

Fig. 41. Vertical distribution of trace constituents of the Õ- and H̃-systems and of neutral hydrogen in the mesosphere. (a) full lines, daytime equilibrium value ($z = 30°$); (b) dashed lines, daytime nonequilibrium values after Hesstvedt [81] (atmosphere at rest); (c) dotted lines, daytime nonequilibrium values under the influence of vertical mixing after Hesstvedt [90]. (A), Õ-system, O and O_3 computed from the classical theory are given for comparison by dash-dotted lines. (B), H_2O, H_2, and H. (C), OH, HO_2, and H_2O_2. The curves are not strictly comparable because the solar elevation, dissociation rates for H_2O (f_1) and mixing ratio of neutral hydrogen to air is not exactly the same in (a) as in (b) and (c). No explanations for the appearance of the interrelated secondary maxima and minima of most of the trace substance in case (b) are given in Hesstvedt's paper [90]. n_2 is given on all three figures for comparison shifted by the indicated power of ten.

The deviations from photochemical equilibrium values produced by transport processes leads to distinct source or sink regions for certain constituents. Such a system is discussed in some details for ozone in Section 4, where the source is mainly in the middle stratosphere and the sink at the earth's surface; the flux between the two is, however, not only vertical but combined with large scale horizontal transport.

We have no knowledge about possible similarly complicated systems in the mesosphere; also as mentioned above there is considerable indication that vertical turbulent mixing is near the mesopause much higher than in the stratosphere [114,115]; therefore the studies made by Hesstvedt are combining photochemical processes only with vertical eddy flux and no other motions.

The sources and sinks so produced are shown in Fig. 42. Water vapor is transported upward to the mesopause and slightly above, where it is destroyed by photodissociation (reaction [1*]). In the region between 70 and 85 km the conversion is finally to molecular hydrogen by reaction [28]. The upper mesosphere and the layer just above the mesopause become thus a source for H_2 from where this gas is diffused up and down by turbulence. The thermosphere above 85 km is a source of atomic hydrogen produced from the imported water vapor (by photodissociation) and H_2 (by reaction [22]) a production which is not balanced at these levels by the removal through reaction [28] due to increased content of H_2 and H_2O and as a consequence of the H loss by turbulent flux. The upper limit of a substantial contribution of water vapor to the upward flux of neutral hydrogen is hereby rather sensitive to the intensity of the mixing. The turbulent atomic hydrogen flux would have a downward direction also above 100 km; however, at these levels molecular diffusion makes a major contribution and this flux has upward direction for the light hydrogen; the source for the H flux through the mesopause does thus not extend much above 100 km.

Vertical motions of a few centimeters per second produce transport effects of the same order of magnitude and might shift the limits between sinks and sources given by Fig. 42 to some extent and alter the magnitude of the effects; they will, however, hardly change the qualitative picture drastically. Further research on this subject is needed.

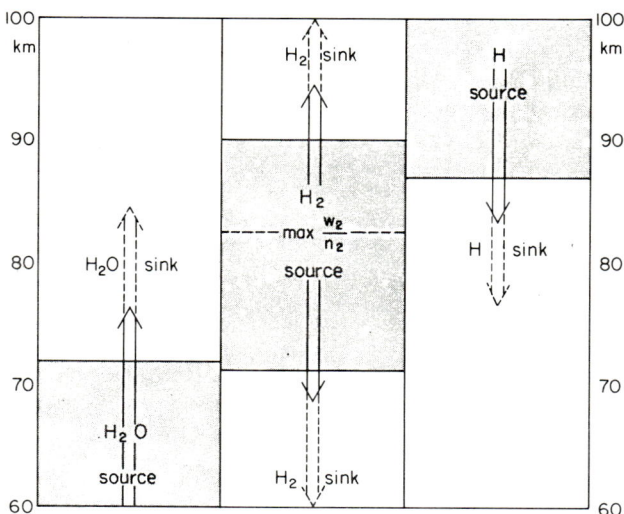

FIG. 42. Source and sink regions of H_2O_2, H_2, and H in the mesosphere and lower thermosphere. Arrows indicate direction of transport by vertical mixing.

Hesstvedt's model on which the foregoing description is based and from which Fig. 42 is derived has an artificial lower limit at 65 km by assuming photochemical equilibrium of H_2 below that level. Photochemical action is, however, too slow to destroy the H_2 flowing downward through that boundary, this should increase the H_2 concentration considerably above equilibrium content. Further, the H_2 equilibrium values are so low in the stratosphere, that considering the concentration measured in the troposphere, there has to be an upward flux through the tropopause (also investigated by Hesstvedt [111]); molecular hydrogen has thus a double source: At the ground and around the mesopause, the stratosphere and the lower mesosphere are the corresponding sink regions; a combined study on the full system has not yet been made. The conversion of H_2 to H_2O in the lower mesosphere and the stratosphere might, together with the oxygenation of CH_4 initiated by reaction [36], be responsible for the slow upward increase of H_2O mixing ratio which is indicated in these layers (the net flux through tropopause could thus be by H_2 only, irrespective of the strong gradient of water vapor mixing ratio at that boundary).

The downward flux of atomic oxygen above and through the mesopause produced by vertical mixing is by far the most intensive transport process in the mesosphere. On the other hand, the destruction of such particles by (secondary) photochemical processes is also much higher than in the case of H_2O or H_2 or even of atomic hydrogen above the mesopause. The removal of odd oxygen particles is increased by the redistribution of the hydrogen compounds in the critical region around the mesopause (especially by the higher water vapor content in that layer).

For these reasons Hesstvedt obtains a very sharp drop in atomic oxygen concentration (Fig. 41A) from values which are around 90 km increased considerably above the photochemical level to subequilibrium concentrations around or just below the mesopause (depending on the assumed strength of the mixing process). The daytime ozone distribution is strictly tied to that of atomic oxygen by

(3.43) $$n_3 = n_1 k_2 n_2^2 / (f_3 + k_{11} x)$$

(the second term in the denominator which was neglected in the discussion of the pure photochemical case stays important down to lower altitudes due to the downward transport on atomic hydrogen by the mixing). Because the $n_3 n_1$ equilibrium is rapidly reached, ozone transport by turbulence is negligible in daytime.

The nighttime conditions around 80 km become still more involved than in the pure photochemical case; the fluxes are changed according to the height dependent variations of the constituents during the dark period, which are in turn more pronounced due to the increased contribution of the $\tilde{H}$ system to

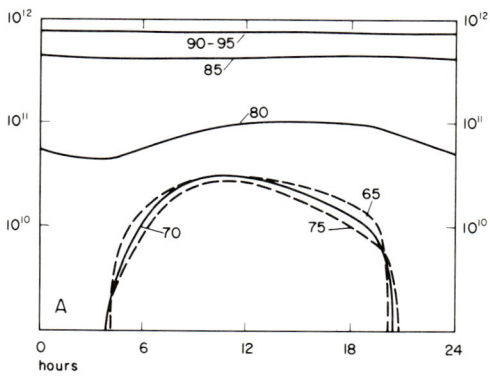

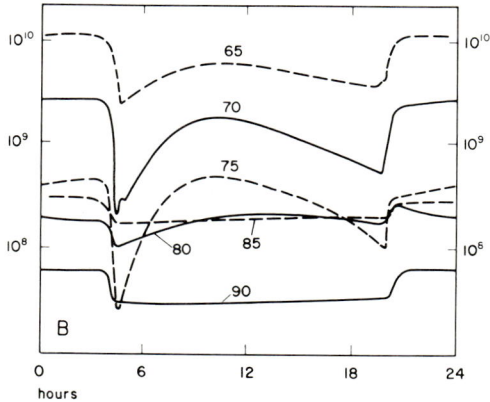

Fig. 43. Diurnal variation of atomic oxygen (A) and ozone concentration (B) in a "moist" mesosphere under the influence of vertical mixing (after Hesstvedt [90]).

odd oxygen destruction. The nighttime minimum of ozone at these levels given by Hesstvedt (Fig. 43) which is not further explained in his paper, is probably a result of such processes,[6] which shall, however, not be discussed here in detail (the daily variations of atomic oxygen and ozone concentration in a mixed mesosphere as calculated by Hesstvedt are shown in Fig. 42).

The depression of odd oxygen particles concentration through the influences of the $\tilde{H}$ system is strongest near the mesopause and is, as mentioned above, still increased through the transport processes by mixing. This leads

---

[6] Such a minimum has been observed by P. B. Hays and R. O. Roble using the satellite occultation technique on hot stars (Personal communication).

to the minimum in absorption of short wave radiational energy—to be expected at the level of minimum temperature—which is not obtained on the basis of the classical photochemical theory.

The nighttime secondary ozone maximum around 70 km predicted by the classical theory is not found anymore (or becomes at least much less pronounced). It is, however, possible that gravitational waves could produce maxima and minima (as observed by Reed [78]) at night when photochemical processes are slow in the lower mesosphere.

Relatively strong horizontal gradients of odd oxygen particles concentration will develop at high latitudes in the winter hemisphere, especially across the polar night border. Thus horizontal transport processes will become of importance for the $\tilde{O}$ distribution in this region and O or O_3 could again as in the lower stratosphere be used as tracers if the necessary techniques for routine measurements at these high levels could be developed.

4. Possible Importance of Nitrogen Oxides to Ozone Photochemistry

The possible contribution of nitrogen oxides to the destruction of O particles has been considered for some time. It was mostly concluded that the concentration of the nitrogen compounds was too low to be of any importance. A recent study by Crutzen [94] based on ideas of Bates and Hays [110] suggests however, that this may not be so.

The additional reactions added to the ozone photochemistry if nitrogen oxides are considered, are

[41] $\qquad NO + O_3 \longrightarrow NO_2 + O_2 \qquad k_{41}$

[6*] $\qquad NO_2 + h\nu \longrightarrow NO + O \qquad f_6 \qquad \lambda_{max} = 3975 \text{ Å}$

and

[42] $\qquad NO_2 + O \longrightarrow NO + O_2$

This leads to an equilibrium between NO and NO_2:

(4.1) $\qquad [NO]\, n_3\, k_{41} = [NO_2]\,[n_1]\, k_{42} + [NO_2] f_6$

and

(4.2) $\qquad \dfrac{[NO]}{[NO_2]} = \dfrac{f_6 + n_1 k_{42}}{n_3 k_{41}} = \dfrac{f_6}{n_3 k_{41}} + \dfrac{f_3}{k_2 n_2{}^2} \dfrac{k_{42}}{k_{41}}$

Both terms are decreasing with decreasing altitude above 25 km, while below that level the first term which is more important at this altitude, increases again. Thus the ratio NO/NO_2 has a minimum in the neighborhood of the level of the ozone maximum.

The Õ-particle destruction by reaction with nitrogen oxides is

(4.3) $$\frac{\partial \tilde{O}}{\partial t} = -k_{41}n_3[NO] - k_{42}n_1[NO_2] + f_6[NO_2]$$

or considering Eq. (4.2)

(4.4) $$\partial \tilde{n}/\partial t = -2k_{42}n_1[NO_2] = -2k_{42}(f_3/k_2 n_2^2)n_3[NO_2]$$

The loss of odd oxygen particles is thus proportional to the first power of the ozone concentration compared to a destruction going with n_3^2 in the classical theory and with $n_3^{3/2}$ for the contribution by $\tilde{H}$ radicals. It is further proportional to the first power of [NO_2], while the destruction by $\tilde{H}$ radicals goes only with the square root of the water vapor mixing ratio. If the nitrogen oxide mechanism were the dominating process, the ozone equilibrium would thus become very sensitive to the nitrogen oxide content of the stratosphere.

Taking into account the additional term discussed above Eq. (3.14) is expanded to

(4.5) $$f_2 n_2 = n_3[An_3 + (B+C)n_3^{1/2} + D]$$

with

(4.6) $$D = (k_{42} f_3/k_2 n_2)[NO_2]/n_2$$

if D should be considerably larger than the other terms in the bracket (as indicated by the numbers given by Crutzen [94]) the ozone concentration becomes:

(4.7) $$n_3 \approx 2(f_2/f_3)(k_2/k_{42})n_2^2(n_2/[NO_2])$$

Under such conditions it is quite directly governed by the nitrogen oxide content of the stratosphere. It becomes inversely proportional to the mixing ratio of NO_2 and is more strongly dependent on the ratio of the dissociation rates f_2 and f_3 than indicated by the classical theory.

The foregoing discussion may not be meaningful in the upper stratosphere where [NO] $\gg$ [NO_2] and where it might thus be more appropriate to express odd oxygen particles destruction through the NO—or the total $\widetilde{NO}$ content ([NO] + [NO_2]) than through that of [NO_2]. However, in the main ozone layer where, with the rate constants indicated by Crutzen [94], [NO_2] is larger than [NO], it should give a useful picture.

However, the present theory of the nitrogen oxide contribution to the ozone photochemistry is certainly incomplete. Following Bates and Hays [110] these compounds would be produced photochemically in the stratosphere from N_2O of initially tropospheric origin (mainly from plant decomposition in the soil). Thus a continuous stratospheric production of NO and NO_2, respectively, is indicated, but no sink. No destruction of these compounds is shown as part of the photochemical system (in contradiction to the behavior of $\widetilde{H}$ particles). It is thus not clear whether the continuous production of nitrogen oxides is balanced by some unknown process within the stratosphere or whether these compounds are lost upward by turbulent mixing, whereby they might be destroyed by some other photochemical processes in the mesosphere, or whether these substances return to the troposphere, after being produced by oxydation of N_2O in the stratosphere, and would then be lost by washout.

The possibility that the thermosphere and upper mesosphere are an additional source for $\widetilde{NO}$-particles has recently been discussed by Strobel et al. [116] and in 1965 by Nicolet [117], who gave a thorough treatment of the photochemistry of $\widetilde{NO}$ and N, however, without discussing a possible influence on odd oxygen particle concentration. It seems thus most probable, that the troposphere is the major sink for $\widetilde{NO}$ particles.

Although there is nothing known about this at present, notable variations of $\widetilde{NO}$ content with latitude and season could be expected. This might considerably alter our present knowledge on the variation of ozone concentration to be expected from pure photochemistry—if (a) $\widetilde{NO}$ concentrations were as high as assumed by Crutzen [94] and if (b) reaction [42] were as fast as indicated [106].

If nitrogen oxides should have the importance suggested by Crutzen's calculations it would certainly be of interest to know the relaxation time of stratospheric NO content, in order to answer the question, could man's increasing activity at these levels augment its level by any appreciable amount? and thus change the ozone content and its distribution, which would then react on stratospheric circulation due to the absorption and emission of radiative energy by O_3. As the troposphere is presumably a sink of $\widetilde{NO}$ (by washout) the relaxation times of this gas in the stratosphere might be similar to that of ozone, and rather high rates of artificial production would thus be needed to alter the overall concentration appreciably.

Recent investigations by Nicolet [118] and by Crutzen [personal communication] indicate that matters are further complicated by an interaction between the $\widetilde{NO}$ and the $\widetilde{H}$ system, which had not been considered so far. The destruction of odd oxygen particles may be further increased by such processes. Also some loss of $\widetilde{NO}$ particles within the stratosphere may occur due to these processes (formation of nitrates, etc.).

Nighttime Destruction of Ozone through the NO Mechanism

After sunset NO is very rapidly converted into NO_2 by reaction [41]. The reconversion to NO is stopped because at stratospheric levels oxygen disappears very rapidly and f_6 becomes zero. Because $[NO] \ll n_3$ only a small percentage of n_3 is destroyed (with the numbers given by Crutzen only about 1% of total ozone, thus an amount which is not observable).

Most of the conversion of NO to NO_2 would actually already occur during dawn; the additional loss of $\widetilde{O}$ particles by the $\widetilde{NO}$ cycle, which is continuing during that time on correspondingly reduced basis below 30 km, is again only a few percent of the total. Thus the nighttime decrease in total ozone resulting from this mechanism is, even if the $\widetilde{NO}$ cycle is as important for the ozone photochemistry as suggested by Crutzen, still smaller than that by the H cycle and thus too unimportant to be observed. No proof of the $\widetilde{NO}$ mechanism can thus be obtained on that basis.

5. Ozone as a Tracer

5.1. Importance of the Photochemical Theory in General Circulation Studies Using Ozone as a Tracer

The discrepancy obtained between theory and observation with respect to the seasonal and latitudinal variation of total ozone had demonstrated 25 years ago that the distribution of the gas was strongly influenced by transport processes and the values computed for the relaxation time in the lower stratosphere substantiated this idea. Ozone could thus be used as a tracer for studying stratospheric circulation.

The information obtained from total ozone observations was not sufficient, however, to allow a decision between the meridional circulation model presented by Brewer [119] and Dobson [120], which largely explains the behavior of ozone as well as the dryness of the stratosphere (but which was criticized on dynamical grounds) and another theory emphasizing the importance of large-scale mixing (Newell [121, 122]).

Only our knowledge on vertical distribution, slowly accumulated during the past decade, brings us in a position to reconsider the problem on a firmer basis. Tracer studies using radioactive debris [123–127] seemed in the meantime to favor the large scale exchange model. However, the results were not completely conclusive, too, mainly because most of the pertinent measurements did not reach much above 20 km leaving still considerable uncertainties.

When we compare the two tracers, radioactive debris have on first sight the advantage of a known source point (or column) and time of injection and a calculable rate of loss (as long as the substance is within the stratosphere and no washout processes thus become effective). Ozone, on the other hand, has world-wide production and a destruction by photochemical processes also within the stratosphere.

This means that except in studies of relatively short-time events (as, for example, exchange of air between stratosphere and troposphere [52]) photochemical processes have to be taken into account, when ozone is used as a tracer in general circulation studies. Ozone has, however, the advantage that in a first approximation production and destruction processes are independent of longitude. While during the time of the big nuclear tests radioactive debris was the preferred tracer, ozone is now rapidly gaining importance for the purpose, also because relatively easy and accurate observational methods have become available.

The remaining uncertainties of the photochemical theory as discussed in the preceeding chapter are at present the biggest difficulty in taking full advantage of ozone tracer studies for stratospheric circulation. The most straightforward procedure in such a tracer study, to compute the three-dimensional field of differences between theoretical and observational distribution and of its variations with time and to derive from such differences the divergence field of the world-wide ozone flux according to Eq. (2.18) or a similar expression including the additional processes discussed in Sections 3 and 4, is not feasible at present, much more so because of the uncertainty of the theoretical results than to that of the observational data.

A presently somewhat more suitable way to solve the problem is the use of observed seasonal variations up to 25–30 km (depending on latitude) and to compute the fluxes, which would be necessary to produce such variations by pure transport, and then to consider whether the changes might be produced almost completely photochemically (as in the middle stratosphere) or what contribution to the full observed change might also at lower levels be of photochemical origin.

Figure 44 shows a completely qualitative representation of the scheme of ozone fluxes, which can be deduced from the observed variations of the three-dimensional ozone field (there is no implication in this representation whether the fluxes represented by the arrows are due to meridional circulation or large-scale mixing).

The main fluxes go from the suggested source region, the tropical stratosphere, to middle and high latitudes of the winter hemisphere; that flux is apparently confined to the period of westerly circulation above 20 km and seems not to exist during the time of the much less disturbed summer easterlies. The downward transport through the lower stratosphere and the transfer into the troposphere where the ozone is finally destroyed, mainly near the ground, however, continues [as is shown by the tropospheric early summer maximum of ozone (Fig. 6)], thus leading to the ozone loss between the tropopause and 20 mb observed from March or April through fall (Fig. 45).

To make full use of this presently only qualitative picture for obtaining insight into the general circulation (i.e., for determining the actual strength of the downward flux into the troposphere and to answer the question whether or not there is a stratospheric backflux of ozone toward lower

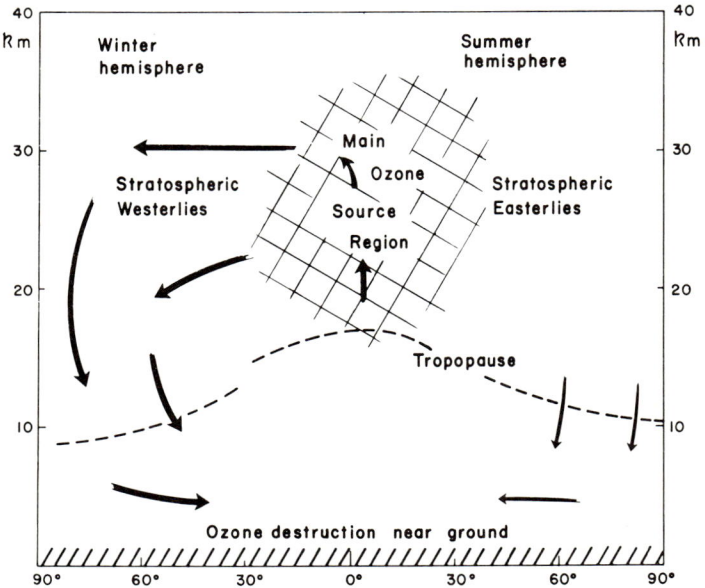

Fig. 44. Tentative model of large scale ozone fluxes and of the seasonal ozone cycle.

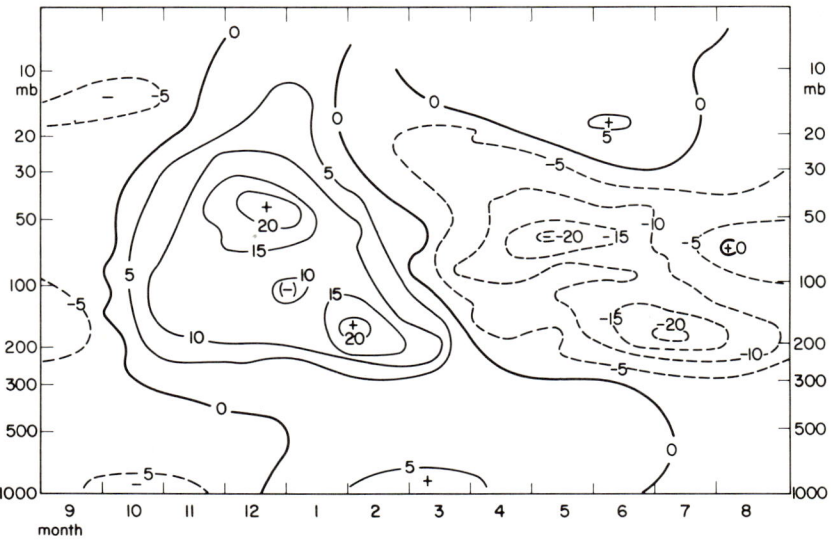

Fig. 45. Time cross section of month-to-month variation of ozone concentration over Switzerland (47°N) in nb.

latitudes in summer, it is necessary to calculate the contribution by photochemical processes to the destruction of the surplus ozone apparently accumulated during the winter, and to a lesser degree to determine to what extent the observed increase during the cold season is retarded photochemically.

Due to the big uncertainty about photochemical relaxation times (see Fig. 37) it is at present not possible to assess the importance of the photochemical contribution, or more precisely, it is not possible to evaluate (from theoretical considerations) down to what level (as a function of latitude and season) the ozone concentration is predominantly determined by photochemical processes and down to what lower level photochemical contribution is yet considerable (under present conditions we may define "considerable" as definitely higher than the observational uncertainty).

If the time of half restoration is less than 15–20 days the influence of air motions on mean ozone content stays small (less than 10% deviation from the photochemically computed value), the procedure indicated above is not useful at these levels; for $\tau > 200$ days the photochemical contribution to the observed variations becomes minor. Figure 37 shows the corresponding levels for different (possible) assumptions on photochemical parameters; the hatched intermediate layer is certainly most important for the interaction between photochemistry and transport.

A rapid determination of the most significant questionable parameters would certainly be the best solution (the biggest uncertainty lies presently in k_{13}, further in k_4 and k_{21}, but even k_3 needs confirmation). Other sources of uncertainty have already been named in Section 2. The possible contribution of the $\widetilde{NO}$ cycle should be intensively studied by obtaining more informations on $\widetilde{NO}$ distribution in the stratosphere and on the pertinent reaction rates.

Although big advances have been made within a few years—the photochemical theory of a "moist" stratosphere looks now much more reasonable than on the basis of the reaction rates compiled by Hunt in 1966 (the best available at that date)—it may still take considerable time until the necessary accuracy is reached. In the meantime the problem might be tackled indirectly by studying with which combination of the questionable parameters a solution can be obtained, which is compatible with the observational facts. As the observed distribution is influenced at least below a certain level by the stratospheric circulation, which we want to study ultimately by using ozone as a tracer in applying a photochemical theory, updated in the indicated manner, one must, however, be very careful to avoid "bootstrap" methods in working along the indicated lines. Additional computations on the availability of the necessary ozone producing quanta in the source region under different assumptions (see Brewer and Wilson [86]) can help to check on the internal consistency of the results.

5.2. Ozone as a Tracer in General Circulation Models

In most tracer studies involving the photochemical theory of ozone which have so far been conducted, certain assumptions on large-scale exchange coefficients or on meridional circulation cells or on both have been introduced into the pure photochemical equations in order to show that a theoretical ozone distribution of a similar general behavior as the observed one can be obtained (mostly using the classical photochemical theory) [65,128,129]. A very close fit has not yet been reached (which is not surprising considering the relative crudness of the circulation models and the neglect of additional reactions in the theory).

The most promising experiment, however, was the introduction of ozone as a tracer in an advanced numerical general circulation model, first by studying the transport by the simulated circulation in case of a given initial distribution (Hunt and Manabe [130]) and in a further step by introducing photochemical processes going along with the model circulation (Hunt [131]). The latter experiment was made in applying independently both, the classical reaction scheme and that of a moist stratosphere[7].

The results indicate very clearly that both large scale mixing and meridional circulation cells contribute to the trace gas transport; in some regions the observed total transport is actually not the sum but rather the difference of the two mechanisms counteracting each other. The meridional circulation produced by this numerical model is, however, not the single cell circulation of the Dobson–Brewer [119,120] model, but rather a double structure with the downward leg over the subtropics and low midlatitudes and upward in equatorial and polar regions (Fig. 46). Somewhat simplified it can be said, that the export of ozone from the main source region, the tropical stratosphere is by the meridional cell (which is an extension of the well-known Hadley cell into the stratosphere); the further transport from the subtropics to higher latitudes, however, is shown to be by large-scale exchange processes, whereby this flux is reduced by the influence of the superimposed meridional cell (Fig. 46).

The experiment shows clearly, that the influence of the general circulation is to reduce ozone concentrations at low and to increase them at high latitudes. As a consequence of this transport a continuous production of ozone results in tropical regions (where its concentration becomes deficient) and a corresponding destruction north of about 30°, where surplus ozone is

---

[7] Some of the relevant reaction rates were considerably altered compared to Hunt's earlier work [71] ($k_4 = 7 \times 10^{-11}$ cm^3 sec^{-1} and $k_{21} = 5 \times 10^{-11}$ cm^3 sec^{-1}). The differences against the set of rates used by Hesstvedt [90] is considerably reduced; the relative contribution of the $\tilde{H}$ reactions is, however, still somewhat higher than in Hesstvedt's computations.

accumulated. After the deviation from pure photochemical equilibrium produced by the general circulation has reached a certain degree the further transport effects are balanced by the photochemical processes. The deviation from pure photochemical equilibrium which must be reached and until the overall balance is attained are much higher in the "dry" than in the "moist" stratosphere, because of the considerably shorter relaxation times in the

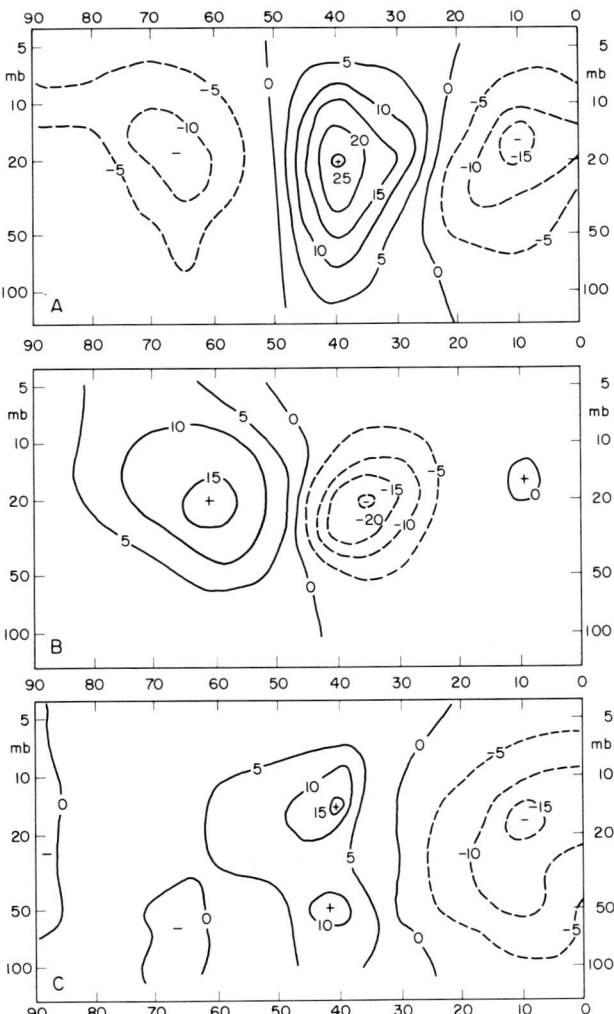

Fig. 46. Change of ozone concentration (in nb per month) produced by the circulation simulated by a 18-layer numerical model after Hunt [131]. A, Change by meridional cells. B, Change by large scale exchange. C, Total change.

latter. The ozone concentrations shown for the classical theory were considerably too high in comparison with the observations (a result which could, however, also be the result of a too low value assumed for the still not-very-well-known rate constant k_3).

A quite promising way of further research is opened by Hunt's experiment. In a next step the seasonal variation of solar elevation must be included in the model atmosphere in order to come to a full understanding of one of the most prominent features of the world-wide ozone distribution—namely, its seasonal variations, which is far from being in phase with that expected from the photochemical theory. Also the relative importance of the ozone transfer into the troposphere and following destruction at the ground as a function of season must be studied in that context.

The study of ozone transport by numerical model atmospheres under inclusion of photochemical processes is not only the probably most promising technique for bringing the ozone problem to a final solution, but the use of ozone as a tracer in such models should on the other hand also be very helpful in their further development and refinement. The prerequisite is, however, that the present uncertainties of the photochemical theory are overcome and also that a considerable further improvement of our knowledge of the three-dimensional ozone distribution is obtained.

6. Interrelation Between Ozone Photochemistry and Stratospheric Dynamics

Ozone is not only a passive tracer, but its radiative properties—strong absorption in the UV, mainly important in the upper and middle stratosphere, relatively weak but broad-band absorption in the intense solar visible (middle and lower stratosphere) and absorption and emission in the ir (9.6 μ band)—let it play a dominant role in determining stratospheric temperature distribution and through this stratospheric dynamics. The mentioned absorption is (at least in the UV and visible), combined with photochemical processes changing the ozone content and therefore in turn again influencing the absorption; variations in the overlaying ozone are at the same time changing the intensity of the radiation; changes in temperature alter some of the reaction rates and thus influence the ozone concentration; they also change the infrared flux; varying temperature distribution influences the field of motion and by this the ozone transport leading especially in the middle and lower stratosphere to changes in the distribution of the trace gas which in turn react on the temperature distribution, etc., a very involved feedback system as indicated by Fig. 47 results.

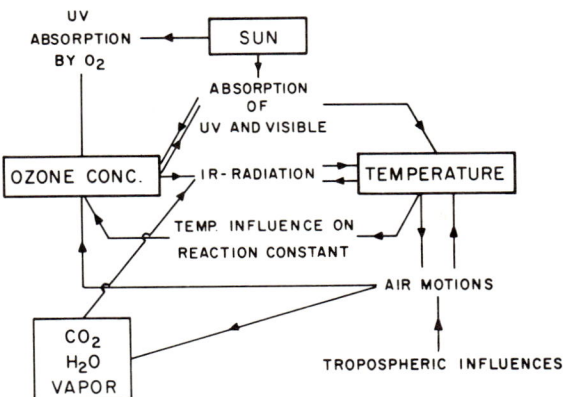

Fig. 47. Schematic representation of the interplay of radiation, photochemistry, and dynamics in the stratosphere.

One of the most interesting features of this system is the rather strong, but for the numerous feedbacks, complicated influence of possible changes in the solar short wave radiation. While the bulk of the solar radiation, which governs tropospheric energetics is known to be rather stable (except for the seasonal variations), such stability between 2000 and 3000 Å in the wavelength region dominating the ozone photochemistry is not yet established. Recent satellite observations actually indicate considerable variability in the 1000–2000 Å region [132]; thus important changes at least in the ozone-producing radiation around 2100 Å seem possible (the question is still open whether the difference between the values of Brewer and Wilson and those of Detwiler et al. given for that region might be real, i.e., demonstrate an actual change with time). Irregular direct solar interference with stratospheric dynamics seems thus possible and a coupling by the ozone layer is still the most reasonable mechanism for the often claimed relation between solar activity and large-scale weather phenomena.

However, before an influence of possible solar variability can be studied the behavior of the normal undisturbed system with all its feedbacks described above must be understood. Work on this problem has only just been started. Lindzen [133–137] obtained some interesting results such as the indication that under certain conditions dynamically stable processes may become unstable in the presence of photochemical reactions. However, in order to solve the problem analytically he had to introduce considerable simplifications.

Again numerical models will presumably become an important tool for studying the feedback system (in the experiments of Hunt [131] quoted in the preceeding chapter the feedback was yet largely excluded by using climatological mean ozone distributions in the computation of the radiation field). Such models should in future also allow to investigate possible solar-weather relationship produced by the ozone layer.

At the same time reliable measurements of the ozone content of the upper stratosphere over long periods are needed—this being the level of most direct solar control having still appreciable ozone concentrations. Umkehr measurements (at present the only method giving data routinely) at Arosa (Fig. 48) indicated a considerable trend apparently related to solar activity [138], but it was also shown, that Umkehr data on the upper stratosphere are very sensitive to slight instrumental changes and the reality of the effect has yet to be proved by parallel observations with two instruments over at least one solar cycle.

It is certainly important to continue every effort in this direction, because it can hardly be doubted that the main interest in ozone research will shift within the next decade to the problem of the dynamical–radiational–photochemical interaction.

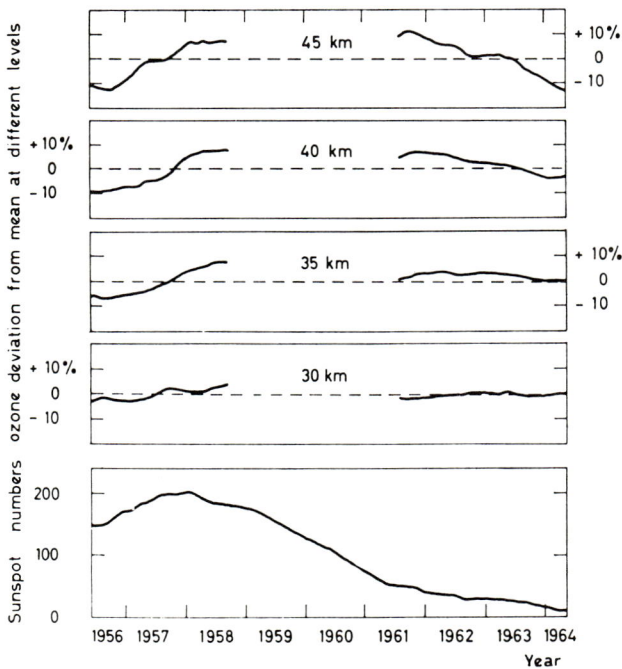

FIG. 48. Secular variation of ozone concentration in the middle and upper stratosphere (12 months overlapping means) compared with the sunspot numbers.

List of Symbols

$\tilde{O}$	Odd oxygen particles
$\tilde{H}$	Active hydrogen compounds (H, OH, HO$_2$)
$\tilde{NO}$	Nitrogen oxides
n_1	Concentration of atomic oxygen [particles cm^{-3}]
n_1^*	Concentration of excited [^{1}D]-oxygen atoms
n_2	Concentration of molecular oxygen
n_{N2}	Concentration of nitrogen
n_3	Concentration of ozone
$\tilde{n}$	$n_1 + n^* + n_3$, concentration of odd oxygen particles
x	Concentration of atomic hydrogen
y	Concentration of OH-radicals
z	Concentration of HO$_2$-radicals
ϕ	$x + y + z$, concentration of $\tilde{H}$-particles
w_1	Concentration of water vapour
w_2	Concentration of molecular hydrogen
w_3	Concentration of hydrogen peroxide
R	z/y
[NO]	Concentration of nitric oxide
[NO$_2$]	Concentration of nitrogen dioxide
[$\tilde{NO}$]	[NO] + [NO$_2$]
k_i	Reaction rates
f_i	Dissociation rates
f_1	Dissociation rate of H$_2$O
f_2	Dissociation rate of O$_2$
f_3	Dissociation rate of O$_3$
f_4	Dissociation rate of H$_2$O$_2$
f_5	Dissociation rate of HO$_2$
f_6	Dissociation rate of NO$_2$
a_2	Fraction of f_2 yielding one excited (^{1}D)-O-atom
a_3	Fraction of f_3 yielding one excited (^{1}D)-O-atom
α_i	Absorption coefficients
d_2	Vertical path length through oxygen in cm NTP above the level under consideration
d_2'	Slant path length of solar beam through oxygen
d_3	Vertical path length through ozone
d_3'	Slant path length through ozone
μ	$d_2'/d_2 = d_3'/d_3$, Bemporad function
H	Scale height
H_0	Height of homogeneous atmosphere (8×10^5 cm)
Radiation 1	Solar energy distribution after Detwiler et al. [85]
Radiation 2	Solar energy distribution after Brewer and Wilson [84] (see p. 268)
τ	Relaxation time (time of "half-restoration") of ozone
τ_1	Relaxation time of $\tilde{n}$ (odd oxygen particles)
τ_c	Relaxation time of night-time conversion of O$_3$ into O
τ_3	Relaxation time of night-time ozone equilibrium around the mesopause

References

1. Dobson, G. M. B. (1931). A photoelectric spectrophotometer for measuring the amount of atmospheric ozone. *Proc. Phys. Soc. (London)* **43**, 324.
2. Dobson, G. M. B. (1957). Observers handbook for the ozone spectrophotometer and the adjustment and calibration of ozone spectrophotometer. *Ann. Int. Geophys. Year* **5**, 46–114.

3. Dütsch, H. U. (1968). Measurement of atmospheric ozone. *Ann. IQSY* **1**, 234–245.
4. Vigroux, E. (1953). Contribution à l'étude expérimentale de l'absorption de l'ozone. *Ann, Géophys.* **8**, 709–762.
5. Vigroux, E. (1967). Détermination des coefficients moyens d'absorption de l'ozone en vue des observations concernant l'ozone atmosphérique à l'aide du spectromètre Dobson. *Ann. Phys. (Paris)* **2**, 209–215.
6. Bojkov, R. D. (1969). Some characteristics of the total ozone deduced from Dobson-spectrometer and filter-ozone-meter-data and their application to a determination of the effectiveness of the ozone station network. *Ann. Géophys.* **25**, 293–299.
7. Guschin, C. P. (1970). On the methods of measurements of the total content of atmospheric ozone. Results of researches of the International Geophysical Projects. *Meteorol. Res. (Moscow)* **17**, 51–57.
8. Dütsch, H. U. (1969). Atmospheric ozone and ultraviolet radiation. "World Survey of Climatology," Vol. 4, Chapter 8, pp. 383–432. Elsevier, Amsterdam.
9. London, J. (1963). The distribution of total ozone in the northern hemisphere. *Beitr. Phys. Atmos.* **36**, 254–263.
10. Khrgian, A. Kh., Kuznetzov, G. I., and Kondratjev, A. B. (1965). Atmospheric ozone. *IGY—Meteorol. Program (Moscow)* **8**, 90.
11. Kuznetzov, G. I., and Khrgian, A. Kh. (1968). Distribution of atmospheric ozone between IGY and IQSY. *Meteorol. Hydr.* **3**, 24–38.
12. Guschin, C. P. (1970). On certain results of atmospheric ozone studies at the Main Geophysical Observatory. Results of Researches of the International Geophysical Projects. *Meteorol. Res. (Moscow)* **17**, 58–71.
13. Petrnko, N. A., and Khrgian, A. Kh. (1970). Results of observations of the total content of atmospheric ozone on the world network of ozonometric stations, 1937–1964.
14. Bugajew, W. A., and Uranowa, L. A. (1967 and 1970). Maps of total ozone content. I. Jan. 1964–July 1966. II Aug. 1966–Dec. 1967. *Annex. Synop. Bull. Hemisphere, Moscow*.
15. Sticksel, P. R. (1970). The annual variation of total ozone in the Southern hemisphere. *Monthly Weather Rev.* **98**, 787–788.
16. London, J. (1965). The geographic distribution of the annual variation of total ozone. *Proc. Ozone Symp. Albuquerque*, 1964 pp. 51–53. WMO, Geneva.
17. Gebhart, R., Bojkov, R. D., and London, J. (1970). Stratospheric ozone: A comparison between observed and computed models. *Beitr. Phys. Atmos.* **43**, 209–227.
18. Kulkarni, R. N., and Garnham, G. L. (1970). Longitudinal variation of ozone in the lower middle latitudes of the Southern hemisphere. *J. Geophys. Res.* **75**, 4174–4176.
19. Prabhakara, C. (1969). Feasibility of determining atmospheric ozone from outgoing infrared radiation. *Monthly Weather Rev.* **97**, 307–314.
20. Prabhakara, C., Conrath, B. J., and Hanel, R. A. (1970). Remote sensing of atmospheric ozone using the 9.6 μ band. *J. Atmos. Sci.* **27**, 689–697.
21. Dave, J. V., and Mateer, C. L. (1967). A preliminary study on the possibility of estimating total atmospheric ozone from satellite measurements. *J. Atmos. Sci.* **24**, 414–427.
22. Götz, F. W. P. (1931). Zum Strahlungsklima des Spitzbergensommers. Strahlungs- und Ozonmessungen in der Königsbucht 1929. *Gerl. Beitr. Geophys.* **31**, 119–154.
23. Mateer, C. L., and Dütsch, H. U. (1964). Uniform evaluation of "Umkehr" observations from the world network. I. Proposed Standard Umkehr Evaluation Technique, Nat. Center Atm. Res., Boulder Colorado.

24. Dütsch, H. U. (1964). Uniform evaluation of Umkehr observations from the world network. III. World-wide ozone distribution at different levels and its variations with season from "Umkehr" observations. Nat. Center Atm. Res., Boulder, Colorado.
25. Bojkov, R. D. (1968). Planetary features of total and vertical ozone distribution. I and II. *Időjárás* **72**, 140–152 and 233–242.
26. Johnson, F. S., Purcell, J. D., and Tousey, R. (1954). Studies of the ozone layer above New Mexico. In "Rocket Exploration of the Upper Atmosphere" (R. L. F. Boyd, M. J. Seaton, and H. S. W. Massey, eds.), pp. 189–199. Pergamon Press, Oxford.
27. Rhandava, J. S. (1970). Results of a rocket experiment designed to measure diurnal variations of atmospheric ozone. *Monthly Weather Rev.* **98**, 402–405.
28. Krueger, A. J. (1969). Rocket measurements of ozone over Hawaii. *Ann. Géophys.* **25**, 225–229.
29. Hilsenrath, E. (1969). An ozone measurement in the mesosphere and stratosphere by means of a rocket sonde. *J. Geophys. Res.* **74**, 6873–6880.
30. Regener, E., and Regener, V. H. (1934). Aufnahme des ultravioletten Sonnenspektrums in der Stratosphäre und vertikale Ozonverteilung. *Phys. Z.* **35**, 788–793.
31. Kulcke, W., and Paetzold, H. K. (1957). Ueber eine Radiosonde zur Bestimmung der vertikalen Ozonverteilung. *Ann. Meteorol.* **8**, 47–53.
32. Vassy, A. (1958). Radiosonde spéciale pour la mesure de la répartition verticale de l'ozone atmosphérique. *J. Sci. Météorol.* **10**, 63–75.
33. Regener, V. H. (1960). On a sensitive method for the recording of atmospheric ozone. *J. Geophys. Res.* **65**, 3975–3977.
34. Regener, V. H. (1964). Measurements of ozone with the chemiluminescent method. *J. Geophys. Res.* **69**, 3795–3800.
35. Brewer, A. W., and Milford, J. R. (1960). The Oxford-Kew sonde. *Proc. Roy. Soc. (London)* A256, 470–496.
36. Komhyr, W. D. (1969). Electrochemical concentration cells for gas analysis. *Ann. Geophys.* **25**, 203–210.
37. Kobayashi, J., and Toyama, Y. (1966). On various methods of measuring the vertical distribution of atmospheric ozone. III. Carbon-iodine type chemical ozone sonde. *Pap. Meteorol. Geophys.* **17** (2), 113–126.
38. Attmannspacher, W., and Dütsch, H. U. (1970). International ozone sonde intercomparison at the observatory Hohenpeissenberg 19.1.–5.2.1970. *Ber. Deut. Wetterdienstes* No. 120.
39. Komhyr, W. D., Grass, R. D., and Proulx, R. A. (1968). Ozone sonde intercomparison tests. *Final Rep. U.S. Navy*, IPR 19–64–805–WEPS.
40. Mateer, C. L., and Heath D. F. (1969). Nimbus D experiment to determine the total amount and vertical distribution of atmospheric ozone. *Symp. Ozone Atmos.*, Monaco pp. 237–238. Centre Nat. Rech. Sci., Paris.
41. Hering, W. S., and Borden, Th. R. (1967). Ozone sonde observations over North America. Vol. 4. *AFCRL, Environ. Res. Pap.* **279**, 365 pp.
42. Dütsch, H. U., Züllig, W., and Ling, Ch. C. (1970). Regular ozone observations at Thalwil, Switzerland and at Boulder, Colorado. *Lab. Atmos. Phys. Eidgen. Tech. Hochsch.*, Zürich, 279 pp.
43. Regener, V. H., and Aldaz, L. (1969). Turbulent transport near the ground as determined from measurements of the ozone flux and the ozone gradient. *J. Geophys. Res.* **74**, 6935–6942.

44. Aldaz, L. (1969). Flux measurements of atmospheric ozone over land and water. *J. Geophys. Res.* **74**, 6943–6946.
45. Fabian, P., and Junge, C. E. (1970). Global rate of ozone destruction at the earth's surface. *Arch. Meteorol. Geophys. Biokl. A* **19**, 161–172.
46. Dütsch, H. U. (1966). Two years of regular ozone soundings over Boulder, Colorado. *NCAR, Boulder, Colorado, Tech. Note* **10**.
47. Pittock, A. B. (1968). Seasonal and year-to-year ozone variations from soundings over South-Eastern Australia. *Quart. J. Roy. Meteorol. Soc.* **94**, 563–575.
48. Komhyr, W. D., and Grass, R. D. (1968). Ozone sonde observations 1962–1966 (2). *Environ. Sci. Serv. Adm. Tech. Rep.* ERL 80—APCL 3.
49. Shimizu, M. (1969). Vertical ozone distribution at Syowa Station, Antarctic in 1966. *Jap. Antarctic Res. Exp. Sci. Rep. Ser. B., Meteorol.* **1**, 38 pp.
50. MacDowall, J. (1960). Some observations at Halley Bay in seismology, glaciology and meteorology, 12. Ozone soundings. *Proc. Roy. Soc. (London)* **A256**, 175–192.
51. Figueira, M. F. (1969). Synoptic analysis of ozone and artificial radioactivity. *Symp. Ozone Atmos. Monaco* pp. 271–272. Centre Nat. Rech. Sci., Paris.
52. Piaget, A. (1971). Utilisation de l'ozone atmosphérique comme traçeur des échanges entre la troposphère et la stratosphère. *Veröff. Schweiz. Meteorol. Zentralanstalt* **21**, 71 pp.
53. Berggren, R. (1965). The vertical distribution of ozone over Arosa on 16, April 1962 and the synoptic situation. *Tellus* **17**, 180–193.
54. Reed, R. J. (1965). The present status of the 26-months oscillation. *Bull. Amer. Meteorol. Soc.* **46**, 374–387.
55. Ramanathan, K. R. (1965). The 26-months-cycle and atmospheric ozone. *Proc. Ozone Symp., Albuquerque, 1964* pp. 1–6. WMO, Geneva.
56. Chapman, S. (1930). A theory of upper atmospheric ozone. *Mem. Roy. Meteorol. Soc.* **3**, 103.
57. Chapman, S. (1930). On the annual variation of upper-atmospheric ozone. *Phil. Mag.* **10**, 345–352.
58. Chapman, S. (1930). On ozone and atomic oxygen in the upper atmosphere. *Phil. Mag.* **10**, 369–383.
59. Maeda, K. (1963). Auroral dissociation of molecular oxygen in the polar mesosphere. *J. Geophys. Res.* **68**, 185–197.
60. Maeda, K., and Aikin, A. C. (1965). Ozone in the polar stratosphere. *Proc. Ozone Symp., Albuquerque, 1964* pp. 37–38. WMO, Geneva.
61. Wulf, O. R., and Deming, L. S. (1936). The theoretical calculation of the distribution of photochemically-formed ozone in the atmosphere. *Terr. Magn.* **41**, 299–310.
62. Wulf, O. R., and Deming, L. S. (1936). The effect of visible solar radiation on the calculated distribution of atmospheric ozone. *Terr. Magn.* **41**, 375–378.
63. Wulf, O. R., and Deming, L. S. (1937). The distribution of atmospheric ozone in equilibrium with solar radiation and the rate of maintainance of the distribution. *Terr. Magn.* **42**, 195–202.
64. Schröer, E. (1949). Theorie der Entstehung, Zersetzung und Verteilung des atmosphärischen Ozons. *Ber. Deut. Wetterdienstes (US-Zone)* **11**, 13–23.
65. Dütsch, H. U. (1946). Photochemische Theorie des atmosphärischen Ozons unter Berücksichtigung von Nichtgleichgewichtszuständen und Luftbewegungen 113 pp. Thesis, Univ. of Zürich.
66. Craig, R. A. (1950). The observations and photochemistry of atmospheric ozone and their meteorological significance. *Meteorol. Monogr.* **1**, No. 2, 50 pp.

67. Paetzold, H. K. (1953). Die vertikale Verteilung des atmosphärischen Ozons nach dem photochemischen Gleichgewicht. *Geofis. Pura Appl.* **24**, 71.
68. Leovy, C. (1964). Radiative equilibrium of the mesosphere. *J. Atmos. Sci.* **21**, 238–248.
69. Bates, D. R., and Nicolet, M. (1950). The photochemistry of atmospheric water vapour. *J. Geophys. Res.* **55**, 301–327.
70. Hampson, J. (1964). "Photochemical Behaviour of the Ozone Layer," 280 pp. Can. Arm. Res. Develop. Estab., Tech. Note 1627 164, Valcartier, Quebec.
71. Hunt, B. G. (1966). Photochemistry of ozone in a moist atmosphere. *J. Geophys. Res.* **71**, 1385–1398.
72. Crutzen, P. J. (1969). Determination of parameters appearing in the "dry" and the "wet" photochemical theories for ozone in the stratosphere. *Tellus* **21**, 368–388.
73. Dütsch, H. U. (1970). Atmospheric ozone—A short review. *J. Geophys. Res.* **75**, 1707–1712.
74. Heilpern, W. (1952). Der Fremdgaseinfluss auf die Lichtabsorption des Sauerstoffs als Funktion vom Druck. *Helv. Phys. Acta* **25**, 753–772.
75. Ditchburn, R. W., and Young, P.A. (1962). The absorption of molecular oxygen between 1850 and 2500 Å. *J. Atmos. Terr. Phys.* **24**, 127–139.
76. Dütsch, H. U. (1950). Einfluss der Temperatur auf das atmosphärische Ozon. *Arch. Meteorol. Geophys. Biokl. A* **2**, 386–400.
77. Reed, E. J., and Scolnik, R. (1965). A nighttime measurement of atmospheric ozone above 40 km. *Proc. Ozone Symp., Albuquerque 1964* p. 27. WMO, Geneva.
78. Reed, R. J., (1968). A nighttime measurement of mesospheric ozone by observation of ultraviolet airglow. *J. Geophys. Res.* **73**, 2951–2957.
79. Kellogg, W. W. (1961). Chemical heating above the polar mesopause in winter. *J. Meteorol.* **18**, 373–381.
80. Young, C., and Epstein, E. S. (1962). Atomic oxygen in the polar winter mesosphere. *J. Atmos. Sci.* **19**, 435–443.
81. Hesstvedt, E. (1968). On the effect of vertical eddy transport on atmospheric composition in the mesosphere and lower thermosphere. *Geofis. Publ.* **27**, No. 4, 35.
82. Campbell, E. S., and Nudelmann, C. (1960). Reaction kinetics, thermodynamics and transport properties in the ozone-oxygen-system. Rep. AFOFR TN-60-502, Dept. Chem., N.Y. Univ.
83. Dütsch, H. U., and Ginsburg, Th. (1968). Parametric studies on ozone photochemistry. *Pure Appl. Geophys.* **72**, 204–213.
84. Brewer, A. W., and Wilson, A. W. (1965). Measurements of solar ultraviolet radiation in the stratosphere. *Quart. J. Roy. Meteorol. Soc.* **91**, 452–461.
85. Detwiler, C. R., Garret, D. L., Purcell, J. D., and Tousey, R. (1961). The intensity distribution in the ultraviolet solar spectrum. *Ann. Géophys.* **17**, 263–272.
86. Brewer, A. W., and Wilson, A. W. (1968). The regions of formation of atmospheric ozone. *Quart. J. Roy. Meteorol. Soc.* **94**, 249–265.
87. Inn, E. C. Y., and Tanaka, Y. (1953). Absorption coefficients of ozone in the ultraviolet and visible regions. *J. Opt. Soc. Am.* **43**, 870–873.
88. Hunt, B. G. (1966). The need for a modified photochemical theory of the ozonosphere. *J. Atm. Sci.* **23**, 88–95.
89. Craig, R. A. (1965). "The Upper Atmosphere," 419 pp Academic Press, New York
90. Hesstvedt, E. (1969). A photochemical atmosphere model containing oxygen, hydrogen and nitrogen, 31 pp. Rep. Inst. Geophys., Univ. Oslo, December.
91. De More, W. B., and Raper, D. F. (1964). Deactivation of $O(^1D)$ in the atmosphere. *Astrophys. J.* **139**, 1381–1383.

92. Hunt, B. G. (1965). Influence of metastable oxygen molecules on atmospheric ozone. *J. Geophys. Res.* **70**, 4990–4991.
93. Schumacher, H. J., and Beretta, U. (1932). Photokinetik des Ozons. I, II. *Z. Phys. Chem. Abt. B* **17**, 405–416, 417–428.
94. Crutzen, P. J. (1970). The influence of nitrogen oxides on the atmospheric ozone content. *Quart. J. Roy. Meteorol. Soc.* **96**, 320–325.
95. McGrath, W. O., and Norrish, G. W. (1960). Studies of the reactions of excited atoms and molecules produced in the flash photolysis of ozone. *Proc. Roy. Soc. (London)* **A254**, 317–326.
96. Benson, S. W., and Axworthy, A. E. (1965). Reconsiderations of rate constants from the thermal decomposition of ozone. *J. Chem. Phys.* **42**, 2614–2615.
97. Reeves, R. R., Manella, G., and Harteck, P. (1960). Rate of recombination of oxygen atoms. *J. Chem. Phys.* **32**, 632–633.
98. Schiff, H. I. (1969). Neutral reactions involving oxygen and nitrogen. *Can. J. Chem.* **47**, 1903–1916.
99. Fitzsimmons, R. V., and Bair, E. J. (1964). Distribution and relaxation of vibrationally excited oxygen in the flash photolysis of ozone. *J. Chem. Phys.* **36**, 2681–2692.
100. Kaufman, F. (1964). Aeronomic reactions involving hydrogen, a review of recent laboratory studies. *Ann. Géophys.* **20**, 106–114.
101. Pettersen, H. L., and Kretchmer, C. B. (1960). Kinetics of recombination of atomic oxygen at room temperature. *U.S. Dept. Commerce*, Office Tech. Service, A.D. 283044.
102. Larkin, F. S., and Thrush, B. A. (1964). Recombination of hydrogen atoms in the presence of atmospheric gases. *Discuss. Faraday Soc.* **37**, 113.
103. Foner, S. N., and Hudson, R. L. (1962). Mass spectrometry of the HO_2 free radical. *J. Chem. Phys.* **36**, 2681–2692.
104. Clyne, M. A. A., and Thrush, B. A. (1963). Rates of elementary processes in the chain reaction between hydrogen and oxygen. Kinetics of the reaction of hydrogen atoms with molecular oxygen. *Proc. Roy. Soc. (London)* **A275**, 559–566.
105. Crutzen, P. J. (1971). Personal communication.
106. Schofield, K. (1967). An evaluation of kinetic rate data for reactions of neutrals of atmospheric interest. *Planet. Space Sci.* **15**, 643–670.
107. Dütsch, H. U. (1968). The photochemistry of stratospheric ozone. *Quart. J. Roy. Meteorol. Soc.*, **94**, 483–497.
108. Leovy, C. B. (1969). Atmospheric ozone; an analytical model for photochemistry in the presence of water vapour. *J. Geophys. Res.* **79**, 417–424.
109. Kaufman, F. (1969). Neutral reactions involving hydrogen and other minor constituents. *Can. J. Chem.* **47**, 1917–1927.
110. Bates, D. R., and Hays, P. B. (1967). Atmospheric nitric oxide. *Planet. Space Sci.* **15**, 643–676.
111. Hesstvedt, E. (1968). On the photochemistry of ozone in the ozone layer. *Geofis. Publ.* **27**, No. 5, 16.
112. Hesstvedt, E. (1964). On the water vapour content of the high atmosphere. *Geofis. Publ.* **25**, No. 3, 1–18.
113. Hampson, J. (1966). Chemiluminescent emissions observed in the stratosphere and mesosphere. *In* "Les problèmes météorologiques de la stratosphère et de la mesosphère." Presse Univ. France, Paris.
114. Blamont, J. E., and De Jager, C. (1961). Upper atmospheric turbulence near the 100 km level. *Ann. Géophys.* **17**, 134–164.
115. Johnson, F. S., and Wilkens, E. M. (1965). Thermal upper limit on eddy diffusion in the mesosphere and lower thermosphere. *J. Geophys. Res.* **70**, 1281–1284.

116. Strobel, D. F., Munten, D. M., and McElroy, M. B. (1970). Production and diffusion of nitric oxide. *J. Geophys. Res.* **75**, 4307.
117. Nicolet, M. (1965). Nitrogen oxides in the chemosphere. *J. Geophys. Res.* **70**, 679–689.
118. Nicolet, M. (1970). Ozone and hydrogen reactions. *Ann. Géophys.* **26**, 531–546.
119. Brewer, A. W. (1949). Evidence for a world circulation provided by the measurements of helium and water vapour distribution in the stratosphere. *Quart. J. Roy. Meteorol. Soc.* **75**, 351–363.
120. Dobson, G. M. B. (1956). Origin and distribution of poly-atomic molecules in the atmosphere. *Proc. Roy. Soc. (London)* **A236**, 187–193.
121. Newell, R. E. (1963). Transfer through the tropopause and within the stratosphere. *Quart. J. Roy. Meteorol. Soc.* **89**, 167–204.
122. Newell, R. E. (1961). The transport of trace substances in the atmosphere and their implications for the general circulation in the stratosphere. *Geofis. Pura Appl.* **49**, 137–158.
123. Feely, H. W., and Spar, J. (1960). Tungsten-185 from nuclear bomb tests as a tracer for stratospheric meteorology. *Nature* **188**, 1062–1064.
124. Machta, L. (1960). Meteorology and radiative fallout. WMO Bull., **9**, No. 2, April, 64–70.
125. Machta, L., List, R. J., and Telegadas, K. (1962). A survey of radioactive fallout from nuclear tests. *J. Geophys. Res.* **67**, 1389–1401.
126. Martell, E. G. (1959). Atmospheric aspects of strontium-90 fallout. *Science* **129**, 1197–1206.
127. Martell, E. G. (1968). Tungsten radio-isotope distribution and stratospheric transport processes. *J. Atmos. Sci.* **25**, 113–125.
128. Prabhakara, C. (1963). Effects of non-photochemical processes on the meridional distribution and total amount of ozone in the atmosphere. *Monthly Weather Rev.* **91**, 411–431.
129. Gebhart, R. (1968). Photochemical, radiative and turbulent effects on the meridional distribution of ozone. A study of the time-dependent problems. *Arch. Meteorol. Geophys. Biokl.* **A7**, 301–335.
130. Hunt, B. G., and Manabe, S. (1968). Experiments with a stratospheric general circulation model. II. Large scale diffusion of tracers in the stratosphere. *Monthly Weather Rev.* **96**, 503–539.
131. Hunt, B. G. (1969). Experiments with a stratospheric circulation model. III. Large scale diffusion of ozone including photochemistry. *Monthly Weather Rev.* **97**, 287–306.
132. Prag, A. G., and Morsel, F. E. (1970). Variations in the solar ultraviolet flux from July 13 to August 9, 1968. *J. Geophys. Res.* **75**, 4613–4621.
133. Lindzen, R., and Goody, R. (1965). Radiative photochemical processes in mesospheric dynamics. I. Models for radiative and photochemical processes. *J. Atmos. Sci.* **22**, 341–348.
134. Lindzen, R. S. (1965). The radiative-photochemical response of the mesosphere to fluctuations in radiation. *J. Atmos. Sci.* **22**, 469–478.
135. Lindzen, R. S. (1966). Radiative and photochemical processes in mesospheric dynamics. II. Vertical propagation of long period disturbances at the equator. *J. Atmos. Sci.* **23**, 334–343.
136. Lindzen, R. S. (1966). Radiative and photochemical processes in meospheric dynamics. III. Stability of a zonal vortex at mid-latitudes to axially symmetric disturbances. *J. Atmos. Sci.* **23**, 344–349.

137. Lindzen, R. S. (1966). Radiative and photochemical processes in mesospheric dynamics. IV. Stability of a zonal vortex at mid-latitudes to baroclinic waves. *J. Atmos. Sci.* **23**, 350–359.
138. Dütsch, H. U., Boller, P., and Ling, Ch. C. (1969). Simultaneous Umkehr observations with two instruments. *Ann. Géophys.* **25**, 215–218.

AUTHOR INDEX

Numbers in parentheses are reference numbers and indicate that an author's work is referred to, although his name is not cited in the text. Numbers in italics show the page on which the complete reference is listed.

A

Adams, J. A. S., 17(1), 19(1), 25(1), 29(1), 30(1), 33(1), *54, 56*
Aditya, P. K., *56*
Ahrens, L. H., 11, 30(2), *54*
Aikin, A. C., 245(60), 264(60), *318*
Aldaz, L., 288(43, 44) *317, 318*
Alfvén, H., 137(6, 7, 8), 139(8), *208*
Aller, L. H., 3(3), 17(3), *54*
Amyx, J. W., 33, *54*
Anderson, C. C., *54*
Anderson, K. A., 146(29), 155(60, 63), 162, *209, 211*
Andronikashuili, E. L., *56*
Arens, J. F., 167, 182(143), 197(143), 198, 199(143), 200(143), 206(143), *211, 214*
Armstrong, A. H., 168(76), 169, 173(116), *211, 213*
Armstrong, J. C., 177(126), 178(126), 180(126), 181(126), 182(126), *214*
Armstrong, T. P., 169, *212*
Arnoldy, R. L., 203(186), 206(186), *216*
Asbridge, J. R., 154(55), *210*
Attmannspacher, W., 224(38), *317*
Austin, M. M., 173(119), 174(119), 175(119), *213*
Axford, W. I., *54*, 169(86), 193(171), 201(171), 203(189), *212, 216, 217*
Axworthy, A. E., 272(96), *320*

B

Badhwar, G. D., 32, *54*
Bair, E. J., 272(99), *320*
Bame, S. J., 154, *210*
Baranov, V. I., *54*
Barrow, G. M., 47, *55*
Barsukov, O. A., *55*
Bass, D. M., 33, *54*

Bates, D. R., 245(69), 271, 272(69), 273(69), 285(110), 290, 303, 304, *319, 320*
Battan, L. J., 64(8), 65(8), 126(8), *133*
Beall, D. S., 177(126), 178(126, 137), 180(126), 181(126), 182(126), 197(137), 198, *214*
Beebe, B. W., 2(11), 45(11), *55*
Beggs, W. C., 206(202), *217*
Benson, S. W., 272(96), *320*
Beretta, U., 271(93), *320*
Bergeron, T., 97, *134*
Berggren, R., 235(53), *318*
Berkner, L. V., 30(8a), 32(8a), *54*
Bezrukikh, V. V., 154(52), *210*
Bieri, R. H., 32(12, 13), 33(12), *55*
Billings, K., *56*
Bird, R. B., 39(35), 40(35), *55*
Blake, J. B., 164(70), 165(70), 167(70), 168(79), 173(113), 178(135), *211, 212, 213, 214*
Blamont, J. E., 297(114), *320*
Blanchard, R. C., 172(108), *213*
Bojkov, R. D., 220(6,17), 221(17), 223(17, 25), *316, 317*
Boller, P., 314(138), *322*
Borden, Th. R., 225(41), *317*
Bostrom, C. O., 171(96, 97), 173, 177, 178(126, 134, 137), 180(126), 181(126), 182, 197(137), 198(137), *212, 213, 214*
Bracken, P. A., 186(161), *215*
Braham, R. R., Jr., 63(7), 64(9), 65(9), 66, 119(9), 131(9), *133*
Brewer, A. W., 224(35), 249, 250, 268 (84, 86), 285, 306, 309, 310, *317, 319, 321*
Brice, N. M., 203(190, 191), 206, *217, 218*
Brier, G. W., 111(28), *134*
Brown, H. A., 111(31), *134*
Brown, W. L., 146(42), 173, 175(114), 176, 183, 200, 206(200), *210, 213, 215, 217*
Bugajew, W. A., 220(14), *316*

C

Burrows, J. R., 191(167, 168), 193, 195(167, 178, 179), 196, 197, 198 (179, 181), 200(181), 203(167, 168), 206(201), *216, 217*

Cahill, L. J., 186(160), 206(200), *215, 217*
Campbell, E. S., 265, *319*
Canfield, E. H., 171(93), 175(120, 121), *212, 213*
Carman, P. C., *55*
Carpenter, D. L., 154(59), *211*
Chan, K. W., 203(186), 206(186), *214*
Chang, D. B., 184(149), *215*
Changnon, S. A., Jr., 97, 124(34), 129, *134*
Chapman, S., 137(4, 5), *208*, 244 (56, 57, 58), *318*
Chappell, C. F., 63(6), *133*
Cherdyntsev, V. V., 17(15), 29(15), 30(15), *55*
Christofilos, N. C., 147(16), 206(16), *209*
Chudakov, E. A., 146(21), 170(89), *209, 212*
Cladis, J. B., 181(141), 183, *214*
Clafin, E. S., 175(122), *214*
Clarke, F. W., 30(16), *55*
Clyne, M. A. A., 273(104), *320*
Conrad, V., 79, 91(13), *133*
Conrath, B. J., 222(20), *316*
Cook, G. A., 12(17), 32(17), *55*
Craft, B. C., *55*
Craig, R. A., 245(66), 270(89), *318, 319*
Craven, J. D., 191(166), 196(166), 203 (166), *216*
Crutzen, P. J., 245(72), 252(72), 265(72), 271, 273(105), 274(72) 277, 285, 303, 304, 305, *319, 320*
Cunningham, R. M., 111(30), *134*
Cupp, E. L., 171(99), *212*
Curtiss, C. F., 39(35), 40(35), *55*

D

Damon, P. E., 2(19), 32(20, 46, *55*
Dave, J. V., 222(21), *316*
Davis, L., Jr., 184(149), *215*
Davis, L. G., 111(29), *134*
Davis, L. R., 159, 160, 188, 190, 206(66, 198, 200), *211, 217*
Davis, T. N., 206(202), *217*
Day, F. H., 19(21), *55*
Decker, W. L., 131(15), *134*
DeJager, C., 297(114), *320*
Deming, L. S., 245(61, 62, 63), *318*
De More, W. B., 270(91), 272(91), *319*
Deney, C. L., 32(7), *54*
Dessler, A. J., 146(22), *209*
Detwiler, C. R., 249, 250, 268, *319*
Ditchburn, R. W., 248, 268, *319*
Dobbin, C. E., 13, *55*
Dobson, G. M. B., 219(1, 2), 306, 310, *315, 321*
Dragt, A. J., 171(105), 172(109), 173, 174, 175, *213*
Dungey, J. W., 184(152), *215*
Dütsch, H. U., 219(3), 220(8), 223(23, 24), 224(38), 227(42), 228(42, 46), 242(46), 245(65, 73), 251(76), 252(65), 265(83), 266(65), 267(65), 268(83), 274(107), 310, 314(138), *316, 317, 318, 319, 320, 322*

E

Elliott, H., 146(25), *209*
Emerson, D. E., 37, *55*
Epstein, E. S., 263(80), 289(80), *319*
Evans, D. S., 193, *216*
Evans, R. D., 9(24), *55*

F

Fabian, P., 228(45), *318*
Fairfield, D. H., 151, 152, 153, 154, 155, *210*
Falthammar, C. G., 184(154, 155, 156), 186(154, 156), *215*
Farley, T. A., 146(26), 176, *209, 214*
Faul, H., 9, 14(25), 17(25), 20(25), 29(25), 37(25), 45, *55*
Feely, H. W., 306(123), *321*
Felthauser, H. E., 154(55), *210*
Fenton, K. B., 168(78), *212*

AUTHOR INDEX

Ferguson, E. E., 32(26), *55*
Ferraro, V. C. A., 137(4, 5), *208*
Figueira, M. F., 234(51), *318*
Fillius, R. W., 171(107), 173(107, 111), *213*
Filz, R. C., 171(101, 103), 172(101, 103, 110), 173(103), 174(103), *213*
Fitzsimmons, R. V., 272(99), *320*
Flamm, E. J., 175(120, 121), *213*
Fleischer, M., 14(27), 19(27), 30, 46(27), *55*
Flueck, J. A., 63(7), *133*
Foner, S. N., 273(103), *320*
Foppl, H., 206(205), *218*
Fowler, W. A., 9(28), 12(28), *55*
Fox, W. F., *55*
Frank, L. A., 155(61, 62), 162, 163(67), 167(61), 186 (159, 162, 163), 187, 188, 191(61, 166), 196(166, 180), 200, 203 (166), 206(162, 163), *211*, *215*, *216*
Freden, S. C., 170(90, 91), 171, 172, 173, 174, 175(91), 176, 177(91), 178(135), *212*, *213*, *214*
Freeman, J. W., Jr., 154, 167(54), 196(54), *210*
Fritz, T. A., 168(80), 169, 193(173, 174), 195, 202, 203(174), *212*, *216*
Funkhouser, J. C., 30, *55*

G

Gabbe, J. D., 173, 175(114), 176, *213*
Gabriel, K. R., 64(10), *133*
Garnham, G. L., 221(18), *316*
Garret, D. L., 249(85), 250(85), 268(85), *319*
Gebhart, R., 220, 221(17), 223(17), *316*, 310(129), *321*
Gerling, E. K., 46(31), *55*
Ginsburg, Th. 265(83), 268(83), *319*
Glass, M., 111(31), *134*
Goldman, D. T., *55*
Goldschmidt, V. M., 10, 14(33), 30(33), 31, 45, *55*
Goody, R., 313, *321*
Gorchakov, E. V., 163(68), *211*
Götz, F. W. P., 223(22), *316*
Grant, L. O., 63(6), *133*
Grass, R. D., 224, 230(48), *317*, *318*
Graybill, F. A., *56*

Grigorov, N. L., 170(89), *212*
Gringauz, K. I., 154, 203, *210*, *211*
Gurnett, D. A., 204(194, 195), *217*
Guschin, C. P., 220(7, 12), *316*

H

Haerendel, G., 171(104), 172, 184(157), 190, 200, 206(204, 205, 207), *213*, *215*, *216*, *218*
Hampson, J., 245(70), 271, 290, *319*, *320*
Hanel, R. A., 222(20), *316*
Harris, H. K., 155(63), *211*
Harrison, F. B., 173(116), *213*
Harteck, P., 272(97), *320*
Hartz, T. R., 203(190), *217*
Haser, L., 206(205), *218*
Haskell, G. P., 154(51), *210*
Hawkins, M. F., *55*
Haymes, R. C., 171(98), *212*
Hays, P. B., 285(110), 303, 304, *320*
Heath, D. F., 225(40), *317*
Heckman, H. H., 168(76), 169, 171(102), 172(102), 173(116), *211*, *213*
Heier, K. S., 14(43), *56*
Heilpern, W., 248, 268, *319*
Heinrich, E. W., 19, 20(34), 25, 28(34), *55*
Hendrickson, R. A., 206(203), *217*
Heppner, J. P., 206(206, 208), *218*
Hering, W. S., 225(41), *317*
Herlofsen, N., 184(148), *215*
Hess, W. N., 146(31), 170, 171(93), 172 (108), 178, 184(152, 153), 206 (132, 202), *209*, *212*, *213*, *214*, *215*, *217*
Hesstvedt, E., 263(81), 270(90), 272(90), 273(90) 279, 282, 284, 288, 289 (81, 90, 112), 290, 297, 299, 301, 302, 310, *319*, *320*
Hills, H. K., 196(180), *216*
Hilsenrath, E., 223(29), *317*
Hinson, H. H., *54*
Hirschfelder, J. O., 39(35), 40(35), *55*
Hoffman, R. A., 186(161), *215*
Holeman, E., 171(103), 172(103, 110), 173(103), 174(103), *213*
Hones, E. W., Jr., 146(38), 154(55), 192, *210*

Howard, R. A., 8(36), *56*
Hudson, H. E., Jr., 129, *135*
Hudson, R. L., 273(103), *320*
Huff, F. A., 60, 61(3, 4), 65(2, 4), 70(12), 79(14), 82(4, 16, 17), 83(16), 84(16, 17), 88(18), 89(17), 91(4, 17), 92(4), 93(12), 96(4), 97(18), 98(17), 101(21), 104(24, 26), 107 (26,27), 108(27), 110(4), 111(4), 114(32), 115(4), 126(2, 3, 14), 127(3, 12, 26), 128(4, 21, 26, 27), 129(4, 36), 130 (4, 12), 131(4, 12), 132(4), *133*, *134*, *135*
Hunt, B. G., 245(71), 269, 270(88, 92), 272(71), 279, 282, 290, 310, *319*, *320*, *321*

I

Ifedili, S. O., 171(99, 100), *212*, *213*
Imhof, W. L., 180(139, 140), 183(140, 144), 198(182), 199(182), 204(182), *214*, *215*, *216*
Inn, E. C. Y., 268, *319*
Ivanenko, I. P., 170(89), *212*

J

Jenkins, R. W., 171, *212*, *213*
Johnson, F. S., 223(26), 297(115), *317*, *320*
Junge, C. E., *56*, 228(45), *318*

K

Kaplon, M. F., 32(7), *54*
Kassander, A. R., 64(8), 65(8), 126(8), *133*
Katz, L., 206(197), *217*
Kaufman, F., 272(100), 273(100), 277, 284(109), *320*
Keesom, W. H., 12(38), 13, 29, *56*
Kellman, S., 175(120, 121), *213*
Kellogg, P. J., 177(129, 130), 184(146), *214*, *215*
Kellogg, W. W., 263(79), 289(79), *319*
Kennel, C. F., 195, 200, 201, 202, 204, 205, *216*, *217*
Kesebir, M., *56*
Khrgian, A. Kh., 220(10, 11,13), *316*
King, J. H., 146(37), *209*
Klinkenberg, L. J., 33(40), *56*

Kniffen, D. A., 168(77), 169, 173(115), *211*, *213*
Kobayashi, J., 224(37), *317*
Kohl, J. W., 199(183), 204(183), *216*
Kohler, M. A., 104, *134*
Komhyr, W. D., 224(36, 39), 230(48), *317*, *318*
Kondratjev, A. B., 220(10), *316*
Konradi, A., 206(199), *217*
Kretchmer, C. B., 272(101), *320*
Krimigis, S. M., 168(75, 80, 81), 169(84, 85), *211*, *212*
Krueger, A. J., 223(28), *317*
Krumbein, W. C., *56*
Kulcke, W., 224(31), *317*
Kulkarni, R. N., 221(18), *316*
Kulp, J. L., 2(19), 46, *55*
Kurt, V. G., 154(53), *210*
Kuznetzov, G. I., 220(10, 11), *316*
Kuznetsov, S. N., 146(43), 163(68), *210*, *211*

L

Lambert, I. B., 14(43), *56*
Lanzerotti, L. J., 146(42), 182(143), 197(143), 198, 199(143), 200(143), 206 (143), *210*, *214*
Larkin, F. S., 272(102), 273(102), *320*
Lebedinskii, A. I., 170 (89), *212*
Lencheck, A. M., 146(24), 170(92), 171 (24, 92), 175(92, 117, 118), 177(131), *209*, *212*, *213*, *214*
Leovy, C. B., 245(68), 274(108), *319*, *320*
Leverson, A. I., 51(44), *56*
Lezniak, T. W., 146(41), *210*
Lifshits, E. M., *56*
Light, P., 104, *134*
Lindstrom, P. J., 171(102), 172(102), *213*
Lindzen, R. S., 313, *321*, *322*
Ling, Ch. C., 227(42), 228(42), 314(138), *317*, *322*
Lingenfelter, R. E., 171(93, 95), 175(120, 121), *212*, *213*
Linsley, R. K., 104, *134*
List, R. J., 306(125), *321*
Lockwood, J. A., 171(99, 100), *212*, *213*
Logachev, Yu. I., 146(43), 163(68), *210*, *211*

AUTHOR INDEX 327

London, J., 220(9, 17), 221(9, 16, 17), 222(9), 223(17), *316*
Lopez, M. E., 117, *134*
Lord, H. C., *56*
Lowder, W. M., 17(1), 19(1), 25(1), 29(1), 30(1), 33(1), *54*
Lucero, A. B., 146(36), 164(69), *209, 211*
Ludwig, G. H., 137(12, 13), 138(13), 145, 147(17), 170(12), *208, 209*
Lust, R., 206(204, 205, 207), *218*

M

Macagno, E., 155(61), 162, 167(61), 191(61), *211*
McCormac, B. M., 146(33, 34, 35), *209*
McDiarmid, I. B., 57, 191(167, 168), 193, 195(167, 178, 179), 196, 197, 198(179, 181), 200(181), 203(167, 168), 206(201), *216, 217*
MacDonald, G. J. F., *56*, 66(11), 78(11), *133*
McDonald, J. E., 60, *133*
MacDowall, J., 230(50), *318*
McElroy, M. B., 305(116), *321*
McEntire, R. W., 206(203), *217*
McGrath, W. O., 271, *320*
McGuiness, J. L., 104, *134*
Machta, L., 306(124, 125), *321*
McIllwain, C. E., 137(12), 142(18), 143, 145, 147(17), 160, 166, 167(72), 170(12), 171(106), 173, 175(123), 176, 177(127), 178(133), 206 (127, 200, 210), *208, 209, 211, 213, 214, 217, 218*
Macy, W. W., 171(103), 172(103), 173, 174, *213*
Maeda, K., 245(59, 60), 264(59, 60), *318*
Maier, E. J. R., 206(202), *217*
Manabe, S., 310, 311, 313, *321*
Manella, G., 272(97), *320*
Marshall, L. C., 30(8a), 32(8a), *54*
Martell, E. G., 306(126, 127), *321*
Mason, B., 9(48), 10(48), 19(48), 29(48), *56*
Mateer, C. L., 222(21), 223(23), 225(40), *316, 317*
Mayne, K. I., 29(49), 30, 31, *56*
Mead, G. D., 142(20), 145, 146(20, 32, 39), 149(20, 48), 153(20), 176, 184(125),

185(39, 125), 188, 189, 191(20), 192, 195, 206(202) *209, 210, 214, 217*
Mead, J. M., 206(202), *217*
Melzner, F., 206(205), *218*
Meyer, B., 206(205), *218*
Meyer, T. O., 37, *55*
Mielke, P. W., Jr., 63(6), *133*
Milford, J. R., 224(35), *317*
Miller, R. D., *56*
Miyake, Y., *56*
Moroz, V. I., 154(53), *210*
Morsel, F. E., 313(132), *321*
Munnerlyn, R. D., *56*
Munten, D. M., 305(116), *321*
Murphy, B. L., 206(202), *217*
Murray, E. G., *56*
Murzin, V. S., 170(89), *212*

N

Nakada, M. P., 176, 184(125, 215), 185, 188, 189, *214, 215*
Nakano G. H., 171(102), 172(102), *213*
Nason, C. K., 117, *134*
Naughton, J. J., 30, 45, *55*
Naugle, J. E., 168(77), 169, 173(115), *211, 213*
Neill, J. C., 104, 124(34), 129, *134*
Ness, N. F., 148(44, 47), 149(44, 47), 167(44), *210*
Neuss, H., 206(205), *218*
Newell, R. E., 306, *321*
Newkirk, L. L., 171(94), 183, 200, 201, 203, *212, 215, 216*
Nicolet, M., 245(69), 271, 272(69), 273(69), 290, 305, *319, 321*
Norrell, G. P., *56*
Norrish, G. W., 271, *320*
Northrup, T. G., 138(15), *209*
Nudelmann, C., 265, *319*

O

O'Brien, B. J., 142(19), 146(30), 167(19), 177(128), 191(19), 193, 194, 197(172), 203(172), *209, 214, 216*
Owens, H. D., 186, 187, 188, *215*
Ozerov, V. D., 154(52), *210*

P

Paetzold, H. K., 224(31), 245(67), *317*, *319*
Palmer, J. A., 164(70), 165(70), 167(70), *211*
Palmer, W. F., 167(71), 191(71), *211*
Paoli, R. J., 155(63), *211*
Parker, E. N., *215*
Parker, L. W., 206(202), *217*
Paulikas, G. A., 164, 165, 167(70), 168(79), 173(113), 178(135), *211*, *212*, *213*, *214*
Petrnko, N. A., 220(13), *316*
Petschek, H. E., 204, *217*
Pettersen, H. L., 272(101), *320*
Pfitzer, K. A., 146(41), 181(142), 182(142), 203(187), *210*, *214*, *217*
Piaget, A., 234(52), 238(52), 254(52), 306, *318*
Pieper, G. F., 173(112), *213*
Pierce, A. P., 13, 28(54), 41, 45, *56*
Pittock, A. B., 229(47), 242(47), *318*
Pizzella, G., 171(106), *213*
Pollak, L. W., 79, 91(13), *133*
Prabhakara, C., 222(19, 20), 310(128), *316*, *321*
Prag, A. G., 313(132), *321*
Proulx, R. A., 224, *317*
Purcell, J. D., 223(26), 249(85), 250(85), 268(85), *317*, *319*
Puri, P. K., *56*

R

Rabben, H., 206(205), *218*
Ragland, P. C., *56*
Ramanathan, K. R., 242(55), 245(55), *318*
Randall, B. A., 169(84, 85), *212*
Rankama, K. 4(58, 59), 8, 19(60), 29, 30, 33(59), *56*
Raper, D. F., 270(91), 272(91), *319*
Ray, E. C., 137(12), 145, 170(12), *208*
Razdan, H., 171(100), *213*
Reagan, J. B., 183(144), *215*
Reed, R. J., 242(54), 259(77, 78), 303, *318*, *319*
Reeves, R. R. 272(97), *320*
Regener, E., 224(30), *317*
Regener, V. H., 224(30, 33, 34), 228(43), *317*
Reid, R. C., *56*

Rhandava, J. S., 223(27), 259(27), *317*
Rieger, E., 206(204, 205), *218*
Roberts, C. S., 146(42), 204, 206(200), *210*, *217*
Roberts, W. J., 129, *135*
Roederer, J. G., 146(40), 148(45), 149(40), 150, 153(40), *210*
Rogers, G. S., 13, 45, *56*
Rosen, L., 173(116), *213*
Rosser, W. G. V., 146(27), *209*
Roy, R. G., *57*
Runcorn, S. K., 11(2), 30(2), *54*
Rybchinsky, R. Ye., 154(52), *210*

S

Sahama, T. G., 14(60), 29, 30, *56*
Schickedanz, P. T., 61(4), 65(4), 82(4, 17), 84(17), 89(17), 91(4, 17), 92(4), 96(4), 98(17), 110(4), 111(4), 115(4), 128(4), 129(4), 130(4), 131(4, 15), 132(4), *133*, *134*
Schiff, H. I., 272(98), *320*
Schofield, K., 273(106), 305(106), *320*
Schröer, E., 245(64), *318*
Schumacher, H. J., 271(93), *320*
Scolnik, R., 259(77), *319*
Scott, M. R., 28(64), *57*
Serlemitsos, P., 154(50), *210*
Shabansky, V. P., 146(28), *209*
Sherwood, T. K., *56*
Shimizu, M., 230(49), *318*
Shipp, W. L., 79(14), 82, 84(17), 89(17), 91(17), 98(17), 101(21), 126(14), 128(21), *133*, *134*
Shklovsky, I. S., 154(53), *210*
Simpson, J., 111(28), *134*
Simpson, R. H., 111(28), *134*
Singer, S. F., 137(9, 10, 11), 146(24), 170(87, 88), 171(24), 175(118), 177(131), *208*, *209*, *212*, *213*, *214*
Smales, A. A., *57*
Smith, F. G., 28(66), *57*
Smith, R. V., 180(139, 140), 183(140, 144), *214*, *215*
Söraas, F., 159, 160, 188, 189, 190, 206 (66), *211*, *215*
Sosnovets, E. N., 146(43), 163(68), *210*, *211*
Spar, J., 126, 131, *135* 306(123), *321*

AUTHOR INDEX

Spitzer, L., 138(14), *209*
Stall, J. B., 61, *133*
Sticksel, P. R., 220(15), 221(15), *316*
Stocker, J., 206(205), *218*
Stoffregen, W., 206(205), *218*
Stolarik, J. D., 206(206, 208), *218*
Stolpovsky, V. G., 146(43), 163(68), *210, 211*
Störmer, C., 137(1, 2, 3), *208*
Strobel, D. F., 305, *321*
Strong, I. B., 154(55), *210*
Stroud, L., 37, *55*
Suess, H. E., 33(67), *57*

T

Tanaka, Y., 268, *319*
Taylor, H. E., 146(38), 192, 193, *210, 216*
Taylow, W. W. L., 204(195), *217*
Telegadas, K., 306(125), *321*
Thrush, B. A., 272(102), 273(102, 104), *320*
Tiratsoo, E. N., 32(68), *57*
Tomassian, A. D., 176(124), *214*
Tousey, R., 223(26), 249(85), 250(85), 268(85), *316, 319*
Toyama, Y., 224(37), *317*
Trichel, M. C., 206(202), *217*
Tverskoy, B. A., 184(150, 151), 186, 189, *215*

U

Uranowa, L. A., 220(14), *316*
Urey, H. C., 11(2), 30(2), *54*

V

Vakulov, P. V., 146(43), *210*
Van Allen, J. A., 137(12, 13), 138(13), 145, 146(23), 147(17), 155(61), 162, 163(67) 167(61), 168, 169(84, 85), 170, 191(61, 166), 196(166, 180), 203(166), *208, 209, 211, 212, 213, 216*
Vassy, A., 224(32), *317*
Vasyliunas, V. M., 154, 203, *210, 211*
Vernor, S. N., 146(21, 43), 163(68), 170(89), 209, *210, 211, 212*
Verzariu, P., 169, *212*

Vette, J. I., 146(36), 156, 157, 164(69), 169(82, 83), 175(64), 176, *209, 211, 212*
Vigroux, E., 220, 268, *316*

W

Wager, L. R., *57*
Walt, M., 176(124), 178(136), 180(136, 138), 183, 200, 201, 203, *214, 215, 216*
Weinstein, A. I., 111(29), *134*
Wentworth, R. C., 177(131), *214*
Wescott, E. M., 206(206, 208), *218*
Whalen, B. A., *57*
Wheeler, H. P., *57*
White, R. S., 171(101, 102), 172(101, 103), 173(103, 119), 174(103, 119), 175(119, 122), *213, 214*
Whiting, R. L., 33, *54*
Wilkens, E. M., 297(115), *320*
Williams, D. J., 142(20), 145, 146(20, 32), 147, 148(44, 46, 47), 149(20, 44, 47), 153(20), 167(44, 46, 71, 73, 74), 171(96, 97), 178(134, 137), 182(143), 184(158), 191(20, 71), 192, 195, 197(137, 143), 198 (137, 143), 199(143, 183), 200(143), 204 (183), 206(143), *209, 210, 211, 212, 214, 215 216*
Williamson, J. M., 206(198), *217*
Wilson, A. W., 249, 250, 268(84, 86), 285, 309, *319*
Wilson, M. D., 206(201), *217*
Winckler, J. R., 146(41), 181(142), 182 (142), 203(187, 188), 206(203), *210, 214, 217*
Wright, J. A., 146(36), *209*
Wulf, O. R., 245(61, 62, 63), *318*

Y

Yoshida, S., 137, 138, *208*
Young, C., 263(80), 289(80), *319*
Young, P. A., 248, 268, *319*

Z

Zartman, R. E., 46(71), 51(71), 52, *57*
Zmuda, A. J., 173(112), *213*
Züllig, W., 227(42), 228(42) *317*

SUBJECT INDEX

A

Alpha radioactivity, 4, 8–9
Argon, ratio to helium in natural gas, 46–47

C

Charged particles in earth's magnetic field, 137–218
  geomagnetic trapping of, 137–146
  magnetic field characteristics, 146–156
  particle survey, 156–169
    alpha particles, 168–169
    electrons, 160–167
    protons, 156–160
  sources, losses, and transport of, 170–206
Cloud seeding, natural precipitation variability and, 64–70

E

Earth's magnetic field, charged particles trapped in, 137–218
Elements
  origin and abundance of, 9–11
  equilibrium theory, 9–11

F

Fluids, helium-generating potential of, 25–28

H

Helium
  accumulation or entrapment of, 41–46
  in atmosphere, 32
  cosmogenic, 12–13
  in crustal rocks, 30–32
  definition and uses of, 11–12
    theories, 12
  distribution and occurrence of, 29–32
    extraterrestrial, 29
    terrestrial, 29–30
  geochemistry and geology of, 1–57
  in hydrosphere, 32
  migration of, 32–40
  in natural gases, 13–14, 46–52
  origin of, 11–14
  potential for generation by rocks and fluids, 14–28

M

Magnetic field, of earth, see Earth's magnetic field

N

Natural gas, helium in, 46–52
Nitrogen oxides, effects on ozone photochemistry, 303–306
Noble gases, in natural gases, 51

O

Ozone (atmospheric) photochemistry, 219–322
  classical theory of, 244–271
    equilibrium theory, 245–252
    nonequilibrium theory, 252–256
  distribution of ozone, 219–244
    seasonal variation, 227–240
    vertical distribution, 222–227
    year-to-year variations, 240–244
  mesospheric ozone, 258–264
  in a moist atmosphere, 271–303
    influence of vertical mixing, 297–303
    nighttime processes in, 288–289
  motion effects and, 256–257
  nitrogen oxide effects in, 303–306
  in ozone use as a tracer, 306–316
  stratosphere dynamics and, 312–314
  uncertainties in, 264–271

SUBJECT INDEX

P

Precipitation records
  climatological studies relevant to, 92–97
  diurnal distribution of storm precipitation, 91–92
  downwind seeding effects on, 119–123
  measurement requirements for, 97–111
    correlation of storm mean precipitation, 102–103
    sampling errors, 103–107
    spatial correlation of point precipitation, 98–101
    sequential variability and lag correlations in, 91
  sources of, 61–64
  space and time variability relationships in, 78–92
  statistical evaluation techniques for, 111–119
    area-depth curves, 114–115
    in seeding experiments, 111–114
  storm-precipitation detection for, 107–110
    areal extent in fixed sampling areas, 110
  time distribution models in, 88–91
  use in weather modification experiments, 59–135

R

Radioactive nuclides
  types of, 4
  natural, 5–7
Radioactivity, general principles and types of, 3–9
Rainfall, *see* Precipitation
Rocks, helium-generating potential of, 14–28

S

Space Prober, For a (poem), 136

W

Weather modification
  design of experiments for, 123–133
  downwind seeding effects on, 119–123
  experiment length in, 131–133
  precipitation climatology and, 70–78
  sampling design, 126–129
  site selection, 124–126
  time-of-day factors in, 130
  time-of-year factors in, 129
  use of precipitation records in, 59–135
  weather types in, 130–131

QC
806
A3
v.15
1971

FEB 29 1972